Einstein Studies

Editors: Don Howard John Stachel

Published under the sponsorship
of the Center for Einstein Studies,
Boston University

Volume 1: Einstein and the History of General Relativity
Don Howard and John Stachel, editors

Volume 2: Conceptual Problems of Quantum Gravity
Abhay Ashtekar and John Stachel, editors

Volume 3: Studies in the History of General Relativity
Jean Eisenstaedt and A. J. Kox, editors

Volume 4: Recent Advances in General Relativity
Allen I. Janis and John R. Porter, editors

Volume 5: The Attraction of Gravitation: New Studies
in the History of General Relativity
John Earman, Michel Janssen, and
John D. Norton, editors

John Earman
Michel Janssen
John D. Norton
Editors

The Attraction of Gravitation: New Studies in the History of General Relativity

Birkhäuser

Boston • Basel • Berlin

John Earman
Department of History
and Philosophy of Science
University of Pittsburgh
Pittsburgh, PA 15260

Michel Janssen
Department of History
and Philosophy of Science
University of Pittsburgh
Pittsburgh, PA 15260

John Norton
Department of History
and Philosophy of Science
University of Pittsburgh
Pittsburgh, PA 15260

Library of Congress Cataloging In-Publication Data

The Attraction of gravitation : new studies in the history of general relativity / edited by John Earman, Michel Janssen, John D. Norton.
p. cm. -- (Einstein studies : v. 5)
Includes bibliographical references and index.
ISBN 0-8176-3624-2 (alk. paper). -- ISBN 3-7643-3624-2 (alk. paper)
1. General relativity (physics) -- History. I. Earman, John. II. Janssen, Michel, 1953- . III. Norton, John D., 1960- . IV. Series.
QC173.6.A85 1993 93-30748
530.1'1--dc20 CIP

Printed on acid-free paper.

The Einstein Studies series is published under the sponsorship of the Center for Einstein Studies, Boston University.

ISBN 0-8176-3624-2
ISBN 3-7643-3624-2

Typeset in TEX by TEXniques, Inc., Newton, MA.
Printed and bound by Quinn-Woodbine, Woodbine, NJ.
Printed in the U.S.A.

9 8 7 6 5 4 3 2 1

Contents

Preface to Volume Five

The attraction of gravitation is universal. Over the last few decades it has led to a resurgence of interest in Einstein's general theory of relativity, our best theory of gravitation. In the mid-1980s, this interest began to extend to the history of general relativity, which is now enjoying international attention of unprecedented vigor and intensity. This volume represents the latest outcome of this new interest. Most of the papers began as presentations at the Third International Conference on the History and Philosophy of General Relativity and, after considerable development and revision, have been brought to their present form. The conference was held at the University of Pittsburgh at Johnstown, Pennsylvania (U.S.A.), June 27–30, 1991. Members of the local organizing committee were John Earman, Al Janis, Michel Janssen, Ted Newman, John Norton, and Alan Walstad (University of Pittsburgh) and Clark Glymour (Carnegie–Mellon University, Pittsburgh). Members of the National and International Committee were Jean Eisenstaedt (Institut Henri Poincaré, Paris), Hubert Goenner (University of Göttingen), Joshua Goldberg (Syracuse University), Don Howard (University of Kentucky), A.J. Kox (University of Amsterdam and Einstein Papers, Boston), Jürgen Renn (Einstein Papers, Boston), and John Stachel (Boston University).

This is the third volume in the *Einstein Studies* series to be devoted to the history of general relativity. There are now sufficiently many scholars working in the area to support a series of conferences and volumes of research articles explicitly devoted to the history of general relativity. John Stachel was the first to tap into this interest when he organized the first international conference on the history of general relativity at Osgood Hill, Massachusetts (U.S.A.), May 8–11, 1986. He and Don Howard founded the series *Einstein Studies* and edited its first volume, *Einstein and the History of General Relativity* (Birkhäuser Boston, 1989), which contained papers from the Osgood Hill conference and elsewhere. Following the success of the first conference, Jean Eisenstaedt organized the Second International Conference on the History of General Relativity, which was held at the International Center of Mathematical Research (CIRM) at Luminy, France, September 6–8, 1988. He and A.J. Kox edited a proceedings volume, *Studies in the History of General Relativity*, which appeared as *Einstein Studies*, Volume Three (Birkhäuser Boston, 1992).

The quality and diversity of papers in this volume demonstrate the ever growing vitality of research in the history of general relativity. We have divided the volume into five sections. The first group of papers deals with disputes between Einstein and other figures in the history of general relativity. These papers remind us that science is a collaborative enterprise, even in the case of general relativity, whose genesis is celebrated almost exclusively as the work of just one person. The papers show us how disputes might sometimes further the interests of science and other times not. John Norton's paper recounts how the prospects of a Lorentz covariant gravitation theory were explored within an extended exchange between Einstein and Nordström at the time that Einstein was laying down the foundations of general relativity. Don Howard and John Norton's paper recalls the final dark months of Einstein's struggle with general relativity, when he still remained convinced through the hole argument that general covariance was physically uninteresting. They conjecture that Paul Hertz at Göttingen communicated a serviceable escape from the hole argument to Einstein—which he misunderstood and brusquely rejected. The main focus of Carlo Cattani and Michelangelo De Maria's paper is the debate over the correct formulation of conservation laws in general relativity. They show how Einstein tenaciously defended his formulation against criticism from various authors, foremost among them Tullio Levi-Civita. Peter Havas' paper portrays an accommodating Einstein entering a dispute with Ludwik Silberstein over the two-body problem in general relativity. We follow the dispute as it grows from a simple disagreement into an acrimonious quarrel that surfaced in the popular press.

While general relativity is not celebrated for its intimate contact with an empirical base, the second group of papers examines some episodes related to the empirical evidence supporting the theory. John Earman and Michel Janssen analyze Einstein's perihelion paper of November 1915, which was the work of only one week. They ask if Einstein achieved this speed by sacrificing mathematical rigor. A.J. Kox discusses Pieter Zeeman's little-known experiments on the equality of inertial and gravitational mass, drawing on the recently discovered Zeeman *Nachlass*.

The mathematical complexity of general relativity stimulated considerable research into the development of new and useful mathematical perspectives on general relativity. This is illustrated by two papers in the third section, "Variational Principles in General Relativity." In the first, S. Kichenassamy gives an overview of the early use of variational principles in general relativity, carefully distinguishing the different notions of variation employed. Carlo Cattani's paper on Palatini reveals that Palatini's contribution to general relativity is not exhausted by the celebrated variational principle to which his name is attached. The reader may find it helpful to read these two papers in conjunction with Cattani and De Maria's paper in the first section.

The largest group of papers in the volume addresses the reception and development of general relativity. Karin Reich investigates the American reception and development of the theory of differential invariants, the branch of mathematics essential to the historical foundation of general relativity and to its further development. Hubert Goenner dissects a less happy episode in the reception of Einstein's work, the malicious 1931 denunciation *A Hundred Authors against Einstein.* Goenner exposes the often murky background and motivations of the volume's contributors. Silvio Bergia gives an extensive survey of attempts to formulate unified field theories along the lines suggested by general relativity. Bergia evaluates these attempts with a carefully chosen set of criteria, articulated at the time of the attempts, thus minimizing the danger of anachronism in his survey.

Gennady Gorelik recounts the life of one of the foremost Russian relativists, Vladimir Fock, revealing a fascinating and complex figure who negotiated controversy within his home country and internationally with dignity and principle. Kameshwar Wali explains why Chandrasekhar's entry into active research in general relativity was delayed until the 1960s. He then reviews Chandra's substantial contributions from the 1960s to the 1990s, starting with relativistic instabilities and post-Newtonian approximations and continuing through rotating stars and black holes.

In the final section, papers by Jean Eisenstaedt and by George Gale and John Urani explore the ever fertile interaction of cosmology and general relativity. Eisenstaedt shows how Lemaître's interest in cosmology was crucial for his important contribution to the modern interpretation of the Schwarzschild solution. Gale and Urani maintain that E.A. Milne's "kinematic relativity" was not merely a dead-end curiosity to be relegated to a footnote in the history of 20th century philosophy. They argue that Milne's program not only helped shape the debate about the nature of cosmology but also played a direct role in the development of the Robertson–Walker metric.

John Earman
Michel Janssen
John Norton

Fall 1993

Acknowledgments

The editors gratefully acknowledge the support, assistance, and encouragement of many people and organizations, their officers, and staff: the Center for Einstein Studies, Boston University; the Center for Philosophy of Science, University of Pittsburgh; the Collected Papers of Albert Einstein Project, Boston University; the Department of History and Philosophy of Science, University of Pittsburgh; the Department of Philosophy, Carnegie–Mellon University; the Franklin J. Matchette Foundation, New York; the University Center for International Studies, University of Pittsburgh; Adam Bryant, Suzanne Durkacs, *Einstein Studies* Series Editors Don Howard and John Stachel; Sara Fleming for preparing the index; and the staff of Birkhäuser Boston.

A Note on Sources

In view of the frequent citations of unpublished correspondence or other items in the Einstein Archive, we have adopted a standard format for such citations. For example, the designation "EA 26-107" refers to item number 26-107 in the Control Index to the Einstein Archive. Copies of the Control Index can be consulted at the Jewish National and University Library (The Hebrew University), Jerusalem, where the Archive is housed; and at Mudd Manuscript Library, Princeton University, and Mugar Memorial Library, Boston University, where copies of the Archive are available for consultation by scholars.

Part I

Disputes with Einstein

Einstein and Nordström: Some Lesser-Known Thought Experiments in Gravitation

John D. Norton

Late in 1907, Einstein turned his attention to the question of gravitation in his new theory of relativity. It was obvious to his contemporaries that Newton's theory of gravitation required only minor adjustments to bring it into agreement with relativity theory. Einstein's first published words on the question (Einstein 1907b, part V), however, completely ignore the possibility of such simple adjustments. Instead he looked upon gravitation as the vehicle for extending the principle of relativity to accelerated motion. He proposed a new gravitation theory that violated his fledgling light postulate and related the gravitational potential to the now variable speed of light. Over the next eight years, Einstein developed these earliest ideas into his greatest scientific success, the general theory of relativity, and gravitation theory was changed forever. Gravitational fields were no longer pictured as just another inhabitant of space and time, like electric and magnetic fields. They were part of the very fabric of space and time itself.

In light of this dazzling success, it is easy to forget just how precarious were Einstein's early steps toward his general theory of relativity. These steps were not based on novel experimental results. Indeed, the empirical result Einstein deemed decisive—the equality of inertial and gravitational mass—was known in some preliminary form as far back as Galileo. Again, there were no compelling theoretical grounds for striking out along the path Einstein took. In 1907, it seemed that any number of minor modifications could make Newtonian gravitation theory compatible with Einstein's new special theory of relativity. One did not have to look for the relativistic salvation of gravitation theory in an extension of the principle of relativity. Einstein himself would later label the motivations for his new approach "epistemological" (Einstein 1916, section 2).

Through the years of his struggle to develop and disseminate general relativity, one of Einstein's greatest strengths was his celebrated mastery of thought experiments. If you doubted that merely uniformly accelerating your coordinates could create a gravitational field, Einstein would have you visualize drugged physicists awakening trapped in a box as it was uniformly accelerated through gravitation-free space (Einstein 1913, pp. 1254–1255). Would not all objects in the box fall just as though the box were unaccelerated but under the influence of a gravitational field? Was not a state of uniform acceleration fully equivalent to the presence of a homogeneous gravitational field?

As vivid and compelling as Einstein's thought experiments proved to be, they still could not mask the early difficulties of Einstein's precarious speculations. Even a loyal supporter, Max von Laue, author of the earliest textbooks on special and general relativity, had objected to Einstein's idea that acceleration could produce a gravitational field. How could this be possible, he complained, since this gravitational field would have no source masses.[1] Einstein's evolving theory had to compete with a range of far more conservative and more plausible approaches to gravitation, and it was to these that physicists such as von Laue looked for a relativistic treatment of gravitation.

We must ask, therefore, about Einstein's own attitude toward these alternatives. In particular, what of the possibility of a small modification to Newtonian gravitation theory in order to render it Lorentz covariant and thus compatible with special relativity? Had Einstein considered this possibility? What reasons could he give for turning away from this conservative but natural path? It turns out that Einstein had considered and rejected this conservative path in the months immediately prior to his first publication of 1907 on relativity and gravitation. He felt such a theory must violate the equality of inertial and gravitational mass. He was forced to revisit these considerations in 1912 with the explosion of interest in relativistic gravitation theories. He first continued to insist that a simple Lorentz covariant gravitation theory was not viable. In the course of the following year, however, he came to see that he was wrong and that there were ways of constructing Lorentz covariant gravitation theories compatible with the equality of inertial and gravitational mass.

After an initial enchantment and subsequent disillusionment with Abraham's theory of gravitation, Einstein found himself greatly impressed by a Lorentz covariant gravitation theory due to the Finnish physicist Gunnar

Nordström. In fact, by late 1913, Einstein had nominated Nordström's theory as the only viable competitor to his own emerging general theory of relativity (Einstein 1913). This selection came, however, only after a series of exchanges between Einstein and Nordström that led Nordström to significant modifications of his theory.

Einstein's concession to the conservative approach proved to have a silver lining; under continued pressure from Einstein, Nordström made his theory compatible with the equality of inertial and gravitational mass by assuming that rods altered their length and clocks their rate upon falling into a gravitational field so that the background Minkowski space-time had become inaccessible to direct measurement. As Einstein and Fokker showed in early 1914 (Einstein and Fokker 1914), the space-time actually revealed by direct clock and rod measurement had become curved, much like the space-times of Einstein's own theory. Moreover, Nordström's gravitational field equation was equivalent to a geometrical equation in which the Riemann–Christoffel curvature tensor played the central role. In it, the full contraction, the curvature scalar, is set proportional to the trace of the stress-energy tensor. What is remarkable about this field equation is that it comes almost two years *before* Einstein recognized the importance of the curvature tensor in constructing field equations for his own general theory of relativity! In this regard, the conservative approach actually anticipated Einstein's more daring approach.

Einstein now had an answer to the objection that general relativity introduced an unnecessarily complicated mechanism for treating gravitation, the curvature of space-time. He had shown that the conservative path led to this same basic result: Gravitational fields come hand-in-hand with the curvature of space-time.

Elsewhere, I have given a more detailed account of Einstein's response to the conservative approach to gravitation and his entanglement with Nordström's theory of gravitation (Norton, 1992). My purpose in this chapter is to concentrate on one exceptionally interesting aspect of the episode. As in Einstein's better-known work on his general theory of relativity, the episode was dominated by a sequence of compelling thought experiments.[2] These experiments concentrate the key issues into their simplest forms and present them in a way that makes the conclusions emerge convincingly and effortlessly. In this chapter I will review this sequence of thought experiments as it carries us through the highlights of the episode.

In particular, we will see how one of the more arcane areas of special relativistic physics proved decisive to the development of relativistic gravitation theory. It emerged from the work of Einstein, von Laue, and others that stressed bodies behave in strikingly nonclassical ways in relativity theory. For example, a moving body can acquire energy simply by being subjected to stress, even though it may not be deformed elastically by the stress. Nonclassical energies such as these provided Einstein with the key for incorporating the equality of inertial and gravitational mass into relativistic physics.

1. First Thought Experiment: Masses Falling from a Tower

The bare facts of Einstein's initiation into the problem of relativizing gravitation theory are known. In late September 1907, Einstein accepted a commission from Johannes Stark, editor of *Jahrbuch der Radioaktivität und Elektronik*, to write a review article on the principle of relativity.[3] That review (Einstein 1907b) was submitted a little over two months later, on December 4, 1907. Its concluding part contained the earliest statement of what came to be the principle of equivalence and of the bold conjectures about gravitation that followed from it. What we know only from later reminiscences by Einstein is that, in this brief period between September and December, he considered and rejected a conservative Lorentz covariant theory of gravitation.[4]

Einstein recalled that he knew how one could take Newton's theory of gravitation and render it Lorentz covariant with small modifications to its equations. Newton's theory is given most conveniently in the usual Cartesian coordinates (x, y, z) by the field equation

$$\nabla^2\phi = \left(\frac{\partial^2}{\partial x^2} + \frac{\partial^2}{\partial y^2} + \frac{\partial^2}{\partial z^2}\right)\phi = 4\pi G\rho \tag{1}$$

for the gravitational field potential ϕ generated by a mass density ρ, where G is the gravitational constant, and by the force equation

$$\mathbf{f} = -m\nabla\phi \tag{2}$$

for the gravitational force $\mathbf{f}$ on a body of mass m. The adaptation to special relativity of the field equation to which Einstein alluded was obvious. One simply replaces the Laplacian operator ∇^2 of (1) with the manifestly Lorentz covariant d'Alembertian $\Box^2$ to recover

$$\Box^2\phi = \left(\nabla^2 - \frac{1}{c^2}\frac{\partial^2}{\partial t^2}\right)\phi = 4\pi G\nu, \tag{3}$$

where ν is an invariant mass density and t the time coordinate. An analogous modification of (2) would also be required. Einstein (1933, pp. 286–287) continued to explain that the outcome of his investigations was not satisfactory.

> These investigations, however, led to a result which raised my strong suspicions. According to classical mechanics, the vertical acceleration of a body in the vertical gravitational field is independent of the horizontal component of its velocity. Hence in such a gravitational field the vertical acceleration of a mechanical system or of its center of gravity works out independently of its internal kinetic energy. But in the theory I advanced, the acceleration of a falling body was not independent of its horizontal velocity or the internal energy of the system.
>
> This did not fit with the old experimental fact that all bodies have the same acceleration in a gravitational field. This law, which may also be formulated as the law of the equality of inertial and gravitational mass, was now brought home to me in all its significance. I was in the highest degree amazed at its existence and guessed that in it must lie the key to a deeper understanding of inertia and gravitation. I had no serious doubts about its strict validity even without knowing the results of the admirable experiments of Eötvos, which—if my memory is right—I only came to know later. I now abandoned as inadequate the attempt to treat the problem of gravitation, in the manner outlined above, within the framework of the special theory of relativity. It clearly failed to do justice to the most fundamental property of gravitation.

The result that troubled Einstein in the theory he advanced came from the relativistic adaptation of the force law (2). As Einstein pointed out in his reminiscences, this adaptation could not be specified so unequivocally. We can proceed directly to the result, however, if we use four-dimensional methods of representation not available to Einstein in 1907. The natural adaptation of (2) is

$$F_\mu = m\frac{\mathrm{d}U_\mu}{\mathrm{d}\tau} = -m\frac{\partial\phi}{\partial x_\mu}, \tag{4}$$

where F_μ is the gravitational four-force acting on a body of rest mass m with four-velocity U_μ; τ is the proper time.[5] We can now apply (4) to the special case of a body whose three-velocity $\mathbf{v}$ has, at some instant of time, no vertical component in a static gravitational field. If the gravitational field at that instant at the mass acts along the z-axis of coordinates, so that the z-axis is the vertical direction in space, then it follows from (4) that the vertical acceleration of the mass is given by

$$\frac{\mathrm{d}v_z}{\mathrm{d}t} = -\left(1 - \frac{v^2}{c^2}\right)\frac{\partial\phi}{\partial z}. \tag{5}$$

We see immediately that this vertical acceleration is reduced as the horizontal speed v is increased, illustrating Einstein's claimed dependence of the rate of fall on horizontal velocity.

The "old experimental fact," which this result contradicts, surely belongs to the famous fable in which Galileo drops various objects of different weights from a tower. Einstein and Infeld (1938, pp. 37–38) certainly identify this story when they wrote:

> What experiments prove convincingly that the two masses [inertial and gravitational] are the same? The answer lies in Galileo's old experiment in which he dropped different masses from a tower. He noticed that the time required for the fall was always the same, that the motion of a falling body does not depend on the mass.

We can combine these ingredients to make explicit the thought experiment suggested by Einstein's analysis. Masses are dropped from a high tower, some with various horizontal velocities and some with none. According to (5), the masses with greater horizontal velocity fall slower, contradicting Einstein's expectation and the familiar classical result that they should all fall alike. See Figure 1.

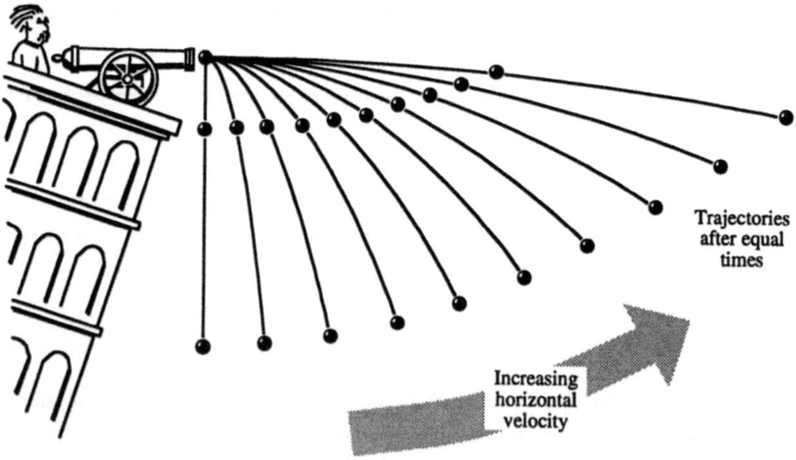

Figure 1. Vertical fall slowed by horizontal velocity in a Lorentz covariant theory of gravitation.

2. Second Thought Experiment: Spinning Tops and Heated Gases

It is not so obvious why Einstein found the outcome of this first thought experiment to be so troubling that he felt justified in abandoning the search for a Lorentz covariant theory of gravitation. The dependence is a minute effect, second order in v/c. Indeed, one might well wonder how even the most ingenious experimentalist could compare the rate of fall of a mass with that of another whizzing past at a horizontal velocity close to the speed of light. Even if this were possible, the experiment had surely not been done in 1907. How could Einstein reject this minute effect as incompatible with an "old experimental fact" whose traditional origins lay with Galileo?

The answer resides in the fact that Einstein derived the dependence of vertical acceleration on the "horizontal velocity or the *internal energy* of the system." What Einstein meant by this was made clear in 1912 when the Finnish physicist Gunnar Nordström published the first of a series of papers on a Lorentz covariant, scalar theory of gravitation (Nordström 1912). The essential assumptions and content of Nordström's theory were contained in equations (3) and (4) above. Nordström did correct, however, a problem with (4). It turns out that this force law can only hold for a mass moving so that the rate of change of the gravitational potential along its world line is zero.[6] (This condition holds instantaneously for the special case used to derive [5].) Thus the force law (4) requires modification if it is to apply to masses along whose trajectories ϕ is not constant. Nordström found two suitable modifications. He favored the one in which the rest mass m of the body is assumed to vary with the gravitational potential ϕ. In particular, he readily derived the dependence

$$m = m_0 \exp\left(\frac{\phi}{c^2}\right), \tag{6}$$

where m_0 is the value of m when $\phi = 0$.

By October 1912, when Nordström sent his paper to *Physikalische Zeitschrift*, Einstein's novel ideas on gravitation had become a matter of public controversy. In July, Einstein found himself immersed in a vitriolic dispute with Max Abraham, who saw in Einstein's admission of a variable speed of light a "death blow" to relativity theory (Abraham 1912). In his response, Einstein (1912, pp. 1062–1063) published his 1907 grounds for abandoning Lorentz covariance in the most general form he could manage. In any Lorentz covariant gravitation theory, he argued, be it a four-vector or six-vector theory, gravitation would act on a moving body with a strength that would vary with velocity. Any such theory was unacceptable, since it violated the requirement of the equality of inertial and gravitational mass.

Therefore it is not at all surprising that Nordström attracted Einstein's attention when he published just such a theory. Einstein's reaction was so swift that Nordström was able to mention it in an addendum to his original paper! The addendum began (Nordström 1912, p. 1129):

> *Addendum to proofs.* From a letter from Herr Prof. Dr. A. Einstein I learn that he had already earlier concerned himself with the possibility

> used above by me for treating gravitational phenomena in a simple way. He however came to the conviction that the consequences of such a theory cannot correspond with reality. In a simple example he shows that, according to this theory, a rotating system in a gravitational field will acquire a smaller acceleration than a non-rotating system.

Einstein's reflection on the acceleration of fall of a spinning system is actually only a slight elaboration of the situation considered in the first thought experiment above. Each element of a suitably oriented spinning body in a gravitational field has a horizontal velocity. Thus, according to (5), which obtains in Nordström's theory, each element will fall slower than the corresponding element without that velocity. What is true for each part holds for the whole. A spinning body falls slower than the same body without rotation.

This example now makes clear Einstein's remark about internal energy. When the body is set into rotation, its parts gain kinetic energy, so its overall energy and its inertia are increased. However, through (5), there is a decrease in the gravitational force acting on it, so that its acceleration of fall is decreased. That is, its rate of fall decreases as the internal energy and inertia increases. Presumably Einstein thought the spinning body just one example of a general effect of this type. In much later reminiscences, Einstein used the example of a kinetic gas.[7] As the gas is heated, each molecule moves faster and thus falls more slowly. Thus the aggregate of molecules, the heated gas, falls more slowly than a colder gas. These two examples comprise the second thought experiment. See Figure 2.

Einstein's result in this form is a far greater threat to Lorentz covariant theories of gravitation such as Nordström's, for it points to effects that might well be experimentally testable. Perhaps the effect might transcend detection by a Galileo-like timing of the fall of spinning tops or hot gases, but would it escape an apparatus similar to that of the Eötvos experiment? Nordström seemed to think so, for he continued his appendix by dismissing Einstein's argument on the basis of the effect being "too small to yield a contradiction with experience." This dismissal depended on a rather bold assumption: that there are no common systems of matter in which a great part of the internal energy, and thus inertia, is due to the kinetic energy of internal motions. Such systems, if they existed, would fall markedly slower than others according to Nordström's theory. Nordström may well have been right that no measurable effect would arise from the spinning of a body, but could he be sure that the energy of commonplace matter did not already have a significant kinetic component? The fundamental theory of matter was then in a state of turmoil and scarcely able to assure him either way. A more prudent Einstein was unwilling to take the risk. Should it turn

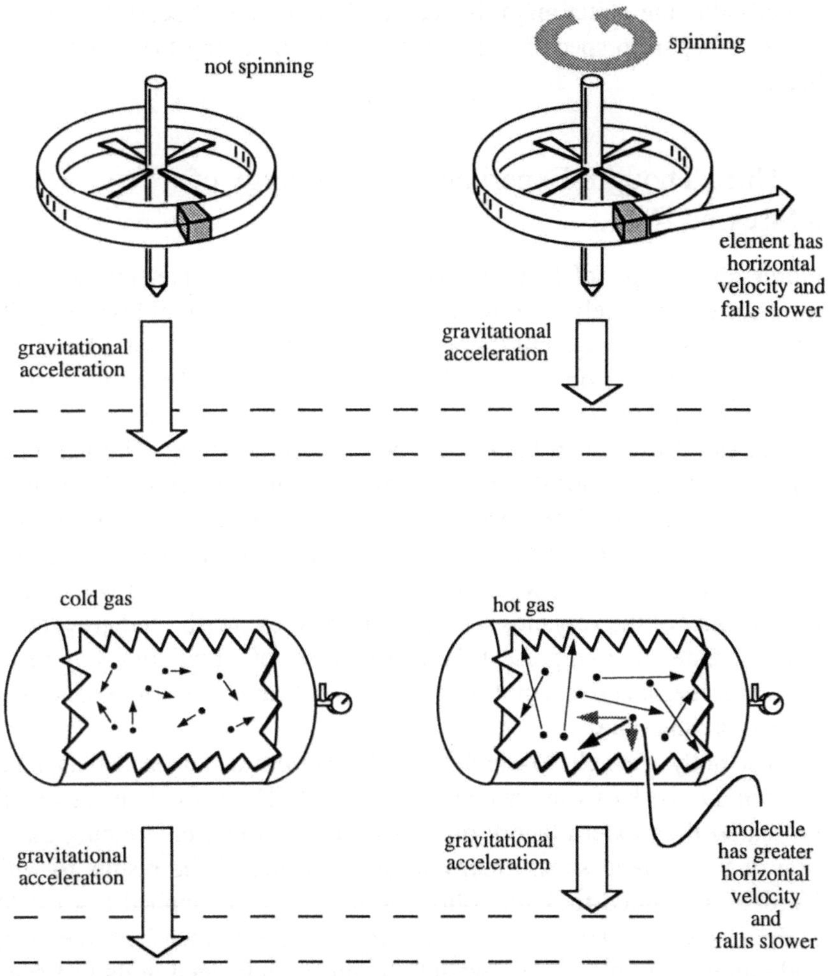

Figure 2. Spinning bodies fall slower than when not spinning. Hot gases fall slower than cold gases, in Nordström's theory.

out that a significant part of the total energy of various types of ordinary matter was due, in different proportion, to an internal kinetic energy, then Nordström's theory might well be refuted by simple observations of the fall of different substances from a tower.

By the time of submission of his next paper on the theory in January 1913, Nordström had become more wary (Nordström 1913a). While still insisting (p. 878) that no observable effect would arise in the case of spinning bodies, he was prepared to raise the question of whether the "molecular motions of a falling body" would influence the rate of fall. He did not state directly that the effect might be measurable, but the effect did worry him, since he began to speculate on a way of incorporating the effect into his theory.

3. Third Thought Experiment: The Energy of a Stressed Rod

Nordström's paper of January 1913 was devoted to a question that would ultimately completely alter the direction of development of his theory. The paper asked which quantity represented the inertial mass of a body. The question was far from trivial. Recent work in the relativistic theory of continua had shown that there were inertial effects that arose when a body was stressed for which there were no classical analogs. Nordström observed (1913a, p. 856) that it had proved possible to ignore this question and develop a complete mechanics of extended bodies without explicitly introducing the concept of inertial mass. This luxury could no longer be afforded, he continued, when one worked in a relativistic gravitation theory, because of the very close connection between inertial and gravitational masses. One had to represent the inertial mass of a body in a way that allowed for inertial effects in stressed bodies that cannot be attributed directly to an individual mass.

The body of results to which Nordström referred had reached its mature form in the work of von Laue (1911a, 1911b). There von Laue essentially presented the modern theory of relativistic continua, introducing the notion of the general stress-energy tensor of matter. The results to which Nordström alluded took the following form. If one applied a stress to a body *without deforming it or setting it into motion*, then both the energy and momentum of the body would remained unchanged in its rest frame. However, if one viewed this same process from a frame of reference in which the body was in motion, then the energy and momentum of the body might change. For example, if the body was influenced by a shear stress[8] p^0_{xy} in its rest frame and then viewed from a frame of reference moving at velocity v in the x direction, then in that frame the body would acquire a momentum in the y direction. The momentum density g_y due to the stress is given by[9]

$$g_y = \gamma \frac{v}{c} p^0_{xy}. \tag{7}$$

If the stress was a normal stress p^0_{xx} in the rest frame, then, when viewed in the relatively moving frame, the body would have acquired both energy and an x-directed momentum. The energy density W and momentum density

g_x acquired is given by

$$W = \gamma^2 \frac{v^2}{c^2} p^0_{xx}, \qquad g_x = \gamma^2 \frac{v}{c^2} p^0_{xx}. \tag{8}$$

These are the effects for which there are no classical analogs. They proved decisive in the relativistic analysis of a number of celebrated thought experiments and real experiments, most notably the Lewis and Tolman bent lever and the Trouton–Noble capacitor.[10]

One of the clearest and earliest analyses of these nonclassical effects is due to a thought experiment of Einstein (1907a, section 1; 1907b, section 12) and was given in the context of his discussion of the inertia of energy. He imagined an extended body at rest carrying a charge distribution. He then imagined that, at some definite instant in its rest frame, the body comes under the influence of an external electromagnetic field. The net external forces are assumed to balance so that the body remains at rest. The effect of the continued action of the forces, however, is to induce a state of stress in the body. Einstein now redescribed this process from a frame in which the body moved uniformly. Because of the relativity of simultaneity, the body does not come under the influence of the external field at one instant. For a brief period, some charge elements are under the influence of the field and some are not. During this period, the external forces exerted by the field do not balance, so that there is a net external force exerted on the body. Work is done on or by the force as the body moves, and there is a net transfer of energy. This energy is the energy described in (8) and associated with the induction of a stressed state in the body.[11]

The beauty of this thought experiment is that it derives the effects of equations (8) directly from the most fundamental, nonclassical effect of special relativity, the relativity of simultaneity. Forces applied simultaneously in one frame of reference need not be seen as applied simultaneously in another. The resulting temporary imbalance leads to an energy and momentum transfer in the latter frame only and these transferred quantities emerge as those of (8). Einstein's analysis is mathematically quite complicated, however, since he considers a body of arbitrary shape and charge distribution. Recapitulating Einstein's analysis for a simpler case is sufficient to reveal the essential physics. That case is a rod of uniform cross section with equal charges at either end. This is the third thought experiment. See Figure 3.

The rod has rest length l, cross-sectional area A, and extends from $x' = 0$ to $x' = l$ in its rest frame (x', t'). At a specific instant $t' = 0$ in its rest frame, the rod comes under the influence of a field that applies equal but

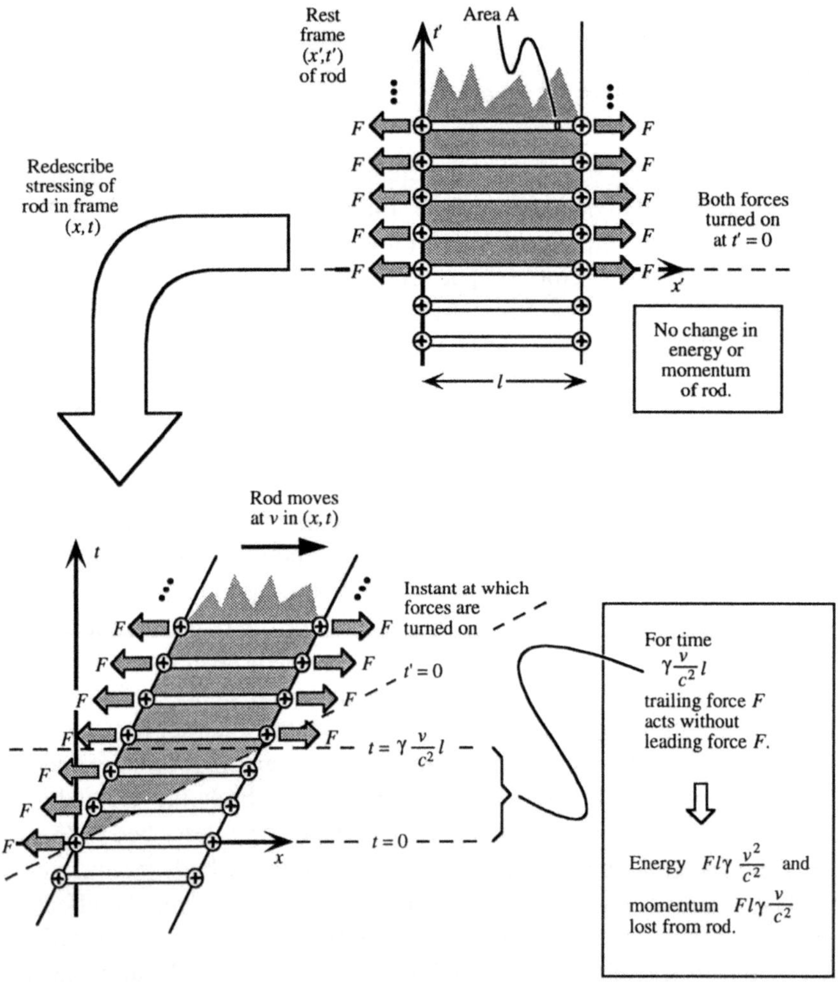

Figure 3. Stressing a moving rod changes its energy and momentum.

oppositely directed forces F to the charges. For concreteness, assume the forces are directed away from the rod along its length. The forces induce a tensile stress on the rod in its rest frame[12]

$$p^0_{xx} = -F/A.$$

If we redescribe this stressing of the rod in a frame (x, t) [13] in which the rod moves at velocity v in the $+x$ direction, we find that the two forces are not activated simultaneously because of the relativity of simultaneity. The force F on the trailing end is activated at a time $\gamma \frac{v}{c^2} l$ earlier than the force F

on the leading end. For this short time period the external force F on the trailing end is not balanced by the other external force. As a result, work is done by the motion of the rod against the force. The resulting loss of energy from the rod is $Fl\gamma\frac{v^2}{c^2}$ and the loss of momentum $Fl\gamma\frac{v}{c^2}$. Recalling the above expression for p^0_{xx} and that the volume of the rod in the frame (x, t) is $V = Al/\gamma$, we recover expressions for the energy E and x-momentum G_x gained by the rod in the process of being stressed:

$$E = \gamma^2 \frac{v^2}{c^2} p^0_{xx} V \quad \text{and} \quad G_x = \gamma^2 \frac{v}{c^2} p^0_{xx} V.$$

Division of these expressions by the volume V yields (8).

4. Fourth Thought Experiment: Radiation in a Massless, Mirrored Box

In his paper (1913a), Nordström had asked the right question. What quantity represents the total inertial mass of a body, including contributions to its inertial properties that arose from stresses? He sought his answer in the form of the source density ν for equation (3), and he looked in the right place for his answer. He expected this density to be a quantity derived from the stress-energy tensor $T_{\mu\nu}$, recently introduced by von Laue. After extensive discussion, he settled upon $1/c^2$ times the rest energy density of the source matter as his source density ν. The rest frame required for this choice was the instantaneous local rest frame of a continuous matter distribution—"dust"—which Nordström assumed contributed to the source matter. We would now express Nordström's choice in manifestly covariant form as

$$\nu = -\frac{1}{c^2} T_{\mu\nu} B_\mu B_\nu, \tag{9}$$

where B_μ is the four-velocity vector field of the continuous distribution of matter.

Nordström's answer was close to the correct answer—but not close enough, as was pointed out by Einstein, in section 7 of his physical part of Einstein and Grossmann (1913).[14] He reported that von Laue himself, also in Zurich but at the University of Zurich, had pointed out to Einstein the only viable choice, the trace of the stress-energy tensor

$$T = T_{\mu\mu}.$$

Einstein proposed to call this scalar "Laue's scalar." What was distinctive about this choice was that it enabled a gravitation theory that employed it to satisfy the requirement of the equality of inertial and gravitational mass, at least "up to a certain degree," as Einstein put it. This degree included examples such as those in the second thought experiment above, as we shall now see.

The key result that enabled satisfaction of this equality was due to von Laue. Von Laue (1911a) had found a single general solution to a range of problematic examples within relativity theory. They all involved systems whose properties appeared to violate the principle of relativity. For example, on the basis of classical electromagnetic theory, Trouton and Noble (1903) believed that a charged, parallel-plate capacitor would experience a net turning couple if it was set in motion with its plates oblique to the direction of motion—although their experiment yielded a celebrated null result. Again, Ehrenfest (1907) had raised the possibility that a nonspherical or nonellipsoidal electron could not persist in uniform translational motion unless forces are applied to it. In both cases the projected behavior would provide an indicator of the uniform motion of the system, violating the principle of relativity.

What these examples had in common was the presence of stresses within the systems and, with the proper treatment of these stresses, the threat to the principle of relativity evaporated. Von Laue noticed that these systems were all what he called "complete static systems," that is, they maintained a static equilibrium in inertial frames of reference without interacting with other systems.[15] The basic result characterizing these systems was that, in their rest frames,

$$\int p^0_{ik}\,\mathrm{d}V^0 = 0, \tag{10}$$

where the integral extends over the rest volume V^0 of the whole body. It follows from (10) that the energy and momentum of a complete static system transforms under Lorentz transformation exactly like the energy and momentum of a point-mass. Since the dynamics of a point-mass was compatible with the principle of relativity, so was the dynamics of a complete static system, and one could not expect a violation of the principle of relativity in the dynamics of these systems.

Von Laue's analysis was very general and powerful because it needed to ask very little of the inner structure of the systems. All one needed to know was whether the system was a complete static system. If it was, one could ignore the further details and simply imagine a black box drawn around the system. Its overall dynamics was now determined.

In effect, what Einstein was able to report in Einstein and Grossmann (1913, section 7) was that von Laue's machinery could be applied directly to the problem of selecting a gravitational mass density. If one chose T as the gravitational mass density, von Laue's result (10) entailed that the total gravitational mass of a complete stationary system in its rest frame was equal to its inertial mass. For, using (10), for such a system we have[16]

$$\begin{aligned}\text{gravitational mass} &= \int T\,\mathrm{d}V^0 = \int (p^0_{11} + p^0_{22} + p^0_{33} + T^0_{44})\,\mathrm{d}V^0 \\ &= \int T^0_{44}\,\mathrm{d}V^0 = \text{total energy} = \text{total inertial mass},\end{aligned} \tag{11}$$

where I follow Einstein in simplifying the analysis by neglecting factors of c^2, so that energy and inertial mass become numerically equal.

The power and subtlety of this rather beautiful result stood out clearly in the example that Einstein employed in his discussion. This example is our fourth thought experiment. The trace T for electromagnetic radiation vanishes. Thus it would seem that electromagnetic radiation can have no gravitational mass.[17] But what of a system of electromagnetic radiation enclosed within a massless box with mirrored walls? Would such a system have any gravitational mass? The radiation itself would not, although that radiation would exert a pressure on the walls of the box. These walls would become stressed and, simply because of this stress, the walls would acquire a gravitational mass. Since it is a complete static system, we need do no direct computation of the distribution of stresses in the walls. The result (11) tells us immediately that the total gravitational mass of the system in its rest frame is given by the system's total inertial mass. See Figure 4.

The same reasoning can essentially be applied to the spinning bodies and heated gases of the second thought experiment, if they are set in a gravitation theory that uses T as its source density. Molecules of gas with horizontal motion will fall slower than those without this motion, thus they do have a smaller effective gravitational mass. They exert a pressure on the walls of the containing vessel, however, which becomes stressed. These stresses alter the value of T and thereby contribute to the gravitational mass. Since (11) applies here, we read immediately from it that the gravitational mass of a gas enclosed in a vessel in its rest frame is given by the inertial mass of the whole system.

Similarly, the individual masses comprising a spinning body do have a smaller effective gravitational mass because of their motion, but the spinning body is stressed by centrifugal forces. We know from (11), without calculation, that the contribution of the stresses to the total gravitational mass exactly compensates for the reduction due the motion of the individual masses. As before, the total gravitational mass is given by the total inertial mass.

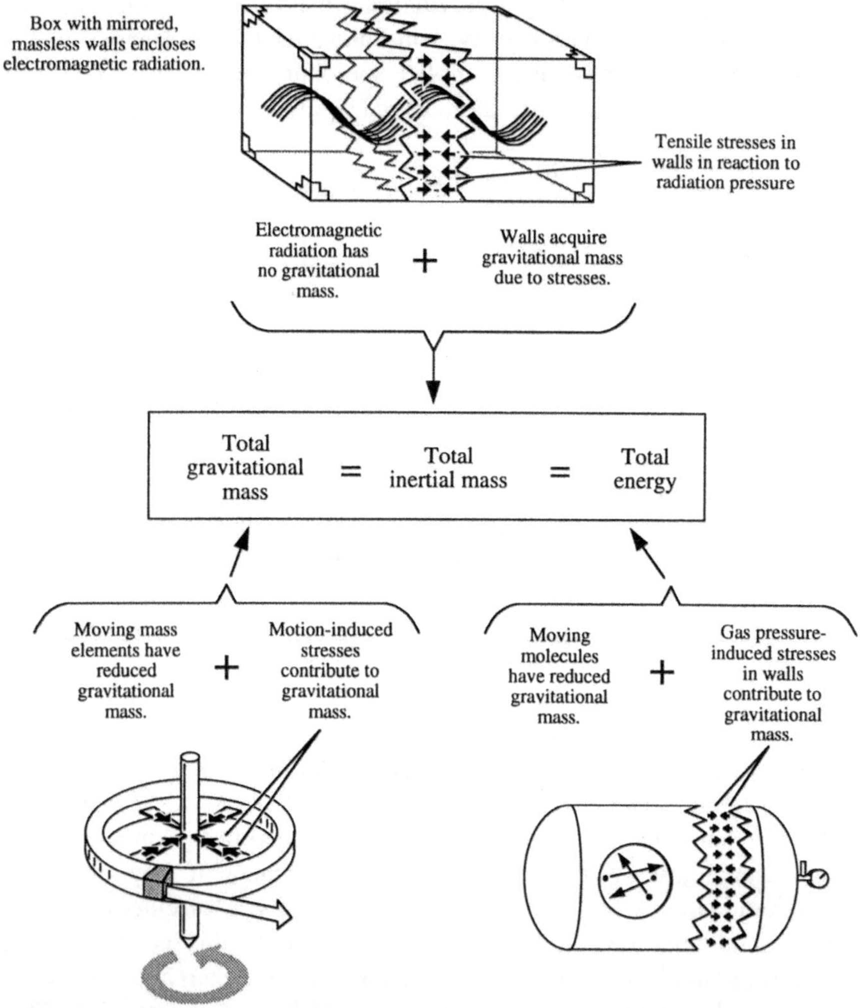

Figure 4. Equality of inertial and gravitational mass for complete stationary systems in a gravitation theory with source density T.

5. Fifth Thought Experiment: Lowering and Raising Radiation

At this point, one might anticipate that Einstein would have to capitulate and cease his opposition to Lorentz covariant gravitation theories. His objection to these theories had been that they failed to satisfy the requirement of equality of inertial and gravitational mass. Most damaging was his conclusion that this equality would fail in the type of cases dealt with in the second thought experiment above. But now his analysis of the choice of T as source density showed how a Lorentz covariant, scalar theory of gravitation could escape Einstein's objection in exactly those most damaging cases.

Einstein was in no mood for retraction, and with good reason. Having presented T as the only viable choice of gravitational source density, he proceeded to argue that the choice was a disaster. A theory that employed T as the gravitational source density must violate the law of conservation of energy. Einstein's argument was presented within a thought experiment—our fifth thought experiment—and it was beguilingly simple. See Figure 5. He imagined electromagnetic radiation trapped in a mirrored, massless box. We shall assume it cubic in shape for simplicity. The system is lowered into a gravitational field. Since it has gravitational mass, an amount of energy proportional to this mass is extracted.

Einstein now introduced another apparatus to raise the radiation. He imagined a mirrored shaft extending out of the gravitational field. Within the shaft are two mirrored, massless baffles, firmly fixed together. The radiation is introduced into the space between the baffles and is raised out of the gravitational field as the baffles are raised. We shall again assume for simplicity that the space between the baffles is cubic.

We have already seen that the gravitational mass of the mirrored box used to lower the radiation is due entirely to the stresses in its walls. It now follows immediately that the system of radiation and baffles has only one-third the gravitational mass of the radiation/box system, for in elevating the radiation trapped between the baffles, one need move only one-third as many stressed members.[18] Only one-third as much energy need therefore be supplied to raise the radiation in the baffle apparatus as is released when the radiation is lowered in the box. Since no energy is involved in raising and lowering the massless box and baffles when devoid of radiation, a complete cycle of raising and lowering the radiation yields a net gain of energy. This violates the law of conservation of energy.

Einstein must have been very pleased with this outcome. In a single blow, it ruled out not just Lorentz covariant, scalar theories of gravitation, but any relativistic gravitation theory that employed a scalar potential. Thus the "undeniable complexity" (Einstein and Grossmann 1913, part 1, section 7) of Einstein's second-rank tensor theory seemed unavoidable.

6. Sixth Thought Experiment: Lowering and Raising a Stressed Rod

Einstein's triumph was short lived. In July 1913, Nordström (1913b) sub-

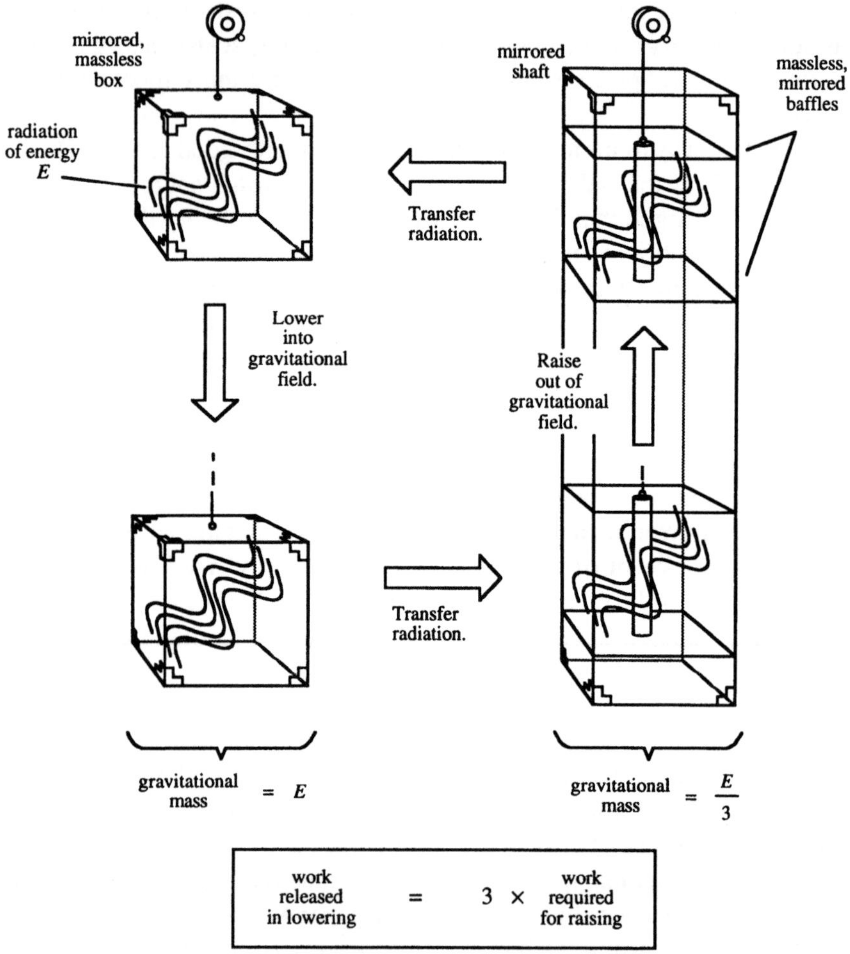

Figure 5. Trace T as source density violates energy conservation.

mitted his so-called "second" theory to *Annalen der Physik.* This theory used the trace T as its gravitational source density and fully exploited the opportunities it provided for enabling the equality of inertial and gravitational mass. Moreover, it was able to incorporate an escape from Einstein's attack on all relativistic scalar theories of gravitation.

The basic equations of the theory remained (3) and (4), except that the four-force F_μ was replaced by a four-force density K_μ:

$$\frac{\partial^2\phi}{\partial x^2}+\frac{\partial^2\phi}{\partial y^2}+\frac{\partial^2\phi}{\partial z^2}+\frac{\partial^2\phi}{\partial u^2}=g(\phi)\nu,$$

$$K_\mu = -g(\phi)\nu\frac{\partial\phi}{\partial x_\mu},$$

where $u = ict$.

The major alteration was the inclusion of the gravitation factor $g(\phi)$. Its purpose was to allow for the fact that the total inertial mass and energy of a system must vary with the gravitational potential, whereas the gravitational mass of the system will be independent of the potential. If a system had inertial mass m when in an external gravitational field of potential ϕ, then its gravitational mass M_g was given by

$$M_g = g(\phi)m. \tag{12}$$

If we now considered a matter distribution whose parts lay in regions of differing gravitational potential, the gravitational mass of the whole distribution would be given by a g-weighted integral over its volume

$$M_g = \int g(\phi)\nu\,\mathrm{d}V.$$

At this point, the expressions for both $g(\phi)$ and the source density ν remained undetermined. Nordström now reversed the direction of Einstein's reasoning. Einstein had shown that choosing T as source density enabled the equality of inertial and gravitational mass for complete static systems. Nordström postulated this equality and from it derived Einstein's choice for source density

$$\nu = -\frac{1}{c^2}T$$

and an expression for g

$$g(\phi) = \frac{c^2}{A+\phi}.$$

The constant A could be set arbitrarily as a gauge freedom. Under the natural choice $A = 0$, which yielded the potential ϕ', Nordström's second theory now provided a very simple relationship between the energy E, inertial mass m, and gravitational mass M_g of a complete stationary system

$$E = mc^2 = M_g\phi'.$$

This dependence of the energy and mass of a system on the gravitational potential ϕ' was closer to familiar classical expressions than the corresponding result (6) of Nordström's first theory.

Satisfactory as these results were, they did not yet provide an escape from Einstein's objection to all relativistic scalar theories of gravitation. It is odd that this objection is mentioned nowhere in Nordström's paper, even though a major part of the paper is devoted to developing effects that were able to defeat that objection. These effects emerged from a long series of analyses of different gravitational systems, including Nordström's model of the electron, stressed rods, light clocks, gravitation clocks, and harmonic oscillators. Nordström found that a very wide range of physical quantities would depend upon gravitational potential. These included the lengths of bodies, times of processes, masses, energies, and stresses. When these dependencies were taken into account, it turned out that Einstein's violation of the law of conservation of energy no longer arose.

A simple thought experiment illustrates most simply how the dependence arises in the case of the lengths of bodies and how this dependence defeats Einstein's objection. This is our sixth thought experiment. Nordström attributed the thought experiment to Einstein although Einstein published it nowhere himself. Since Nordström (1913b) was submitted from Zurich, the home of both Einstein and von Laue, this raises the question of precisely who developed the ideas that enable escape from Einstein's objection.

Einstein's thought experiment cuts directly to the heart of the mechanism that allowed a violation of energy conservation in the fifth thought experiment. A body gains gravitational mass upon being stressed. This additional gravitational mass generates energy when the body is lowered into a gravitational field. That gravitational mass disappears when the body is unstressed. If we raise the unstressed body, we create a cycle that yields a net gain in energy. The radiation in the fifth thought experiment actually only plays an incidental role in providing a mechanism for stressing bodies that were to be raised and lowered.

The escape Nordström and Einstein now offered is ingenious. If a stressed body expanded upon being lowered into a gravitational field, then energy would be absorbed as the work required to expand the body against the stresses. Could the expansion be so adjusted that it absorbed exactly all the energy released in the fall of the gravitational mass of the stresses themselves? If so, the construction of an energy-generating cycle would be blocked. Nordström's (1913b, pp. 545–545) account of Einstein's thought experiment shows us that this adjustment is easily achieved (see Figure 6). He wrote:

> Herr Einstein has proved that the dependence in the theory developed here of the length dimensions of a body on the gravitational potential must be a general property of matter. He has shown that otherwise

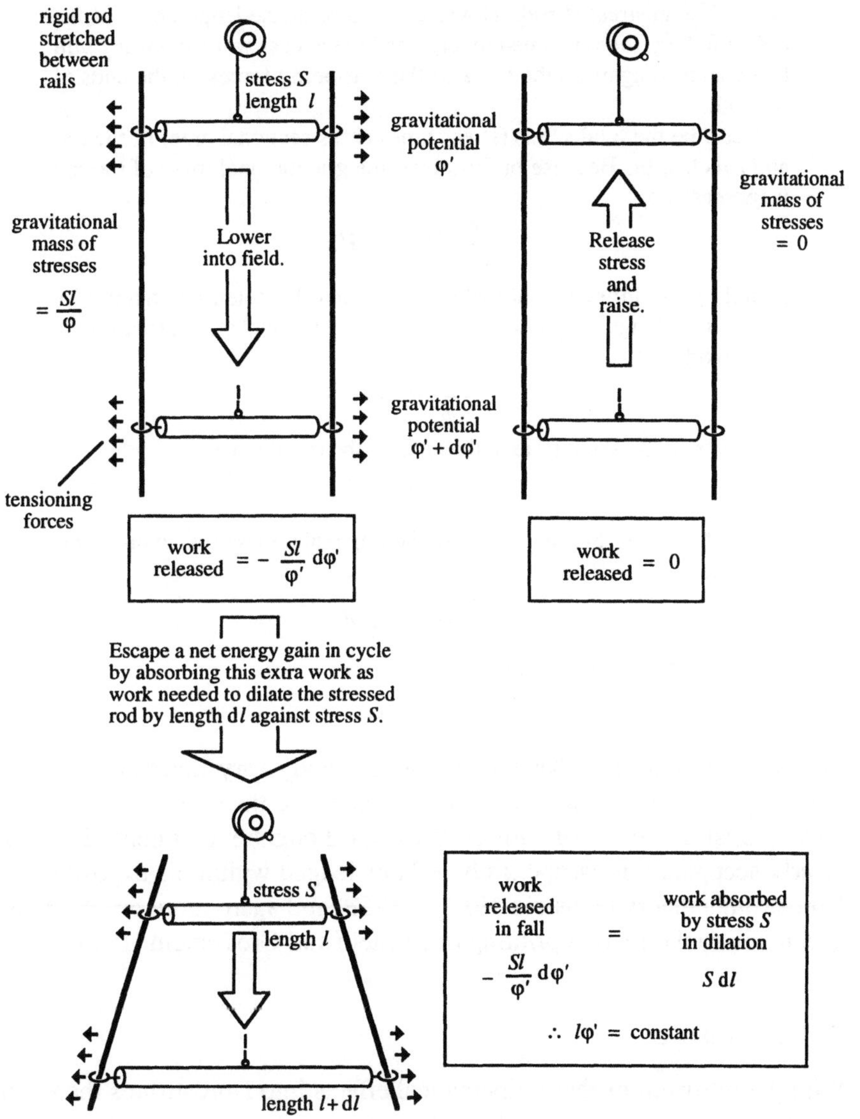

Figure 6. Gravitational potential dependence of length restores energy conservation.

it would be possible to construct an apparatus with which one could pump energy out of the gravitational field. In Einstein's example, one considers a non-deformable rod that can be tensioned movably between two vertical rails. One could let the rod fall stressed, then relax it and raise it again. The rod has a greater weight when stressed than unstressed, and therefore it would provide greater work than would be consumed in raising the unstressed rod. However because of the lengthening of the rod in falling, the rails must diverge and the excess work in falling will be consumed again as the work of the tensioning forces on the ends of the rod.

Let S be the total stress (stress times cross-sectional area) of the rod and l its length. Because of the stress, the gravitational mass of the rod is increased by

$$\frac{g(\phi)}{c^2} Sl = \frac{1}{\phi'} Sl.$$

In falling [an infinitesimal distance in which the potential changes by $d\phi'$ and the length of the rod by dl], this gravitational mass provides the extra work

$$-\frac{1}{\phi'} Sl \, d\phi'.$$

However, at the same time at the ends of the rod the work

$$S \, dl$$

is lost [to forces stressing the rod]. Setting equal these two expressions provides

$$-\frac{1}{\phi'} d\phi' = \frac{1}{l} dl,$$

which yields on integration

$$l\phi' = \text{const}.$$

Thus simply requiring that the length of a body vary inversely with the gravitational potential ϕ' is sufficient to preserve the conservation of energy against the threat of Einstein's earlier thought experiment. Einstein clearly accepted this escape, as he acknowledged within his exposition of Nordström's theory (Einstein 1913, p. 1253) and again more briefly in his addendum to the journal printing of Einstein and Grossmann (1913).

7. Conclusion

With the intrusion of these kinematical effects into Nordström's theory, it ceased to be a conservative, Lorentz covariant theory of gravitation and became more akin to Einstein's own theory, in which gravitation, space, and time were intimately intermingled. Just how close it had come to Einstein's theory was revealed by Einstein and Adriaan D. Fokker in a paper the following February (Einstein and Fokker 1914). Since the times of all processes and the lengths of all bodies were affected equally by the gravitational potential ϕ, the times and spaces of the background Minkowski space-time had ceased to be directly measurable by real rods and clocks. Instead they revealed a non-Minkowskian space-time with the characteristic property that there exist preferred coordinate systems (x, y, z, t) in which the invariant interval is given by

$$ds^2 = \phi^2(dx^2 + dy^2 + dz^2 - c^2\,dt^2). \qquad (13)$$

After postulation of this basic property for space-time, the theory developed in a remarkably similar way to Einstein's theory. The trajectory of a body in free fall in the gravitational field was a geodesic of the space-time. The law of conservation of gravitational and non-gravitational energy-momentum was given by the vanishing of the covariant divergence of the stress-energy tensor. Finally, the field equation of Nordström's second theory proved to be just

$$R = kT,$$

where R is the curvature scalar and k a constant. Einstein was not able to introduce generally covariant field equations based on the Riemann curvature tensor into his own gravitation theory until November 1915.

In 1914, Einstein could not offer decisive grounds for picking between his and this final version of Nordström's theory. The strongest argument he could muster against Nordström's theory was that it failed to satisfy the requirement of the relativity of inertia, a requirement whose essential content would be transformed into Mach's principle. The presence of the preferred coordinate systems (x, y, z, t) in (13) was judged by Einstein as a residual, absolute element that had to be jettisoned if the principle of relativity were to be generalized to accelerated motion.

The three soon-to-be classic tests of general relativity could offer no help in deciding between the two theories. Both Einstein's and Nordström's theory predicted a red shift in light from the sun and of equal magnitude. Unlike Einstein's theory, Nordström's theory predicted no deflection in a beam of starlight grazing the sun. However, the world would still wait five years for Eddington's celebrated expeditions. Finally, accounting for the anomalous motion of Mercury had not yet emerged as a *sine qua non* of any new gravitation theory. Einstein's theory of 1913 actually failed to account for this anomalous motion, a shortcoming that was oddly never mentioned in Einstein's publications of this period. Nordström (1914) analyzed planetary motions according to his theory. He found that it predicted changes in planetary orbits that were very small in comparison with the perturbations due to other planets and thus felt justified in concluding that this theory was "in the best agreement with experience" (p. 1109).

What decisively changed the standards for evaluation of gravitation theories was a result communicated by Einstein (1915) to the Prussian Academy on November 15, 1915. He showed that his gravitation theory, now equipped with generally covariant field equations, was able to account almost exactly for the anomalous advance of Mercury's perihelion. Overnight, the margin of error in astronomical prediction allowed a gravitation theory dropped by at least an order of magnitude. As von Laue noted in his sympathetic review (1917, p. 305), Nordström's theory was no match for Einstein's when it came to Mercury, for Nordström's theory predicted a slight retardation of the planet's perihelion. The failure was now deemed so complete that von Laue did not even bother to report the magnitude of the retardation.

After the excitement of Eddington's eclipse expedition and the public acclaim of Einstein and his theory, the fate of Nordström's theory was sealed. It could offer little competition to the seductive charms of Einstein's theory. By the time of Pauli's authoritative survey (1921, section 50), in less than a paragraph Nordström's theory was dismissed briefly and decisively as a viable gravitation theory.

Notes

[1] M. von Laue to A. Einstein, December 27, 1911, EA 16-008. For further discussion, see Norton (1985, section 4.1).

[2] For philosophical analyses of thought experiments from various perspectives, see Horowitz and Massey (1991), which contains Norton (1986), and see also Brown (1991) and Sorensen (1992).

[3] Einstein to J. Stark, September 25, 1907, EA 22-333.

[4] One of the most informative is Einstein (1933, pp. 286–287).

[5] Here and henceforth, Greek indices will vary over 1, 2, 3, 4 and Latin indices over 1, 2, 3. I will employ the coordinate system $(x_1, x_2, x_3, x_4) = (x, y, z, u = ict)$ as was common in four-dimensional physics in the early 1910s. Summation over repeated indices will be implied.

[6] From the orthogonality of four-velocity U_μ and four-acceleration $\mathrm{d}U_\mu/\mathrm{d}\tau$, we infer from the contraction of (4) with U_μ that

$$0 = F_\mu U_\mu = -m\frac{\partial\phi}{\partial x_\mu}\frac{\mathrm{d}x_\mu}{\mathrm{d}\tau} = -m\frac{\mathrm{d}\phi}{\mathrm{d}\tau},$$

so that $\mathrm{d}\phi/\mathrm{d}\tau = 0$.

[7] In a lecture given on April 14, 1954, according to notes taken by Wheeler (1979, p. 188).

[8] p^0_{ik} is the (three-dimensional) stress tensor.

[9] $\gamma = 1/\sqrt{1 - v^2/c^2}$.

[10] See Norton (1992, section 9), and Janssen (manuscript).

[11] Einstein's analysis did not consider the corresponding exchange of momentum associated with the temporary imbalance of external forces, which would lead to the momentum expression in (8). I add this to my analysis below since it is a trivial and obvious extension of Einstein's original thought experiment.

[12] I follow Einstein in assuming that we are treating a case in which the forces between the charges on the body are small compared with the external forces and can be neglected.

[13] As usual, we have $t = \gamma(t' + (v/c^2)x')$ and $x = \gamma(x' + vt')$, where $\gamma = 1/\sqrt{1 - v^2/c^2}$.

[14] One obvious problem with (9) that Einstein did not mention is that it is ill-defined for source matter that, unlike dust, has no natural rest frame.

[15] Von Laue's (1911a, section 5) definition was unnecessarily restrictive and did not include bodies rotating uniformly about their axes of symmetry. Nordström (1913b, pp. 534–535) quietly extended the analysis to "complete *stationary*" systems, which did include such rotating bodies.

[16] Under Nordström's choice of coordinate system, with $x_4 = ict$, $T_{44} = -$(energy density), whereas under Einstein and Grossmann's (1913) choice of metrical signature $(-, -, -, +)$, $T_{44} = +$(energy density). I have also followed Einstein in simplifying the analysis by ignoring the fact that the total energy of a system must vary with gravitational potential, whereas its gravitational mass will not. Thus the expression for the proportionality of the inertial and gravitational mass of a system must contain a factor that is a function of the gravitational potential. This effect is explicitly incorporated into Nordström's (1913b) second theory through the factor $g(\phi)$, and the proportionality is expressed as relation (12) of Section 6 below. For the analysis of this section and the following, this g factor can be taken as approximately constant and its effect absorbed into other constants in the equations.

[17] This conclusion holds for free radiation, and for this reason there is no gravitational bending of light in Nordström's (1913b) second theory, since it employs T as its source density.

[18] To see this most clearly, imagine that each pair of opposing walls of the box are held together by a slender rod that carries all the stresses needed to hold the walls against radiation pressure. One set of opposing walls and rods forms the set of baffles. Three identical sets can be fitted together to form the cubical box.

References

Abraham, Max (1912). "Relativität und Gravitation. Erwiderung auf einer Bemerkung des Hrn. A. Einstein." *Annalen der Physik* 38: 1056–1058.

Brown, James R. (1991). *Laboratory of the Mind: Thought Experiments in the Natural Sciences*. London: Routledge.

Ehrenfest, Paul (1907). "Die Translation deformierbarer Elektronen und der Flächensatz." *Annalen der Physik* 23: 204–205.

Einstein, Albert (1907a). "Über die vom Relativitätsprinzip gefordete Trägheit der Energie." *Annalen der Physik* 23: 371–384.

—— (1907b). "Über das Relativitätsprinzip und die aus demselben gezogenen Folgerungen." *Jahrbuch der Radioaktivität und Elektronik* 4: 411–462; 5: 98–99.

—— (1912). "Relativität und Gravitation. Erwiderung auf eine Bemerkung von M. Abraham." *Annalen der Physik* 38: 1059–1064.

——— (1913). "Zum gegenwärtigen Stande des Gravitationsproblems." *Physikalische Zeitschrift* 14: 1249–1262.

——— (1915). "Erklärung der Perihelbewegung des Merkur aus der allgemeinen Relativitätstheorie." *Königlich Preussische Akademie der Wissenschaften* (Berlin). *Sitzungsberichte*: 831–839.

——— (1916). "Die Grundlage der allgemeinen Relativitätstheorie." *Annalen der Physik* 49: 769–822; translated without p. 769 as "The Foundation of the General Theory of Relativity" in *The Principle of Relativity*. Hendrik A. Lorentz, Albert Einstein, Hermann Minkowski, and Hermann Weyl. New York: Dover, 1952, pp. 111–164.

——— (1933). "Notes on the Origin of the General Theory of Relativity." In *Ideas and Opinions*. Carl Seelig, ed. Sonja Bargmann, trans. New York: Crown, 1954, pp. 285–290.

Einstein, Albert and Fokker, Adriaan D. (1914). "Die Nordströmsche Gravitationstheorie vom Standpunkt des absoluten Differentialkalküls." *Annalen der Physik* 44: 321–328.

Einstein, Albert and Grossmann, Marcel (1913). *Entwurf einer verallgemeinerten Relativitätstheorie und einer Theorie der Gravitation*. Leipzig and Berlin: B.G. Teubner (separatum). Reprinted with added "Bemerkungen" by Einstein in *Zeitschrift für Mathematik und Physik* 63: 225–261.

Einstein, Albert and Infeld, Leopold (1938). *The Evolution of Physics*. Cambridge: Cambridge University Press.

Horowitz, Tamara and Massey, Gerald, eds. (1991). *Thought Experiments in Science and Philosophy*. Savage, Maryland: Rowman and Littlefield.

Janssen, Michel (manuscript). "Condensers, Contraction and Confusion: Accounts of the Trouton–Noble Experiment in Classical Electrodynamics."

Nordström, Gunnar (1912). "Relativitätsprinzip und Gravitation." *Physikalische Zeitschrift* 13: 1126–1129.

——— (1913a). "Träge und schwere Masse in der Relativitätsmechanik." *Annalen der Physik* 40: 856–878.

——— (1913b). "Zur Theorie der Gravitation vom Standpunkt des Relativitätsprinzip." *Annalen der Physik* 42: 533–554.

——— (1914). "Die Fallgesetze und Planetenbewegungen in der Relativitätstheorie," *Annalen der Physik* 43: 1101–1110.

Norton, John D. (1985). "What was Einstein's Principle of Equivalence?" *Studies in History and Philosophy of Science* 16: 203–246. Reprinted in *Einstein and the History of General Relativity: Einstein Studies, Vol. 1*. Don Howard and John Stachel, eds. Boston: Birkhäuser, 1989, pp. 3–47.

——— (1986) "Thought Experiments in Einstein's Work," presented at the workshop "The Place of Thought Experiments in Science and Philosophy," Center for Philosophy of Science, University of Pittsburgh, April 17, 1986; published in Horowitz and Massey (1991).

——— (1992). "Einstein, Nordström and the Early Demise of Scalar, Lorentz Covariant Theories of Gravitation." *Archive for History of Exact Sciences* 45: 17–94.

Pauli, Wolfgang (1921). "Relativitätstheorie." In *Encyklopädie der mathematischen Wissenschaften, mit Einschluss an ihrer Anwendung. Vol. 5, Physik*, Part 2. Arnold Sommerfeld, ed. Leipzig: B.G. Teubner, 1904–1922, pp. 539–775. [Issued November 15, 1921]. English translation, *Theory of Relativity*. With supplementary notes by the author. G. Field, trans. London: Pergamon, 1958; reprint New York: Dover, 1981.

Sorensen, Roy A. (1992). *Thought Experiments*. New York: Oxford University Press.

Trouton, Frederick T. and Noble, H.R. (1903). "The Mechanical Forces Acting on a Charged Condensor Moving through Space." *Philosophical Transactions of the Royal Society of London* 202: 165–181.

von Laue, Max (1911a). "Zur Dynamik der Relativitätstheorie." *Annalen der Physik* 35: 524–542.

—— (1911b). *Das Relativitätsprinzip*. Braunschweig: Friedrich Vieweg und Sohn.

—— (1917). "Die Nordströmsche Gravitationstheorie." *Jahrbuch der Radioaktivität und Elektronik* 14: 263–313.

Wheeler, John A. (1979). "Einstein's Last Lecture." In *Albert Einstein's Theory of General Relativity*. Gerald E. Tauber, ed. New York: Crown, pp. 187–190.

Out of the Labyrinth? Einstein, Hertz, and the Göttingen Answer to the Hole Argument

Don Howard and John D. Norton

In his lifetime, Einstein became a living oracle. We are told time and time again of lesser-known scientists grappling with overwhelming problems who made the pilgrimage to consult Einstein, perhaps just for encouragement or endorsement, or perhaps in the hope that he might hand them the thread that would lead them out of their labyrinth. Our paper tells the story of a scientist who had become hopelessly lost in a labyrinth of his own making as he struggled with the most important discovery of his life. A correspondent gives him the thread that could be followed out of the labyrinth, but the scientist impatiently dismisses this gift as a confused distraction, only to discover a similar way out a few months later. What makes our story special is that the scientist was not just anyone—it was Einstein himself—and the discovery was general relativity.

The time was 1915. Einstein's correspondent was Paul Hertz, then a physicist working in Göttingen and taking regular part in the activities of the group centered around David Hilbert. The problem was the so-called hole argument, through which Einstein had convinced himself that no physically acceptable version of his still-incomplete general theory of relativity could be generally covariant. We will conjecture that Hertz provided Einstein with a serviceable and sophisticated escape from this ill-fated conclusion, and that Einstein misunderstood and dismissed it, only to arrive at a similar escape a few months later in the form of his point-coincidence argument. Finally, on the basis of an intriguing similarity in wording and timing, we will suggest that Einstein may have drawn immediate inspiration for the final formulation of his point-coincidence argument from another hitherto unrecognized source.

Our argument for our main conclusion will be somewhat unusual, resting, as it does, upon our conjectural reconstruction of letters from Hertz to Einstein on the basis of Einstein's surviving replies to Hertz. Such an approach raises obvious methodological and historiographical questions about the use of evidence that is as much conjectured as discovered. However, in the absence of more direct evidence, our only alternative is to say nothing at all; but this is an issue too interesting and important to pass over in silence.

1. Background: General Covariance Lost and Regained

In the summer of 1915, when our story is set, Einstein's long struggle toward his general theory of relativity was drawing to a close. Roughly two years earlier, he and Marcel Grossmann had published the first outline of the theory, complete in all essential details excepting the gravitational fields equations offered, which were not generally covariant (Einstein and Grossmann 1913). To make matters worse, Einstein soon suppressed his concern over this lack of general covariance by convincing himself that any generally covariant field equations that one might propose must be physically uninteresting. His principal argument for this surprising conclusion was the "hole argument," published in its final and most complete form in Einstein 1914b, pp. 1066–1067 (see Norton 1987; Stachel 1989).

In the hole argument, Einstein considered a "hole," a region of space-time devoid of "material processes" (the stress-energy tensor $T_{ik} = 0$), and a solution g_{ik}, in a coordinate system x^m, of supposedly generally covariant field equations for the metric tensor g_{ik}, given a matter distribution that is nonvanishing only outside the hole. He then showed that the general covariance of the field equations allowed him to construct a second solution, with components g'_{ik}, in the *same* coordinate system x^m, that agreed with the first solution g_{ik} outside the hole but came smoothly to differ from it within the hole. Einstein found the existence of two such solutions in the same coordinate system unacceptable, for he took it to violate the "principle of causality," which seemed here to amount to the requirement that the field and matter distribution outside the hole should determine uniquely the processes or events within the hole. His presumption, apparently, was that there is a unique, real state of affairs within the hole (and elsewhere) that is supposed to be described, uniquely, by a theory of gravitation (see Howard 1992).

In brief, Einstein constructed these two solutions by means of a transformation from the original coordinate system x^m to a new coordinate system $x^{m'}$ that agreed with the original outside the hole but came smoothly to differ from it within the hole. Under this transformation the first solution g_{ik}, in x^m, becomes g'_{ik}, in $x^{m\prime}$, which general covariance guarantees is also a solution of the field equations. To recover the second solution mentioned above, Einstein looked upon the components g'_{ik} as ten functions of the arguments $x^{m\prime}$ and imagined that these arguments were replaced by numerically identical values of the original x^m without changing the functional form of g'_{ik}. The result is two differing solutions of the field equations in the same coordinate system x^m. (See Figure 1.)

It will be important for later discussion to pause here and note that these two solutions have the following characteristic property, although Einstein did not stress this fact: There exist two coordinate systems x^m and $x^{m\prime}$ that agree outside the hole but come smoothly to differ within the hole, such that the components of the second solution, in the coordinate system x^m, are precisely the same functions of the coordinates as are the components of the first solution, in the second coordinate system $x^{m\prime}$.[1]

For example, in the case of the two-dimensional space-time of Figure 2, if the matrix of values of the second solution is $\begin{bmatrix} 1 & 0 \\ 0 & -2 \end{bmatrix}$ at (1, 1) in the first coordinate system, then the matrix of values of the first solution is also $\begin{bmatrix} 1 & 0 \\ 0 & -2 \end{bmatrix}$ at (1, 1) in the second coordinate system. Notice, however, that if (1, 1) are the coordinates of a point p inside the hole, then, by construction, (1, 1) in the second coordinate system will be the coordinates of a *different* point, p', in the hole.

The hole argument forced Einstein to limit the range of coordinate systems used in his theory in such a way that, for any arbitrarily selected region of space-time, he could not use two coordinate systems that agreed outside but came smoothly to disagree within the region. To see how close the covariance of his 1913 theory came to this limit, Einstein defined the notion of the "adapted coordinate system," analyzed most completely in Einstein 1914b. The coordinate system adapted to a given field was defined by a variation principle so contrived that it selected a single coordinate system from all those that came smoothly to agree on the boundary of any given region of space-time. This entails a result that will become important below: *For any region of space-time, it is impossible for there to be two different adapted coordinate systems that come smoothly to agree at the boundary*. Einstein could also show that his 1913 field equations were covariant under transformations between these adapted coordinate systems, so that while these field equations were not generally covariant, they had at least the maximum covariance permitted by the hole argument.[2]

Einstein's failure to offer generally covariant field equations was a great worry and embarrassment to him. His frequent protestations of the unacceptability of generally covariant field equations, however, such as Einstein 1914a, and his publication in October 1914 of a lengthy review article (Einstein 1914b) of the theory suggested that he felt the theory had achieved some stability in its then non-generally covariant formulation.

In late June and early July of 1915, Einstein visited Göttingen and gave six lectures on his theory to a group including David Hilbert, Felix Klein and, more likely than not, Emmy Noether and Paul Hertz. Einstein described this visit to several correspondents. Thus, on August 16, he wrote to Berta and Wander Johannes de Haas: "To my great delight, I succeeded in convincing Hilbert and Klein completely" (EA 70-420).[3] And one month earlier, on July 15, Einstein had reported enthusiastically to Sommerfeld:

> In Göttingen I had the great pleasure of seeing everything understood, down to the details. I am quite enthusiastic about Hilbert. A man of consequence. (EA 21-381; reprinted in Hermann 1968, p. 30)[4]

That report to Sommerfeld, however, also showed that Einstein was not yet entirely reconciled to his new theory. He wrote Sommerfeld that he would prefer not to include one or two papers on his new theory (Einstein 1911b, 1914b) in the collection *Das Relativitätsprinzip*, since none of the current presentations were "complete."

As it turned out, Einstein had been understood in Göttingen even better than he realized. Hilbert was particularly excited, writing to Karl Schwarzschild on July 17, 1915: "During the summer we had here as guests the following: Sommerfeld, Born, Einstein. Especially the lectures of the last on gravitational theory were an event" (quoted in Pyenson 1979a, p. 193, n. 83). The excitement in Göttingen was tempered, however, by a widely shared belief that Einstein's mathematical abilities might not be up to the task of perfecting the new theory of gravitation. Typical of this attitude are a couple of remarks found in Felix Klein's lecture notes on general relativity from the summer of 1916. Thus, on the first day of the lectures, July 15, 1916, Klein remarked to his audience that, in the popular mind, relativity theory was surrounded by a "fog of mystery" [*Nebel der Mystik*], adding:

> Einstein's own way of thinking is partly to blame for this mystery, for it starts out again from the most general philosophical speculations and is guided, above all, more by strong physical instinct than by clear mathematical insight.[5]

More to the point, however, is a remark later in that same lecture, in the middle of a section entitled "On the Choice of Coordinates Encountered in Einstein." In Einstein's new theory, Klein tells his students, we enter upon the terrain of arbitrary coordinates, "familiar" to us from the work of Lagrange, Gauss, and Riemann, where the $g_{\mu\nu}$ and the ds^2 must be treated according to the rules of Ricci's absolute differential calculus, or "more

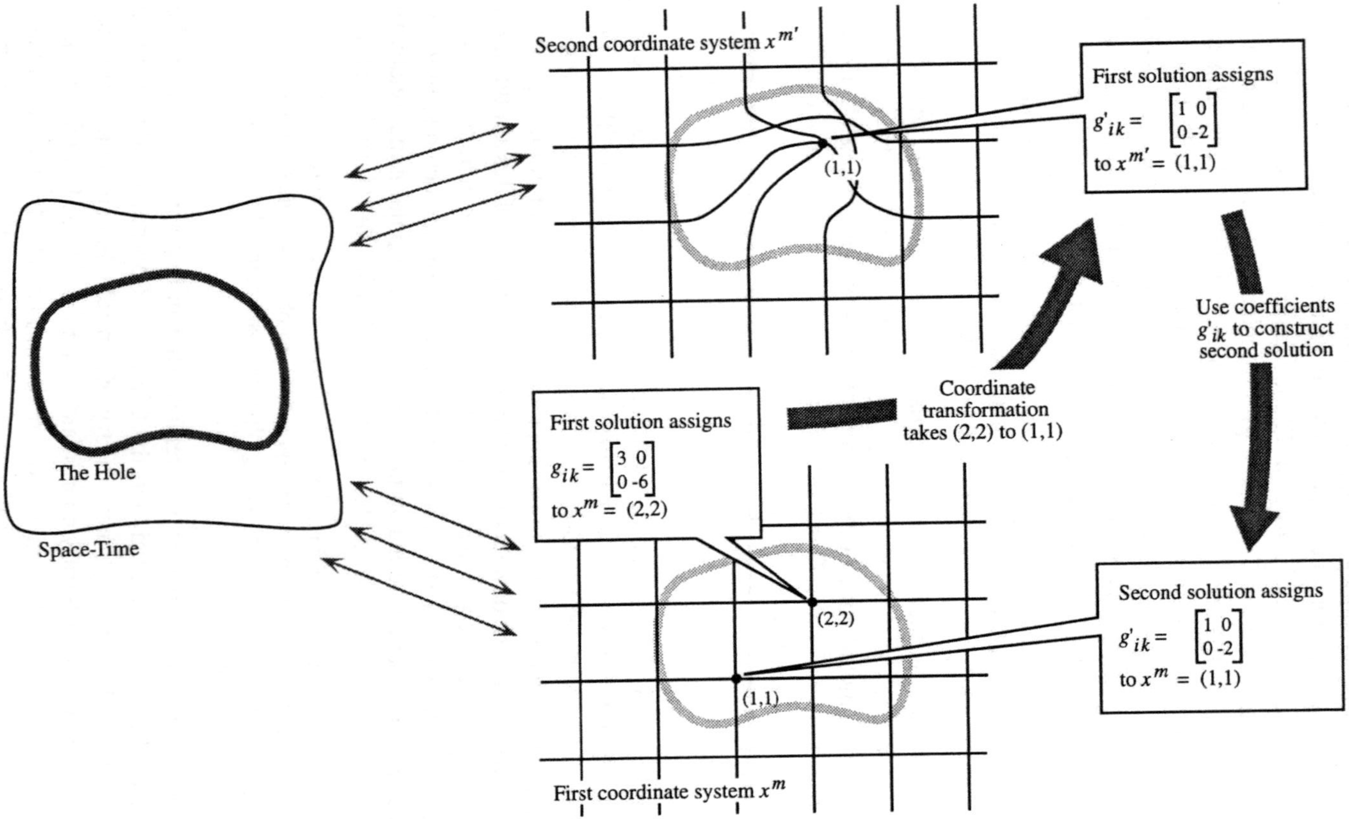

Figure 1. Construction of the two solutions of the hole argument.

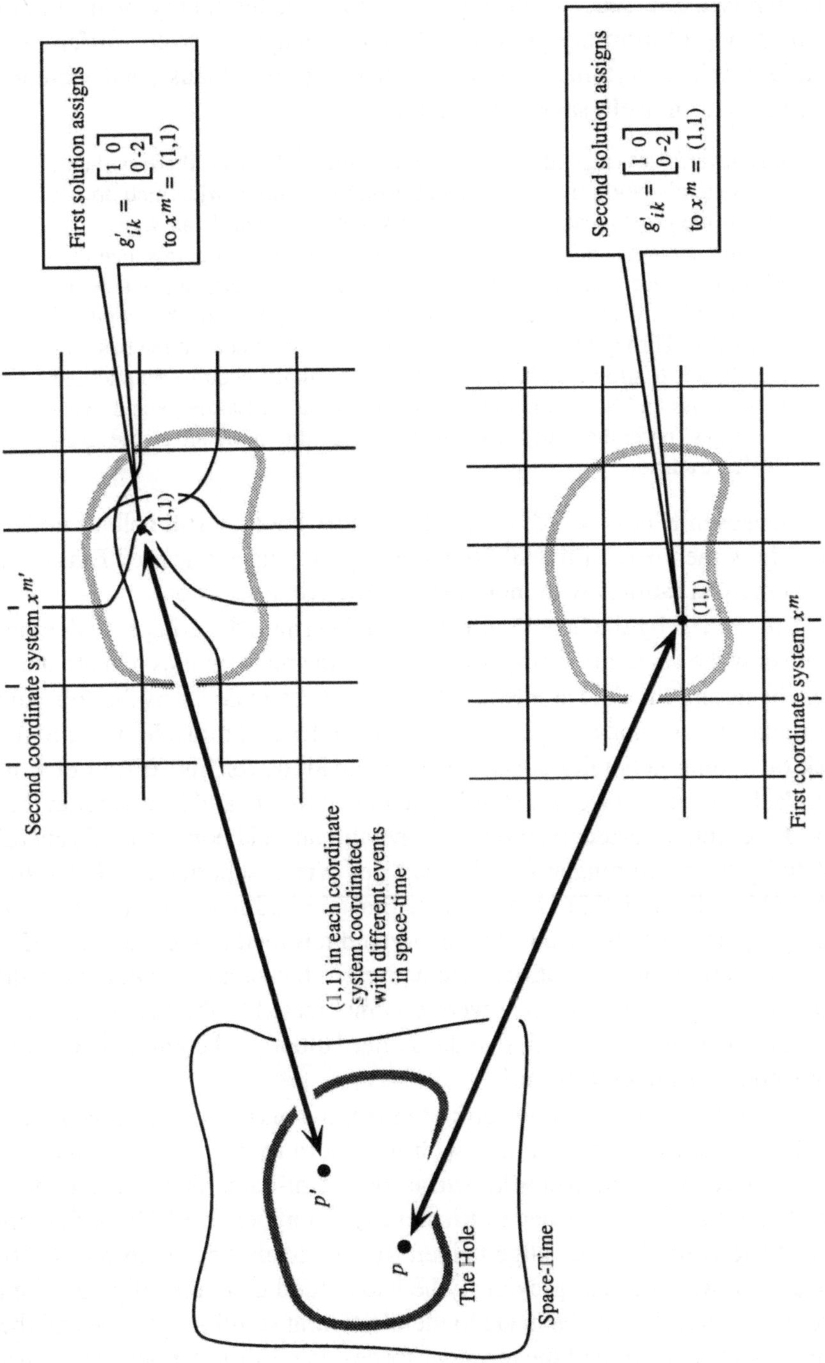

Figure 2. Properties of the two solutions of the hole argument.

objectively expressed," according to the rules of the theory of invariants of the group of arbitrary point transformations applied to the differential invariant ds^2. Everything we learned about Lagrange, Gauss, and Riemann may be clear in itself, says Klein. Still,

> It is nevertheless a good idea to explain it further, because there are here, in Einstein's work, imperfections [*Unvollkommenheiten*], which do not impair the great ideas in his new theory, but hide them from view.
>
> This is connected with the repeatedly mentioned circumstance that Einstein is not innately [*von Hause aus*] a mathematician, but works rather under the influence of obscure [*dunkelen*], physical-philosophical impulses. Through his interaction with Grossmann and on the basis of the Zurich tradition he has, to be sure, gradually become acquainted with Gauss and Riemann, but he knows nothing of Lagrange and over-estimates (parenthetically) Christoffel, under the influence of the local Zurich tradition.[6]

One senses in Klein's words a hint of jealousy, but they still help us understand how members of the Göttingen group may have regarded Einstein's mathematical failings with more than a little condescension.

Undeterred by the hole argument, and determined, perhaps, to demonstrate how the vaunted Göttingen expertise at the mathematics of mathematical physics might yield dividends of a kind not yet achieved by the "obscure physical-philosophical impulses" of Einstein, Hilbert himself turned to the task of finding generally covariant field equations for his version of Einstein's theory, a fusion of Einstein's gravitation theory and Mie's matter theory. He communicated the modern gravitational field equations of general relativity to the Göttingen Gesellschaft der Wissenschaften on November 20, 1915 (Hilbert 1915). Meanwhile, Einstein had lost confidence in the lack of general covariance of his theory and returned to the quest for generally covariant field equations. He arrived at the same gravitational field equations as Hilbert, and they were communicated to the Prussian Academy on November 25, 1915, five days after Hilbert had communicated the same equations in Göttingen.[7]

Einstein soon turned to the task of informing his correspondents of how he reconciled his hole argument with his return to general covariance by means of a consideration now known as the "point-coincidence argument."[8] The latter was first published in Einstein's comprehensive 1916 review article, "Die Grundlage der allgemeinen Relativitätstheorie" (Einstein 1916, pp. 117–118). Whereas previously he had argued that generally covariant equations typically can be made to yield different solutions for one and the same coordinatization of the physical space-time, Einstein now argued that while the two solutions g_{ik} and g'_{ik} may be mathematically distinct, they are not physically distinct, for both solutions catalogue the identical set of space-time coincidences, which exhaust the reality captured by the theory. Thus, Einstein wrote to Paul Ehrenfest on December 26, 1915:

> The physically real in the world of events (in contrast to that which is dependent upon the choice of a reference system) consists in *spatiotemporal coincidences.** Real are, e.g., the intersections of two different world lines, or the statement that they *do not* intersect. Those statements that refer to the physically real therefore do not founder on any univocal [*eindeutige*] coordinate transformation. If two systems of the $g_{\mu\nu}$ (or in general the variables employed in the description of the world) are so created that one can obtain the second from the first through mere space-time transformation, then they are completely equivalent [*gleichbedeutend*]. For they have all spatiotemporal point coincidences in common, i.e., everything that is observable.
>
> *)and in nothing else! (EA 9-363)

An example of these space-time coincidences would be the collision of two point-masses.

We illustrate Einstein's point-coincidence argument in a way that will be suggestive below. Let two point-masses originate at a point-event q outside the hole, separate, and then collide at some point-event within the hole. See Figure 3. According to the second solution, g'_{ik}, the particles will collide at the point[-event] with coordinates (1, 1) in the first coordinate system, x^m. According to the first solution, g_{ik}, the particles will collide at the point with coordinates (1, 1) in the second coordinate system, $x^{m'}$. As illustrated in Figure 2, Einstein had earlier assumed that the two sets of coordinates would represent *different* point[-event]s, p and p', in the physical space-time. He now understands that, on the contrary, they must represent the same point[-event], because the two sets of trajectories agree in all physically significant quantities and thus cannot pick out physically different point[-event]s. For example, measurements of physical time elapsed along the trajectory qap as determined by the first solution g_{ik} would be identical to that along qap' as determined by the second solution g'_{ik}.[9]

2. Letters from Paul Hertz

Einstein later recalled the intense emotions that simmered and boiled within himself through the years of his struggle with general covariance when he wrote of the episode: "But the years of anxious searching in the dark, with their intense longing, their alternations of confidence and exhaustion and final emergence into the light—only those who have experienced it can understand that" (Einstein 1934, pp. 289–290). Into this emotional

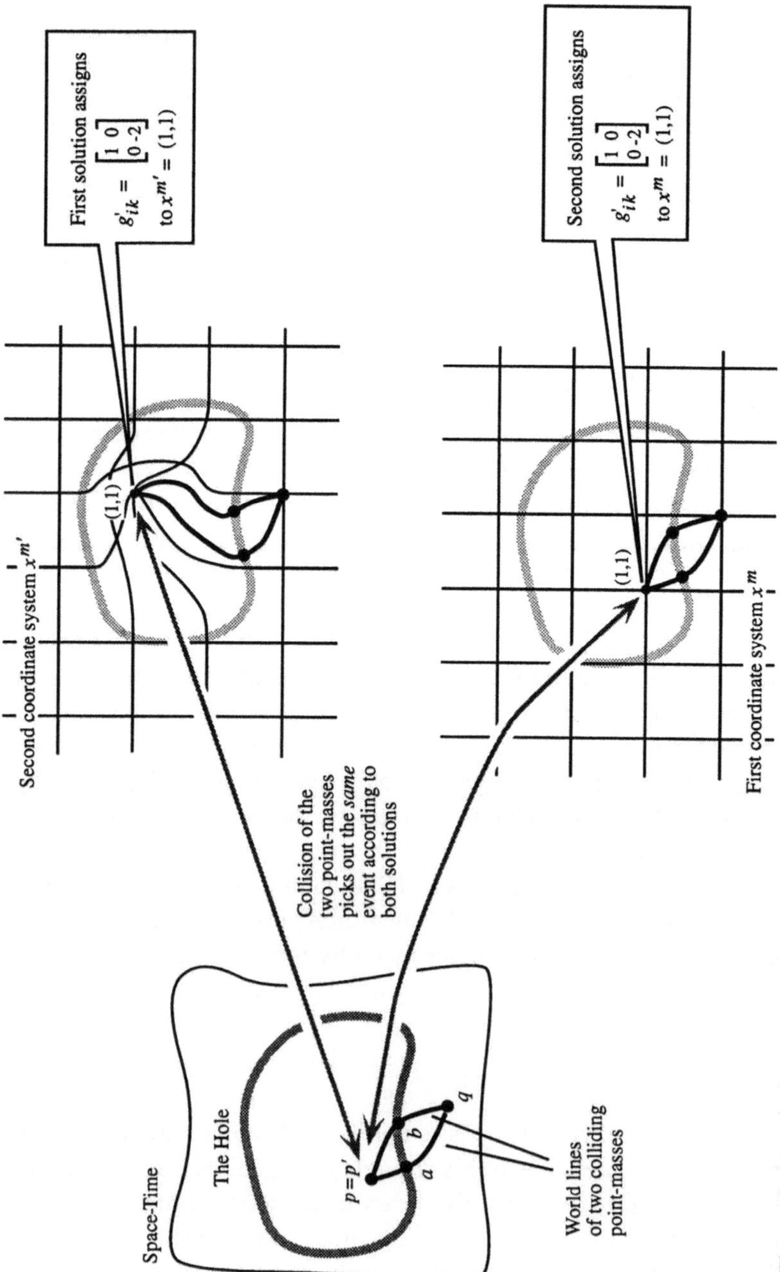

Figure 3. Resolution of the hole argument.

and intellectual cauldron around August 1915 was added an exchange in correspondence with Paul Hertz, just a few months before the struggle drew to its dramatic close that November.

Hertz was born in 1881 in Hamburg. In 1915 he was a *Privatdozent* at Göttingen and a member of the group clustered around Hilbert and Klein. He had taken a degree at Göttingen in 1904 under Max Abraham, with a dissertation on discontinuous movements of an electron (Hertz 1904).[10] After publishing a few additional studies on electron theory, he turned his attention to the foundations of statistical mechanics, an interest that culminated in his seminal 1916 monograph in the *Repertorium für Physik* (Hertz 1916), and also led to his acquaintance with Einstein. This acquaintance was a direct result of Hertz's critical remarks (Hertz 1910) on Einstein's early papers on the subject (Einstein 1902, 1903, 1904), remarks to which Einstein replied in a short note in the *Annalen* in 1911 (Einstein 1911a). They had begun corresponding by August 1910 and had become personally acquainted no later than early September 1910, at a meeting of the Schweizerische Naturforschende Gesellschaft in Basel.[11] Hertz was by this time acquainted with several of Einstein's closer friends and colleagues, most importantly Paul Ehrenfest, who had been a student in Göttingen at the same time as Hertz,[12] and Jakob Laub, another fellow student from Göttingen, who was a colleague of Hertz's in Heidelberg from 1909 to 1911.[13] In 1921, Hertz finally received an appointment as Ausserordentlicher Professor in Göttingen, the same year that he and Moritz Schlick published their influential edition of Helmholtz's epistemological writings (Helmholtz 1921). And in later years, Hertz turned his attention to various topics in the philosophy of science, including pioneering studies, very much in the Göttingen tradition, of the formal axiomatics of scientific theories.[14] Einstein provided a letter of recommendation for Hertz after his emigration to the United States (EA 12-221). He died in Philadelphia in 1940.

We do not know for certain that Hertz was present when Einstein lectured in Göttingen in late June and early July of 1915. Given the nature of the previous relationship between Hertz and Einstein, given Hertz's role in the group around Hilbert, and given the character of Hertz's correspondence with Einstein later that summer, it is more than likely, however, that he was present.

We know of the letters that Hertz wrote to Einstein only because Einstein's replies still exist (EA 12-201 and EA 12-203). Einstein's letter EA 12-203 is dated "22. VIII" (August 22). The content is compatible only with the years 1913–1915. The year must be 1915 because of the mention in a postscript of a coming visit to Zurich ("Aug. 26 to about September 15"), the address of his friend Heinrich Zangger being given for correspondence. Einstein made a visit to Zurich fitting this description in 1915.[15]

Einstein's letter is written in a friendly and encouraging tone. It reflects on the great problems Einstein had faced in finding a way to restrict the coordinate systems of his theory and sketches the difficulties still facing the theory in this area. The letter begins:

> One who has himself poked about so much in the chaos of possibilities can understand very well your fate. You haven't the faintest idea what I, as a mathematical ignoramus, had to go through until I entered this harbor.

And about his specific restriction to "adapted" coordinates, he comments:

> How can one pick out a coordinate system or a group of such? It appears not to be possible in any way simpler than that which I have chosen. I have groped about and tried everything possible.... The coordinate restriction that was finally introduced deserves particular confidence because it can be brought into connection with the postulate of the complete determination of events.

This last remark alludes to the fact that adapted coordinate systems were first introduced by Einstein in order to block the conclusion of the hole argument.

The letter's primary purpose, however, is to respond encouragingly to an idea of Hertz's alluded to in the first paragraph, which presumably concerns the restriction of the coordinate systems. Hertz's idea is presumably also the one that Einstein refers to in both the opening sentence—"A surface-theoretical interpretation of preferred systems would be of very great value"—and the closing sentence of paragraph five—"Perhaps one could get an overview on the question if one succeeded in finding the geometrical interpretation for which you seek"—for such an interpretation is not given or even mentioned by Einstein anywhere else in the letter. And Einstein's other letter, EA 12-201, contains a response to a proposal by Hertz that is cast in the older language of the theory of two-dimensional Gaussian surfaces.[16]

Einstein's EA 12-201 is dated "Berlin, Saturday" but, because of the close similarity of content, it was quite plausibly written at about the same time as EA 12-203. The earliest possible date is August 14, since Hertz's son, Hans, who is mentioned at the end of the letter, was born on Sunday, August 8.[17] The letter was probably written no later than about Saturday, October 9, since it betrays no doubts on Einstein's part about the restricted covariance of the Einstein–Grossmann (1913) theory, whereas by October 12 Einstein is writing to Lorentz that he now realizes that something is amiss with the theory.

The letter responds to another proposal by Hertz, but, as we shall see, it is written in a very different tone. The letter is at times impatient, discouraging and almost hostile—Einstein did not like Hertz's proposal! On the basis of Einstein's reply in EA 12-201, we reconstruct Hertz's proposal to amount to an escape from the hole argument, coupled with a proposal for setting up generally covariant gravitational field equations. The reconstruction that follows is the only one we have found that is compatible with the entirety of Einstein's response.

At this point, some readers might like to scan ahead and read the letter EA 12-201, which is quoted in full in Section 4, in order to see the raw material upon which our reconstruction is based. Readers who like puzzles might even want to try to build their own reconstruction before reviewing the one we offer below in Section 3.

3. Our Reconstruction of Hertz's Proposed Escape from the Hole Argument

Hertz tried to show Einstein that he should not be troubled by the differences between the two solutions considered in the hole argument. He considered the hole argument for the case of a two-dimensional Gaussian surface. We would now write the line element of such a surface in the quadratic differential form $\mathrm{d}s^2 = g_{11}(\mathrm{d}x^1)^2 + 2g_{12}\,\mathrm{d}x^1\,\mathrm{d}x^2 + g_{22}(\mathrm{d}x^2)^2$, where Hertz used the older notation introduced by Gauss, wherein one writes $\mathrm{d}s^2 = E\,\mathrm{d}u^2 + 2F\,\mathrm{d}u\,\mathrm{d}v + G\,\mathrm{d}v^2$. In the case of variable curvature, this geometry seems to allow the defining of a special coordinate system (u, v), whose curves are the curves of constant curvature and of maximum curvature gradient, and are thus adapted to the geometry. We shall call such systems "Hertz-adapted" to avoid confusing them with Einstein's "adapted" coordinate systems. Presumably such coordinates were proposed because they would be defined in terms of invariant features of the surface and because they might be proved to exist for spaces of both positive and negative curvature, unlike isometric coordinates.

Hertz examined the two solutions of the hole argument in the way outlined in Section 1 above. He considered one solution with coefficients E, F, and G in his original coordinate system (u, v) and the other with coefficients E^x, F^x, and G^x in the second coordinate system (u^x, v^x) so that the E, F, and G are the same functions of the variables u and v as the functions E^x, F^x, and G^x are of the variables u^x and v^x.[18] Moreover, Hertz ensured that the coordinate system (u, v) is Hertz-adapted to the geometry represented by E, F, and G, which entails that the coordinate system (u^x, v^x) is also Hertz-adapted to the geometry represented by E^x, F^x, and G^x.

He then asked after the nature of the underdetermination of the geometry revealed by the admissibility under general covariance of the two solutions constructed in the hole. To do so, he asked after the geometry within the hole according to the two solutions at two points that correspond in the sense that the coordinates of the first point in the first coordinate system (u, v) are numerically equal to the coordinates of the second point in the second coordinate system (u^x, v^x). To find the points, one must follow the two coordinate curves corresponding to the coordinate values selected and pursue them until they meet in the hole. Since the two coordinate systems are Hertz-adapted to superficially different geometries, the coordinate curves must diverge upon entering the hole, according to whether the system was adapted to the first or second solution of the field equations. For the coordinate system adapted to the first solution, the curves would meet at the point $P(u, v)$. For the coordinate system adapted to the second solution, the curves would meet at the point $P^x(u^x, v^x)$. See Figure 4, which is our rendering of the diagram Einstein gives in his letter (which is reproduced as Figure 5).

But what are the differences between the two solutions revealed by the construction? Hertz could point to no geometrically significant differences. Spelling out the argument in a way that employs the equations Einstein writes in his letter EA 12-201, the points selected by the construction would have the same coordinate values in each of the geometrically significant Hertz-adapted coordinate systems so that

$$u^x = u \quad \text{and} \quad v^x = v.$$

Moreover, the geometries at each point in the corresponding solutions are the same. For if E, F, and G are the coefficients assigned by the first solution to P, and if E^x, F^x, and G^x are the coefficients assigned by the second solution to P^x, then the geometries at the two points are the same in so far as $E^x = E$, $F^x = F$, and $G^x = G$.[19]

Perhaps Hertz might now have said that the two solutions are geometrically the same in every respect, for these identities would hold for corresponding points covering every point of both solutions. We can think of each solution as representing a different geometric surface. The construction shows how one of them can be mapped into the other by the map that takes point P to point P^x while preserving all geometric properties. In modern language, the two are isomorphic.

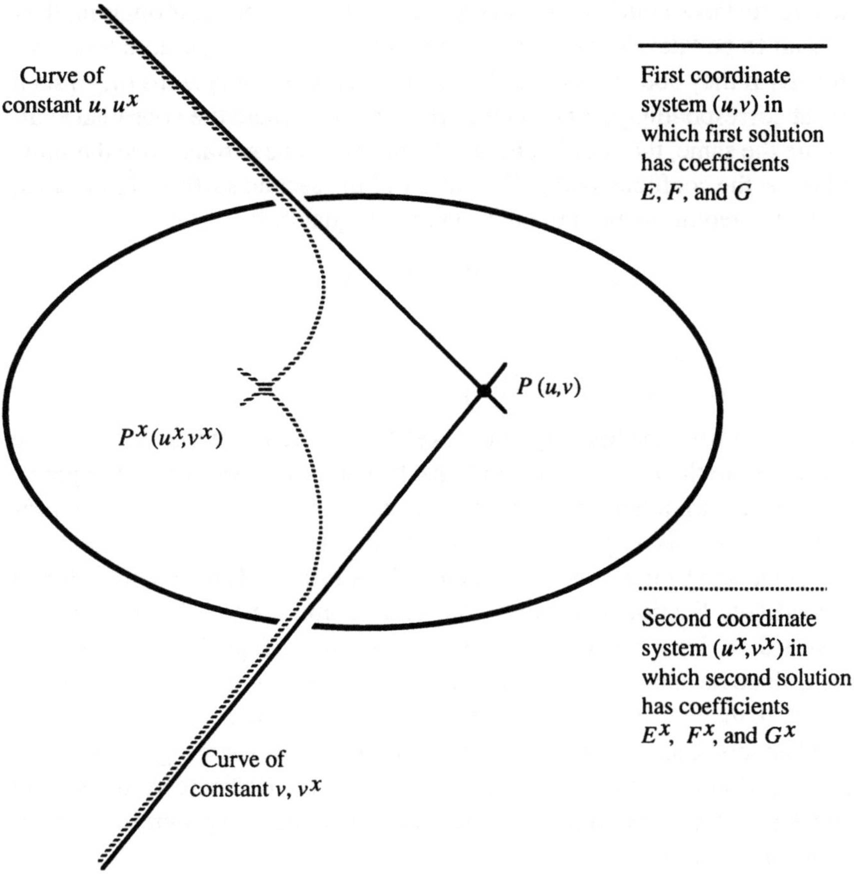

Figure 4. Interpretation of figure in Einstein's letter (cf. Figure 5).

We might rephrase this last point using the only direct quotation Einstein gives of Hertz: Since the two solutions amount to the same surface geometrically, we merely recall that, by the construction, this surface "is developable [i.e., isomorphically mappable] into itself," a clumsy but intelligible way of making the point. This usage of the term "developable" as meaning isomorphically mappable was standard at the time and was even applied to precisely the case Hertz treats using exactly the same set of equations.

Consider, for example, the discussion of two two-dimensional Gaussian surfaces embedded in a three-dimensional space that is found in Johannes Knoblauch's *Grundlagen der Differentialgeometrie* (Knoblauch 1913, pp. 121–124), then regarded as a standard text in Göttingen.[20] If the two surfaces could be laid upon one another without deformation, they are said to be "developable onto one another." The two surfaces have this property if they both admit two-dimensional coordinate systems (u, v) such that at corresponding points on the two surfaces, where the coordinate values are the same, the coefficients E, F, and G of one surface have the same values as the coefficients E_1, F_1, and G_1 of the second surface. Knoblauch wrote this requirement in the now-familiar equations:

$$E_1 = E, \quad F_1 = F, \quad G_1 = G.$$

4. Einstein's Immediate Response

The escape from the hole argument sketched above is obviously very close in strategy to the escape Einstein himself would offer shortly as the point-coincidence argument, but Einstein's immediate response to Hertz's proposal was just a list of protests and complaints. Einstein took Hertz-adapted coordinates to be the same as the adapted coordinates Einstein himself had defined (see Section 1 above). The letter from Einstein began with the protest that Hertz had misrepresented Einstein's adapted coordinate systems, since he had failed to retain the crucial property stressed in Section 1 above, namely that two different (Einstein-)adapted coordinate systems could not come smoothly to agree on the boundary of some region of space-time. And in any case—whether or not the two coordinate systems were adapted—they were supposed to have properties that, in general, could not obtain. Einstein wrote:

> Berlin, Saturday
>
> Dear Herr Hertz,
>
> If I have understood your letter correctly, then you make a completely erroneous representation of that which I call "adapted coordinate systems." How do you come to require that a pair of coordinate systems [Figure 5 = figure from Einstein's letter] should exist, such that for
>
> $$u^x = u$$
> $$v^x = v$$
>
> one has also
>
> $$E^x = E$$
> $$F^x = F$$
> $$\langle G^x = G \rangle\, \phi^x = \phi$$
>
> and over and above this they agree on the boundary of the region?
>
> I am rather convinced that (excepting perh.[aps] quite special fields) *this is never allowed to be possible.* I have never posited the existence of systems equivalent in this sense.[21]

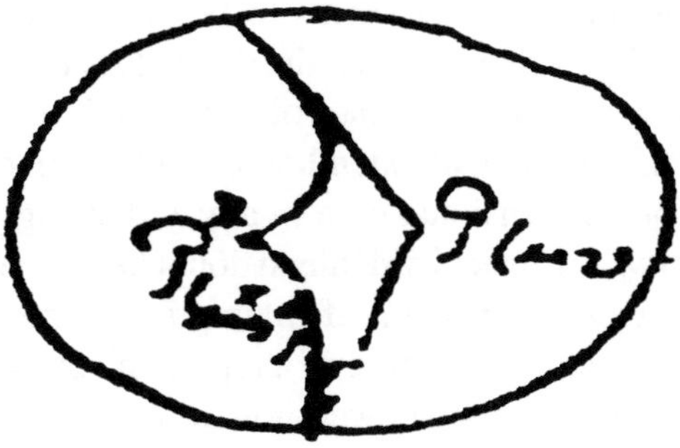

Figure 5. Diagram in Einstein to Hertz, "Berlin, Samstag" [1915] (EA 12-201).

We can only conjecture about how Einstein came to see the adapted coordinates of Hertz's proposal as being the same as the adapted coordinates he himself had defined for his 1913 theory. Both would use the term "adapted" naturally as an appropriate term for coordinate systems that they define in a way that responds to the geometry of the metric field, but it is hard to see that the use of the term alone would be sufficient to lead to this misunderstanding. Recall that in EA 12-203 Einstein had encouraged Hertz in his attempts to find a "surface-theoretic interpretation" of the preferred systems of coordinates of Einstein's theory. If EA 12-203 was written before EA 12-201, we could well imagine Einstein anticipating such a proposal from Hertz when he received EA 12-201. Or, even if EA 12-201 did predate EA 12-203, Hertz himself might have thought his adapted coordinates would serve as the surface-theoretic interpretation of Einstein's adapted coordinates and offered them as such. Finally, a minor factor that might well be crucial in such circumstances: Einstein complains later in the letter that he cannot read Hertz's handwriting on page five of his letter. We might well wonder, then, how clearly written the other pages were.

Einstein's more general complaint about the inadmissibility of the two coordinate systems (u^x, v^x) and (u, v) is readily explicable. All he need assume is that both coordinate systems with their components (E^x, F^x, G^x) and (E, F, G) are coordinate systems and components of the *same* field, not of two different fields as is crucial to both the hole argument and the proposal of Section 3 above. (Perhaps this is already assumed in Einstein's objection that the two systems cannot both be adapted coordinate systems.) As Einstein points out, only quite special fields can be transformed in the way indicated. A coordinate transformation in general produces a quite different set of components for the field that will fail to match in the indicated way.

Einstein continued in what seems to be an attempt further to worry Hertz's proposal. He pointed out that the defined special coordinate system would become degenerate in the case of a space of constant curvature and then mentioned the problem of extending the definition of these coordinate systems to the four-dimensional case in a way that suggested some doubt about its feasibility. If Einstein did intend doubt here, he was shortly proven wrong about the general program of finding four-dimensional coordinate systems that fit the natural structure of a region of space-time, for less than two years later Kretschmann showed how a four-dimensional coordinate system could be constructed in general relativity from curvature invariants (Kretschmann 1917, pp. 592–599).[22] The search for coordinates somehow "adapted" to the intrinsic geometry of the space was, in any case, characteristic of the Göttingen approach to general relativity, as reflected in Hilbert's employment of what he termed "Gaussian coordinates" (Hilbert 1916, pp. 58–59), which are now commonly designated geodesic normal coordinates.[23] The passage quoted above continues thus:

> Independently of this, I understand how you establish a special coordinate system on a two-dimensional manifold by curves of constant curvature and those of maximal curvature gradient. What is problematic [*verdächtig*] about this, however, is that, in regions of constant curvature, the ⟨surfaces⟩ curves (or surfaces) of constant curvature are shifted infinitely far away from one another. The difference, in principle, of the two coordinates that have been introduced is also problematic. You could, nevertheless, attempt to see whether such a thing can be done in a four-dimensional manifold.

Hertz had apparently also coupled his analysis with a proposal for a generally covariant field equation. Einstein replied sharply, asking whether or not Hertz agreed with the need to restrict the covariance of his theory, which again suggests that Hertz had been less than clear in explaining that the proposal, as outlined in Section 3, was intended as an escape from the hole argument. Einstein wrote:

> I have not understood the proposal for the setting-up of a gravitation law, because I cannot read your writing on page 5. After all, I have said in my work that a usable gravitation law is not allowed to be generally covariant. Are you not in agreement with this consideration?

Einstein then returned to his earlier objection about the two coordinate systems that Hertz had introduced and closed with these words:

> So once again: I would not think of requiring that the world should be "developable onto itself," and I do not understand how you require such a dreadful thing of me. In my sense, there is certainly a huge manifold of adapted systems that do *not*, however, agree on the boundary.

With best regards to you, your wife, and your gentleman son, who is already surprisingly affable and fond of writing, I remain, riveted upon your further communications, yours

A. Einstein

Einstein had understood, in effect, that Hertz required the transformation relating the two coordinate systems to be an isometry of the surface, so that he could say that the surface could be developed onto itself by the transformation. As Einstein had pointed out, surfaces admitting such isometries are exceptional and, in any case, the transformation could not be between Einstein's adapted coordinate systems, since such systems would never agree on the boundary of the region in the way Hertz required.

5. Einstein's Eventual Assimilation of the Lessons Hertz Tried to Teach Him

Even though Einstein's immediate response to Hertz was so prickly and defensive, he eventually came to appreciate and advocate Hertz's central point: If a system is developable onto another, the two represent the same reality. This advocacy is nowhere more in evidence than in Einstein's correspondence with Ehrenfest in late December and early January 1916. Ehrenfest was reluctant to accept the generally covariant form of the theory of gravitation announced by Einstein in November 1915, and he pressed his reservations by reminding Einstein, as had other correspondents, of the earlier hole argument. More specifically, in a letter that no longer exists

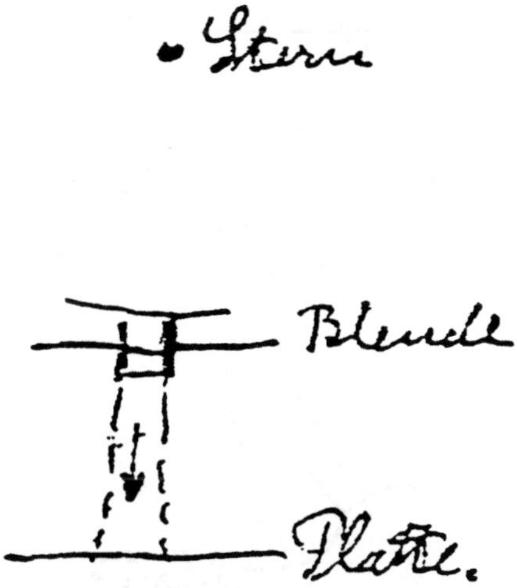

Figure 6. First diagram in Einstein to Ehrenfest, January 5, 1916 (EA 9-372).

from late December 1915, Ehrenfest evidently asked Einstein to consider a situation in which light from a distant star passes through one of Einstein's notorious holes and then strikes a screen with a pinhole in it that directs the light onto a photographic plate.[24] Given that generally covariant equations allow for two different solutions, $g^{A}_{\mu\nu}$ and $g^{B}_{\mu\nu}$, inside the hole, Ehrenfest asks how we can be sure that light from the distant star following different paths through the hole determined by the two different solutions can be guaranteed to strike the same place on the plate.[25]

We quote the relevant section of Einstein's detailed answer in its entirety:

> In the following way you obtain all of the solutions that general covariance brings in its train in the above special case. Trace the little figure above [see Figure 6] on completely deformable tracing paper. Then deform the tracing paper arbitrarily in the paper-plane. Then again make a copy on stationery. You obtain then, e.g., the figure [Figure 7]. If you now refer the figure again to orthogonal stationery-coordinates, then the solution is mathematically a different one from before, naturally also with respect to the $g_{\mu\nu}$. But physically it is exactly the same, because even the stationery-coordinate system is only something imaginary [*eingebildet*]. The same points of the plate always receive light....
>
> What is essential is this: As long as the drawing paper, i.e., "space," has no reality, the two figures do not differ at all. It is only a matter of "coincidences," e.g., whether or not the point on the plate is struck by light. Thus, the difference between your solutions A and B becomes a mere difference of representation, with physical agreement. (EA 9-372)

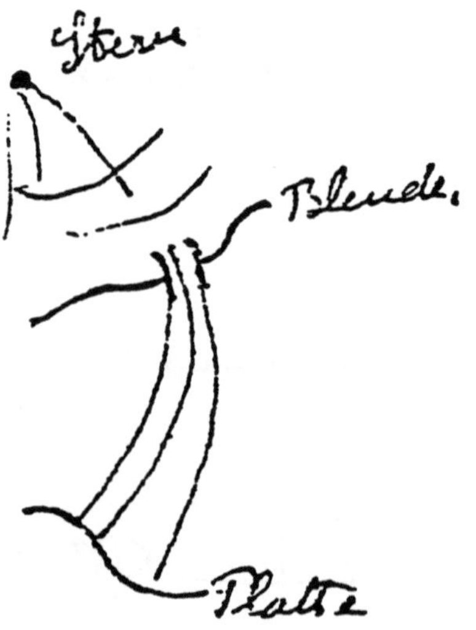

Figure 7. Second diagram in Einstein to Ehrenfest, January 6, 1916 (EA 9-372).

Aside from the talk of "coincidences," Einstein's point here is exactly Hertz's, namely, that one can have two solutions that are mathematically different, while being physically or geometrically (they come to same thing in this context) indistinguishable.

6. Hilbert's Escape from the Hole Argument

The reconstruction of what Hertz wrote to Einstein as conjectured in Section 3 above was based on an analysis of Einstein's letters. We then sought some independent evidence for our conjecture, but the existing documentation provided none. There is additional correspondence between Einstein and Hertz from early October 1915, concerning whether or not Hertz should resign his membership in some society seemingly concerned with political matters. And something that Einstein wrote in this connection so irritated Hertz that he threatened to break off the correspondence, an eventuality that Einstein earnestly sought to avoid.[26] Further communication was no doubt made even more difficult by the fact that Hertz soon found himself in the military, posted to a flight school in Posen.[27]

If we could not confirm independently that Hertz suggested such an escape from the hole argument, then, we asked ourselves, could we at least determine whether or not such an escape was common knowledge in Göttingen at the time so that Hertz was either initiating or reflecting a standard response? To our surprise and pleasure we found—after we had completed the construction of the conjecture of Section 3—that Hilbert had offered almost exactly the escape in the second of his famous papers on general relativity and the foundations of physics (Hilbert 1916).

The relevant remarks are found in Hilbert's somewhat labored discussion of the "causality problem" in general relativity, the designation Einstein often used for the hole argument (Hilbert 1916, pp. 59–63).[28] Hilbert points out that the Cauchy problem is not well posed for his own generally covariant version of general relativity (Hilbert 1915). That theory has fourteen independent variables—the ten gravitational potentials, $g_{\mu\nu}$, and the four electromagnetic field potentials, q_s—but the gravitational field equations and Maxwell's equations provide only ten independent field equations. Hilbert illustrates this underdetermination with a pair of solutions, the first of which represents an electron at rest throughout all time, with the gravitational and electromagnetic fields everywhere time-independent. In a manipulation reminiscent of the hole argument, the second solution is obtained by a coordinate transformation that is the identity for the time coordinate $x_4 \leq 0$, but comes to differ for $x_4 > 0$. In the second solution, the electron adopts a nonvanishing velocity and the fields become time-dependent after $x_4' = 0$. While the possibility of such different solutions at first seems to threaten the principle of causality, however, Hilbert proposes to rescue it by offering a definition of what it means for an object, a law, or an expression to be "physically meaningful." According to Hilbert, something should be regarded as physically meaningful only if it is invariant with respect to arbitrary transformations of the coordinate system. And in this sense, the causality principle is satisfied, since, he asserts, all physically meaningful expressions, which is to say, all invariant expressions, are unambiguously determined by the generally covariant equations.[29]

It is at this point in Hilbert's exposition that his argument converges upon what we believe Hertz proposed to Einstein. Hertz, we believe, exploited a geometrically adapted coordinate system to display the essential agreement between the two solutions E, F, G and E^x, F^x, G^x. Hilbert summarized his basic claim and then promised to prove the claim by exploiting the geometrically adapted Gaussian coordinate system:

> *The causality principle holds in this sense*:
>
> From a knowledge of the 14 physical potentials, $g_{\mu\nu}$, q_s, follow all assertions about them for the future necessarily and uniquely, *insofar as they have physical significance*.
>
> In order to prove this claim, we employ the Gaussian space-time coordinate system. (Hilbert 1916, p. 61; Hilbert's emphasis)

Hilbert begins by noting that the selection of Gaussian coordinates provides the four extra constraints needed to ensure that the fourteen potentials are determined uniquely by fourteen equations. The Gaussian coordinate system is uniquely defined, and, most importantly, the unique assertions then made about the potentials in the Gaussian coordinate system are of invariant character. Thus, the present can uniquely determine the invariant and therefore physically meaningful content of the future and no contradiction with the causality principle remains.

Hilbert proceeded to indicate three ways in which invariant assertions can be given mathematical expression. Reminiscent of our reconstruction of Hertz's proposal, the first two of Hilbert's ways resorted to specially adapted coordinate systems.[30] The first recapitulated the use of invariant coordinate systems, such as what he termed Gaussian (geodetic normal) coordinates, and elaborated on its application to the example of the electron at rest. The second allowed invariant character for an assertion that *there exists* a coordinate system in which some nominated relation holds. As an illustration, he resorted again to the case of the electron and claimed invariant character for the assertion that there exists a coordinate system according to whose x_4 time coordinate the electron is at rest.

That Hertz, as we reconstruct him, and Hilbert, both working in Göttingen, should rely so heavily on specially adapted coordinate systems to reveal the physically significant elements of a theory provides strong evidence for our reconstruction. It also raises the further question of the origin of these ideas. Were they Hertz's own? Or was he acting, in effect, as a spokesperson for Hilbert and the Göttingen group?

7. Other Influences on Einstein's Resolution of the Hole Argument

Hertz's proposal to Einstein—as reconstructed by us—would have provided a serviceable escape from the hole argument. The escape route actually followed by Einstein, however, his point-coincidence argument, differed in crucial ways from that of Hertz and the Göttingen group. The latter was the mathematician's escape, relying principally on the mathematical notion of invariance; the former was the physicist's escape, relying principally on general dicta about physical reality. Was the point-coincidence argument another unprimed outpouring of Einstein's genius? Or can we identify who primed the pump? We believe that there are at least two plausible candidates.

The first of these, chronologically, is Joseph Petzoldt, a Privatdozent at the Technische Hochschule Berlin–Charlottenburg, founder in 1912 of the Gesellschaft für positivistische Philosophie (of which Einstein was a founding member), and author of numerous books and articles promoting a point of view that Petzoldt labeled "relativistic positivism," a mélange of ideas from Mach and Einstein, the chief aim of which was a critique of the traditional metaphysical notion of substance. Petzoldt's most important contribution for the purposes of our discussion was his introduction in 1895 of what he termed "Das Gesetz der Eindeutigkeit" ("The Law of Uniqueness" or "Univocalness") (Petzoldt 1895), according to which, in one of its forms, a theory would be acceptable only if it determined a unique model of the reality it aimed to describe. Petzoldt's "law of uniqueness" and the major discussion stimulated by it form an essential part of the background to Einstein's hole and point-coincidence arguments, since it is this very methodological principle that lies at the root of both.[31]

By 1915, Einstein and Petzoldt were in personal contact with one another. There is evidence that Petzoldt was attending Einstein's lectures on relativity in Berlin in either the winter semester of 1914–1915 or the summer semester of 1915. A postcard from Einstein to Petzoldt in late 1914 or early 1915 makes it clear that Einstein had been reading Petzoldt's work and approved of its general tendency: "Today I have read with great interest your book in its entirety, and I happily infer from it that I have for a long time been your companion in your way of thinking" (EA 19-067); the book was most likely Petzoldt's *Das Weltproblem vom Standpunkte des relativistischen Positivismus aus, historisch-kritisch dargestellt* (Petzoldt 1912b).[32]

Against this background, one may wonder whether Einstein had absorbed the point of view exemplified by a remark in Petzoldt's "Die Relativitätstheorie im erkenntnistheoretischer Zusammenhang des relativistischen Positivismus" (Petzoldt 1912a), which would have appeared early in 1913 in the proceedings of the Deutsche Physikalische Gesellschaft. The relevant remark concerns the way Petzoldt's epistemological perspectivalism is allegedly embodied in special relativity. Petzoldt writes,

> The task of physics becomes, thereby, the unique [*eindeutige*] general representation of events from different standpoints moving relative to one another with constant velocities, and the unique setting-into-relationship of these representations. Every such representation of whatever totality of events must be uniquely mappable onto every other one of these representations of the same[1)] events. The theory of relativity is one such mapping theory. What is essential is that unique connection. Physical concepts must be bent to fit for its sake. We have theoretical and technical command only of that which is represented uniquely by means of concepts.
>
> [1)] Better: representations of events in arbitrarily many of those systems of reference that are uniquely mappable onto one another are representations of "*the same*" event. Identity must be *defined*, since it is not given from the outset. (Petzoldt 1912a, p. 1059)

It is the footnote that grabs one's attention, for it expresses a fundamental presupposition of Einstein's point-coincidence argument. What is interesting about Petzoldt's remark is that this way of talking about identity under a mapping, especially of what are clearly, from context, Minkowskian point-events, was not commonplace in the pre-1915 literature on relativity.

To appreciate the role of the second figure possibly influencing Einstein's formulation of the point-coincidence argument, recall that Einstein's struggle to find generally covariant field equations came to a close with his November 25, 1915 communication to the Prussian Academy (Einstein 1915b). Already in his immediately preceding communication of November 18, 1915, he remarked that through general covariance, "time and space have been robbed of the last trace of objective reality" (Einstein 1915a, p. 831), by which he meant that "the relativity postulate in its most general formulation ... turns the space-time coordinates into physically meaningless parameters" (Einstein 1915b, p. 847). This makes it clear that, at this time, in late November, Einstein was in possession of an answer to the hole argument involving essentially the idea that coordinatizations are not sufficient for the individuation of points in the physical space-time. Curiously, however, when he begins informing his correspondents about these developments in late December, he adds, for the first time, the talk of coincidences so characteristic of the familiar form of the point-coincidence argument.

It seems likely to us that Einstein's immediate inspiration for the point-coincidence talk came from the work of Erich Kretschmann. His 1915 essay, "Über die prinzipielle Bestimmbarkeit der berechtigten Bezugssysteme beliebiger Relativitätstheorien," is a lengthy and labored discussion of the determination of coordinate systems in which the notion of spatiotemporal coincidence plays a prominent role. The paper clearly anticipates essential elements of the point-coincidence argument, as Kretschmann himself seemed to think when, in a later publication, he cited his own 1915 paper "for further details" (Kretschmann 1917, p. 576) on the point-coincidence argument, citing Einstein's version of the argument solely for the introduction of the German term "*Koinzidenzen*," replacing Kretschmann's 1915 "*Zusammenfallen*" (see below).[33]

In his 1915 paper, Kretschmann argues that only what he calls "topological" relations in the form of coincidences have empirical significance, since all observation requires that we bring a part of the measuring instrument into contact with the measured object:

> What is observed here—if we neglect, at first, all direct metrical determinations—is only the completely or partially achieved spatiotemporal coincidence [*Zusammenfallen*] or non-coincidence [*Nichtzusammenfallen*] of parts of the measuring instrument with parts of the measured object. Or more generally: topological relations between spatiotemporally extended objects. (Kretschmann 1915, p. 914)

A similar insistence on the observability of coincidences figures prominently in the best-known of Einstein's statements of the point-coincidence argument, where Einstein writes:

> All our space-time verifications invariably amount to a determination of space-time coincidences [*Koinzidenzen*].... Moreover, the results of our measurings are nothing but verifications of such meetings of the material points of our measuring instruments with other material points, coincidences [*Koinzidenzen*] between the hands of a clock and points on the clock dial, and observed point-events happening at the same place at the same time. (Einstein 1916, p. 117)[34]

There is, to be sure, the one difference noted later by Kretschmann, which is that Einstein uses the term "*Koinzidenzen*," not Kretschmann's "*Zusam-*

menfallen." The former term is more suggestive of the topologist's notion of the intersections of lines at extensionless points, whereas the latter is more suggestive of macroscopic congruences of bodies at the level of observational practice. Thus, Kretschmann can talk more comfortably of "completely or partially achieved coincidences [*Zusammenfallen*]." The similarity is nonetheless striking.

Kretschmann proceeds in the 1915 paper to develop now-familiar ideas concerning coordinate systems. In particular, he urges on the basis of his earlier assertions on coincidences that, "in no case can a soundly based decision be made, through mere observations, between two quantitatively different but topologically equivalent mappings of the world of appearance onto a space-time reference system" (Kretschmann 1915, p. 916). An immediate application of Kretschmann's remark (but *not* offered by Kretschmann) is the case of the two solutions, g_{ik} and g'_{ik} (in the same coordinate system x^m) of the hole argument. They are "two quantitatively different... mappings of the world of appearance onto a [single] space-time coordinate system." Nonetheless, they are "topologically equivalent," since they agree on all point-coincidences, and hence observation allows no soundly based decision between them. But if observation reveals no difference, does there remain any factual difference between them? If we pursue the development of Kretschmann's ideas, we find that whatever differences obtain between the two solutions, g_{ik} and g'_{ik}, must be merely matters of convention: "Insofar as the kinematical assertions of a system of physical laws cannot be reduced to purely topological relations, they are henceforth to be considered as mere—at most methodologically grounded—conventions" (Kretschmann 1915, p. 924).[35]

Of course, there is no reason to think that Kretschmann intended his discussion to be applied to Einstein's hole argument. However, the similarity between Einstein's expositions of the point-coincidence argument and Kretschmann's discussion is so striking that it cannot be (dare we say!) a mere coincidence and must have resulted from some sort of connection between Einstein and Kretschmann. The only question to be resolved is the nature of that connection. What is extremely suggestive is that Kretschmann's paper appeared in an issue of the *Annalen der Physik* that was distributed on December 21, 1915, five days before the earliest of the surviving letters in which Einstein articulates the point-coincidence argument, his letter to Ehrenfest of December 26 (EA 9-363). We are unaware of any similar invocation of point-coincidences in the corpus of Einstein's writings—both published and unpublished—prior to this letter. What is more, when, in a letter of December 14, 1915 (EA 21-610), Einstein informed Moritz Schlick about the exciting developments of November 1915, he remarked only on space and time having lost the last vestige of physical reality, with no mention of point-coincidences. These facts make almost irresistible the conclusion that Einstein read Kretschmann's paper or learned of its content when it appeared, found the ideas on coincidences extremely congenial, and turned to refine and exploit them to explain to his correspondent Ehrenfest where his hole argument had failed.

Other paths of transmission of these ideas between Einstein and Kretschmann are possible, but seem less likely. Kretschmann completed his Ph.D. in 1914 under Max Planck and Heinrich Rubens in Berlin, standing for the Promotionsprüfung on February 5 of that year. But Kretschmann reports that he finished his studies in Berlin in 1912 (see the *Lebenslauf* at the end of Kretschmann 1914), and the manuscript of his 1915 paper was submitted from Königsberg, where he had finished *Gymnasium* in 1906 and where he became a *Privatdozent* in 1920. Were he present in Berlin after Einstein's arrival in April 1914, it is plausible that he might have had some contact with Einstein, through which contact Einstein may have supplied the ideas about coincidences to or learned them from Kretschmann. Whatever contact they may have had in Berlin, however, cannot have been extensive or engaging to Kretschmann as far as Einstein's still incomplete general theory of relativity was concerned. While he was elsewhere rather long-winded, Kretschmann's 1915 paper contains only a brief discussion of Einstein's theory (pp. 977–978), citing just two of the earlier joint publications by Einstein and Grossmann (Einstein and Grossmann 1913, 1914), and omitting the major review article of November 1914 (Einstein 1914b). The discussion is sketchy and fails to make any serious contact with the idea of adapted coordinates, an idea that was a major focus of Einstein's Berlin work on the theory at that time and very relevant to the subject of Kretschmann's paper. Finally, of course, the possibility of such earlier transmission completely fails to explain the extraordinary fact that the point-coincidence argument and mention of space-time coincidences in this general context appear for the first time in a letter of Einstein's of December 26, 1915, only days after the issue of the *Annalen* containing Kretschmann's paper was distributed.[36]

ACKNOWLEDGMENTS. Our sincere thanks are due to Rudolf Hertz, Paul Hertz's son, for his invaluable assistance in our research. Thanks are also due to Peter Havas, for having drawn our attention to the notes from Klein's lectures on general relativity during the summer of 1916. The research for this paper was supported in part by grants from the National Science Foundation (no. SES-8421040, DH), the American Philosophical Association (DH), the Deutscher Akademischer Austauschdienst (DH), and the University of Kentucky Research Foundation (DH). We would like to thank the Hebrew University of Jerusalem, which holds the copyright, for permission to quote from Einstein's unpublished letters, and to thank the Niedersächsische Staats- und Universitätsbibliotek, Göttingen, for permission to quote from Felix Klein's unpublished lectures. Items in the Einstein Archive are cited by their numbers in the Control Index.

Notes

[1] To see this, note that the first solution transformed from x^m to $x^{m'}$ has the functional form g'_{ik} of the coordinates $x^{m'}$, which is the same functional form as the components of the second solution in the coordinate system x^m.

[2] For a summary of the mathematical machinery Einstein used to analyze his adapted coordinates, see Norton (1984, section 6).

[3] This letter is dated on the basis of its place in a sequence of letters discussing the shipment of the de Haas's furniture from Berlin to the Netherlands, the shipment being overseen by Einstein.

[4] For more on this visit, see the discussion in Pais 1982, pp. 250 and 259.

[5] Cod. Ms. Klein 21L, p. 63, Niedersächsische Staats- und Landesbibliothek Göttingen.

[6] Cod. Ms. Klein 21L, p. 69, Niedersächsische Staats- und Landesbibliothek Göttingen.

[7] This timing, the fact that Einstein and Hilbert engaged in an intense correspondence through November 1915 and then had a brief falling out after that correspondence, has raised the possibility that Einstein stole the field equations from Hilbert. We do not take this possibility seriously for the reasons given in Norton (1984, pp. 314–315).

[8] See, for example, Einstein to Paul Ehrenfest, December 26, 1915 (EA 9-363), December 29, 1915 (EA 9-365), and January 5, 1916 (EA 9-372), as well as Einstein to Michele Besso, January 3, 1916 (EA 7-272; reprinted in Speziali 1972, pp. 63–64).

[9] Notice that such magnitudes as "time elapsed" are in turn reducible to space-time coincidences. A crude physical time could be measured by an idealized light clock, which is a small rigidly co-moving rod along whose length a light pulse is repeatedly reflected. The time elapsed is measured by the number of collisions of the light pulse with the mirrored ends of the rod.

[10] Hilbert was the titular director of Hertz's dissertation, but Hertz actually did the work under Abraham, who was then *Privatdozent*; see Pyenson 1979b, p. 76.

[11] See Einstein to Hertz, August 14, 1910 (EA 12-195) and August 26, 1910 (EA 12-198). For more on the beginning of their acquaintance, see Stachel et al. 1989, p. 44, and Klein et al. 1993, p. 315.

[12] See the Hertz–Ehrenfest correspondence in the Ehrenfest scientific correspondence in the Archive for the History of Quantum Physics.

[13] See Pyenson 1990, as well as Laub to Einstein, May 16, 1909 (EA 15-465), Einstein to Laub, May 19, 1909 (EA 15-480), and Einstein to Laub, October 11, 1910 (EA 15-489), November 4, 1910 (EA 15-491).

[14] See, for example, Hertz 1923, 1929a, 1929b, 1930, 1936a, 1936b.

[15] See Clark 1971, p. 184. The chief purpose of Einstein's trip was to meet the novelist Romain Rolland at Vevey, this as part of Einstein's efforts to promote international intellectual cooperation in spite of the barriers raise by World War I. For more on the meeting with Rolland and Einstein's related activities, see Nathan and Norden 1968, pp. 12–18. The year could not be 1913, because Einstein was then still in Zurich, and such a trip would not likely have been undertaken in late August 1914, immediately after the outbreak of the war.

[16] See below. In particular, Hertz uses the older "E, F, and G" notation for what we would now call the components of the metric tensor.

[17] Rudolf Hertz (Paul's son), private communication.

[18] To see the correspondence between our account of the hole argument in Section 1 and Hertz's construction, notice that our second solution, g'_{ik}, in the first coordinate system, x^m, corresponds to Hertz's E, F, G in (u, v), while our first solution, g_{ik}, in the second coordinate system, $x^{m'}$, corresponds to Hertz's E^x, F^x, G^x in (u^x, v^x). Of course, there is the inconsequential change of context. Einstein's argument is formulated in a space-time with an indefinite metric, whereas Hertz's argument is formulated for the space of a two-dimensional Gaussian surface.

[19] Obviously, this construction and the point-coincidence argument have the following in common: They pick out a point in the physical space by the intersection of curves with invariant geometrical properties. In Hertz's case, the curves are curves of constant curvature and maximal curvature gradient; in the case of the point-coincidence argument, they are geodesics.

[20] In his *Vorlesungen über die Entwicklung der Mathematik im 19. Jarhundert* (Klein 1927, pp. 147–148), Felix Klein lists Knoblauch 1913 as one of the "great textbooks" appearing around the turn of the century, along with Darboux's *Leçons sur la théorie générale des surfaces* (Darboux 1914–1915) and Bianchi's *Vorlesungen über Differentialgeometrie* (Bianchi 1910). Although first published in 1927, Klein's lectures were delivered in the years 1915 through 1917.

[21] Einstein's replacing of G, the g_{22} component of the metric, by ϕ is explicable in terms of his 1913 theory. In Einstein's 1913 theory, the $g_{\text{“time” “time”}}$ component of the metric in a static field in a suitably adapted coordinate system represents the single gravitational potential of the field, commonly represented by ϕ. Note that the angle brackets indicate a strikeout in Einstein's original.

[22] In a footnote, Kretschmann comments that the possibility of finding "absolute" coordinates, meaning coordinates picked out uniquely by the geometry of the space being thus coordinatized, had been pointed out to him already in a letter from Gustav Mie in February 1916; see Kretschmann 1917, p. 592, n. 1.

[23] For more on Hilbert's introduction of "Gaussian coordinates," see Stachel 1992, pp. 410–412.

[24] The approximate date of Ehrenfest's letter to Einstein can be determined from his remark, in a letter to Lorentz of December 23, 1915, that he had invited Einstein to spend the holidays in Leiden. Einstein's reply to Ehrenfest's thought experiment is contained in the same letter of January 5, 1916 (EA 9-372), in which he explains that the border's being blocked was the reason why he could not have come to Holland at that time. We thank A.J. Kox for making available transcriptions of

the Ehrenfest–Lorentz correspondence, these from his forthcoming edition of the scientific correspondence of Lorentz.

[25] The reconstruction of Ehrenfest's thought experiment is based upon Einstein's reply of January 5 (EA 9-372) and on the description found in Ehrenfest's letter to Lorentz of January 9, in which he enclosed Einstein's letter, asking for Lorentz's opinion.

[26] See Einstein to Hertz, undated 1915 (EA 12-205), October 1915 (EA 12-206), Hertz to Einstein, October 8, 1915 (EA 12-207), and Einstein to Hertz, October 9, 1915 (EA 12-208). Though the dating of some of these letters is problematic, they seem clearly to form a sequence written over a short period. It should be noted that most of Hertz's are missing, the letter of October 8 having survived because Hertz retained a copy in his files.

[27] See Hertz to Hilbert, February 17, 1916 (Cod. Ms. Hilbert 150, Handschriftenabteilung, Niedersächsische Staats- und Universitätsbibliothek Göttingen).

[28] Hilbert's only footnote in this section of the paper (Hilbert 1916, p. 61) cites Einstein's most complete version (1914b, p. 1067) of the hole argument.

[29] For more on Hilbert and the causality principle in general relativity, see Stachel 1992, pp. 410–412.

[30] The third merely allowed invariant character to a fully covariant law, such as the law of conservation of energy-momentum expressed as the vanishing covariant divergence of the stress-energy tensor.

[31] For more on Petzoldt and a more detailed bibliography of his writings, see Howard 1992.

[32] For the dating of Einstein's postcard to Petzoldt and other details about their relationship, see Howard 1992.

[33] For more on Kretschmann's papers, see Norton 1992, pp. 295–301.

[34] See Howard 1992, n. 25, for a critical discussion of Friedman's (1983, pp. 22–25) interpretation of this passage as anticipating the verificationist theory of meaning that later became popular among the logical positivists.

[35] In a footnote to the word "convention," Kretschmann carefully indicates the precise sense of the word intended. It is to mean that which is not demonstrable through observation, rather than something arrived at by some kind of free agreement.

[36] We might also conjecture that Einstein was asked to review the paper by Planck, the editor of *Annalen*. Kretschmann's paper is dated October 15 and was received on October 21. If it was sent out for review, Einstein would have been the obvious reviewer. The short time between submission and publication, October 21 to December 21, suggests that, even though Kretschmann was a first-time author in the *Annalen*, the manuscript was not sent out for review, since a two-month period between submission and publication was more or less normal for established authors (see Pyenson 1983). This would not be surprising, since Planck had supervised Kretschmann's Ph.D., was presumably confident of Kretschmann's scholarship, and possibly already familiar with the work submitted.

References

Bianchi, Luigi (1910). *Vorlesungen über Differentialgeometrie*, 2nd ed. Max Lukat, trans. Leipzig and Berlin: B.G. Teubner.

Clark, Ronald W. (1971). *Einstein: The Life and Times*. New York and Cleveland: World.

Darboux, Gaston (1914–1915). *Leçons sur la théorie générale des surfaces et les applications géometriques du calcul infinitésimal*, 2nd ed. Paris: Gauthier-Villars.

Einstein, Albert (1902). "Kinetische Theorie des Wärmegleichgewichtes und des zweiten Hauptsatzes der Thermodynamik." *Annalen der Physik* 9: 417–433.

——— (1903). "Eine Theorie der Grundlagen der Thermodynamik." *Annalen der Physik* 11: 170–187.

——— (1904). "Zur allgemeinen molekularen Theorie der Wärme." *Annalen der Physik* 14: 354–362.

——— (1911a). "Bemerkungen zu den P. Hertzschen Arbeiten: 'Über die mechanischen Grundlagen der Thermodynamik'." *Annalen der Physik* 34: 175–176.

——— (1911b). "Über den Einfluß der Schwerkraft auf die Ausbreitung des Lichtes." *Annalen der Physik* 35: 898–908.

——— (1914a). "Prinzipielles zur verallgemeinerten Relativitätstheorie und Gravitationstheorie." *Physikalische Zeitschrift* 15: 176–180.

——— (1914b). "Die formale Grundlage der allgemeinen Relativitätstheorie." *Königlich Preussische Akademie der Wissenschaften* (Berlin). *Sitzungsberichte*: 1030–1085.

——— (1915a). "Erklärung der Perihelbewegung des Merkur aus der allgemeinen Relativitätstheorie." *Königlich Preussische Akademie der Wissenschaften* (Berlin). *Sitzungsberichte*: 831–839.

——— (1915b). "Der Feldgleichungen der Gravitation." *Königlich Preussische Akademie der Wissenschaften* (Berlin). *Sitzungsberichte*: 844–847.

——— (1916). "Die Grundlage der allgemeinen Relativitätstheorie." *Annalen der Physik* 49: 769–822. Reprinted as a separatum Leipzig: Johann Ambrosius Barth, 1916. Page numbers are cited from the English translation: "The Foundations of the General Theory of Relativity." In Hendrik A. Lorentz, Albert Einstein, Hermann Minkowski, and Hermann Weyl, *The Principle of Relativity*. W. Perrett and G.B. Jeffrey, trans. London: Methuen, 1923; reprint New York: Dover, 1952.

——— (1934). "Einiges über die Entstehung der allgemeinen Relativitätstheorie." In *Mein Weltbild*. Amsterdam: Querido, pp. 248–256. Quotations are from the English translation: "Notes on the Origin of the General Theory of Relativity." In *Ideas and Opinions*. Carl Seelig, ed.; Sonja Bargmann, trans. New York: Crown, 1954, pp. 285–290.

Einstein, Albert and Grossmann, Marcel (1913). *Entwurf einer verallgemeinerten Relativitätstheorie und einer Theorie der Gravitation. I. Physikalischer Teil von Albert Einstein. II. Mathematischer Teil von Marcel Grossmann*. Leipzig and Berlin: B.G. Teubner. Reprinted with added "Bemerkungen," *Zeitschrift für Mathematik und Physik* 62 (1914): 225–261.

—— (1914). "Kovarianzeigenschaften der Feldgleichungen der auf die verallgemeinerten Relativitätstheorie gegründeten Gravitationstheorie." *Zeitschrift für Mathematik und Physik* 63: 215–225.

Eisenstaedt, Jean and Kox, A.J., eds. (1992). *Historical Studies in General Relativity*. Einstein Studies, Vol. 3. Boston: Birkhäuser.

Friedman, Michael (1983). *Foundations of Space-Time Theories: Relativistic Physics and Philosophy of Science*. Princeton, New Jersey: Princeton University Press.

Helmholtz, Hermann von (1921). *Schriften zur Erkenntnistheorie*. Paul Hertz and Moritz Schlick, eds. Berlin: Julius Springer.

Hermann, Armin, ed. (1968). *Albert Einstein/Arnold Sommerfeld*. Briefwechsel. Basel and Stuttgart: Schwabe.

Hertz, Paul (1904). "Untersuchungen über unstetige Bewegungen eines Elektrons." Ph.D. dissertation. Göttingen.

—— (1910). "Über die mechanischen Grundlagen der Thermodynamik." *Annalen der Physik* 33: 225–274, 537–552.

—— (1916). "Statistische Mechanik." In *Repertorium der Physik*. Rudolf Heinrich Weber and Richard Gans, eds. Vol. 1, *Mechanik und Wärme*. Part 2, *Kapillarität, Wärme, Wärmeleitung, kinetische Gastheorie und statistische Mechanik*. Rudolf Heinrich Weber and Paul Hertz, eds. Leipzig and Berlin: B.G. Teubner, pp. 436–600.

—— (1923). *Über das Denken und seine Beziehung zur Anschauung*. Part 1, *Über den funktionalen Zusammenhang zwischen auslösendem Erlebnis und Enderlebnis bei elementaren Prozessen*. Berlin: Julius Springer.

—— (1929a). "Über Axiomensysteme beliebiger Satzsysteme." *Annalen der Philosophie und philosophischen Kritik*. 8: 179–204.

—— (1929b). "Über Axiomensysteme für beliebiger Satzsysteme." *Mathematische Annalen* 101: 457–514.

—— (1930). "Über den Kausalbegriff im Makroskopischen, besonders in der klassischen Physik." *Erkenntnis* 1: 211–227.

—— (1936a). "Kritische Bemerkungen zu Reichenbachs Behandlung des Humeschen Problems." *Erkenntnis* 6: 25–31.

—— (1936b). "Regelmäßigkeit, Kausalität und Zeitrichtung." *Erkenntnis* 6: 412–421.

Hilbert, David (1915). "Die Grundlagen der Physik. (Erste Mitteilung)." *Königliche Gesellschaft der Wissenschaften zu Göttingen. Mathematisch-physikalische Klasse. Nachrichten*: 395–407.

—— (1916). "Die Grundlagen der Physik: zweite Mitteilung." *Königliche Gesellschaft der Wissenschaften zu Göttingen. Mathematisch-physikalische Klasse. Nachrichten*: 55–76.

Howard, Don (1992). "Einstein and *Eindeutigkeit*: A Neglected Theme in the Philosophical Background to General Relativity." In Eisenstaedt and Kox 1992, pp. 154–243.

Howard, Don and Stachel, John, eds. (1989). *Einstein and the History of General Relativity*. Based on the Proceedings of the 1986 Osgood Hill Conference, North Andover, Massachusetts, May 8–11, 1986. Einstein Studies, Vol. 1. Boston: Birkhäuser.

Klein, Felix (1927). *Vorlesungen über die Entwicklung der Mathematik im 19. Jahrhundert. Part 2, Die Grundbegriffe der Invariantentheorie und ihr Eindringen in die mathematische Physik*. Richard Courant and S. Cohn-Vossen, eds. Berlin: Julius Springer.

Klein, Martin, Kox, A.J., Renn, Juergen, and Schulmann, Robert, eds. (1993). *The Collected Papers of Albert Einstein*. Vol. 3, *The Swiss Years: Writings, 1909–1911*. Princeton: Princeton University Press.

Knoblauch, Johannes (1913). *Grundlagen der Differentialgeometrie*. Leipzig and Berlin: B.G. Teubner.

Kretschmann, Erich (1914). "Eine Theorie der Schwerkraft im Rahmen der ursprünglichen Einsteinschen Relativitätstheorie." Ph.D. dissertation. Berlin.

—— (1915). "Über die prinzipielle Bestimmbarkeit der berechtigten Bezugssysteme beliebiger Relativitätstheorien." *Annalen der Physik* 48: 907–942, 943–982.

—— (1917). "Über den physikalischen Sinn der Relativitätspostulate, A. Einsteins neue und seine ursprüngliche Relativitätstheorie." *Annalen der Physik* 53: 575–614.

Nathan, Otto and Norden, Heinz, eds. (1968). *Einstein on Peace*. New York: Schocken.

Norton, John (1984). "How Einstein Found his Field Equations: 1912–1915." *Historical Studies in the Physical Sciences* 14: 253–316. Reprinted in Howard and Stachel 1989, pp. 101–159.

—— (1987). "Einstein, the Hole Argument and the Reality of Space." In *Measurement, Realism and Objectivity*. J. Forge, ed. Dordrecht and Boston: D. Reidel, pp. 153–188.

—— (1992). "The Physical Content of General Covariance." In Eisenstaedt and Kox 1992, pp. 281–315.

Pais, Abraham (1982). *'Subtle is the Lord...': The Science and the Life of Albert Einstein*. Oxford: Clarendon; New York: Oxford University Press.

Petzoldt, Joseph (1895). "Das Gesetz der Eindeutigkeit." *Vierteljahrsschrift für wissenschaftliche Philosophie und Soziologie* 19: 146–203.

—— (1912a). "Die Relativitätstheorie im erkenntnistheoretischer Zusammenhang des relativistischen Positivismus." *Deutsche Physikalische Gesellschaft. Verhandlungen* 14: 1055–1064.

—— (1912b). *Das Weltproblem vom Standpunkte des relativistischen Positivismus aus, historisch-kritisch dargestellt*, 2nd ed. Wissenschaft und Hypothese, vol. 14. Leipzig and Berlin: B.G. Teubner.

Pyenson, Lewis (1979a). "Mathematics, Education, and the Göttingen Approach to Physical Reality, 1890–1914." *Europa: A Journal of Interdisciplinary*

Studies 2, no. 2. Quotations are taken from the reprinting in Pyenson 1985, pp. 158–193.

——— (1979b). "Physics in the Shadow of Mathematics: The Göttingen Electron-theory Seminar of 1905." *Archive for History of Exact Sciences* 21: 55–89.

——— (1983). "Physical Sense in Relativity: Max Planck Edits the Annalen der Physik, 1906–1918." In *Proceedings of the Ninth International Conference on General Relativity and Gravitation.* Ernst Schmutzer, ed. Berlin: Akademie-Verlag, pp. 285–302. Reprinted in Pyenson 1985, pp. 194–214.

——— (1985). *The Young Einstein: The Advent of Relativity.* Bristol and Boston: Adam Hilger.

——— (1990). "Laub, Jakob Johann." In *Dictionary of Scientific Biography*, vol. 17, suppl. 2. Frederic L. Holmes, ed. New York: Charles Scribner's Sons, pp. 528–529.

Speziali, Pierre, ed. (1972). *Albert Einstein–Michele Besso. Correspondance 1903–1955.* Paris: Hermann.

Stachel, John (1989). "Einstein's Search for General Covariance, 1912–1915." In Howard and Stachel 1989, pp. 63–100.

——— (1992). "The Cauchy Problem in General Relativity: The Early Years." In Eisenstaedt and Kox 1992, pp. 407–418.

Stachel, John, Cassidy, David C., Renn, Juergen, and Schulmann, Robert, eds. (1989). *The Collected Papers of Albert Einstein.* Vol. 2, *The Swiss Years: Writings, 1900–1909.* Princeton: Princeton University Press.

Conservation Laws and Gravitational Waves in General Relativity (1915–1918)

Carlo Cattani and Michelangelo De Maria

1. Introduction

This chapter deals with two closely related debates in general relativity in 1916–1918, one on gravitational waves, the other on the correct formulation of conservation laws. Both issues involve the definition of a quantity representing the stress-energy of the gravitational field. Such definitions were typically proposed in the context of deriving the gravitational field equations from a variational principle. A proper understanding of the debates on gravitational waves and conservation laws therefore requires some discussion of the rather complicated history of attempts to derive gravitational field equations from a variational principle.[1]

We will trace Einstein's work on gravitational waves and his work on conservation laws during the years 1916–1918 in this more complex network. We will look at objections to Einstein's approach from Levi-Civita, Schrödinger, and Bauer; at alternative approaches suggested by Lorentz and Levi-Civita; and at Einstein's response to all of them. In particular, we will examine the 1917 correspondence between Einstein and Levi-Civita. We will see how Levi-Civita's criticism of Einstein's formulation of conservation laws strengthened Einstein in his conviction that physical considerations force one to adopt a noncovariant formulation of conservation laws for matter plus gravitational field.

2. The Importance of the Conservation Laws in Einstein's 1914 Gravitational Theory

In Einstein and Grossmann 1914 and Einstein 1914, Einstein used a variational method to derive the field equations of limited covariance of his so-called *Entwurf* theory (Einstein and Grossmann 1913). He used conservation of energy-momentum of matter plus gravitational field—the stress-energy of the latter being represented by a pseudotensor rather than a tensor—to define the Lagrangian H for the gravitational field and to restrict the covariance of his theory. Einstein believed he had found a very general argument to fix the Lagrangian for the gravitational field. This Lagrangian leads to the field equations of the *Entwurf* theory.

By substituting the gravitational tensor into the law of conservation of energy-momentum of matter (with stress-energy tensor $\mathcal{T}_\mu{}^\nu$), Einstein was able to derive certain constraints on H that he thought uniquely fixed its form. Imposing conservation of energy-momentum of matter and unaware of the contracted Bianchi identities, he obtained a set of equations to be satisfied by the gravitational field:

$$\frac{\partial}{\partial x^\nu} S_\sigma{}^\nu - \mathcal{B}_\sigma = 0, \qquad (\sigma, \nu, \ldots = 0, 1, 2, 3)$$

where[2]

$$\mathcal{B}_\mu \stackrel{\text{def}}{=} \frac{\partial^2}{\partial x^\sigma \partial x^\alpha}\left(g^{\nu\alpha} \frac{\partial H\sqrt{-g}}{\partial g_\sigma^{\mu\nu}}\right) \tag{1}$$

and

$$S_\sigma{}^\nu \stackrel{\text{def}}{=} g^{\nu\tau}\frac{\partial H\sqrt{-g}}{\partial g^{\sigma\tau}} + g_\mu^{\nu\tau}\frac{\partial H\sqrt{-g}}{\partial g_\mu^{\sigma\tau}} + \tfrac{1}{2}\delta_\sigma^\nu H\sqrt{-g} - \tfrac{1}{2} g_\sigma^{\mu\tau}\frac{\partial H\sqrt{-g}}{\partial g_\nu^{\mu\tau}}. \tag{2}$$

Then Einstein showed that both $\mathcal{B}_\mu$ and $S_\sigma{}^\nu$ must vanish:

$$\mathcal{B}_\mu = 0, \qquad S_\sigma{}^\nu = 0, \tag{3}$$

and used these conditions to define the form of H. He finally obtained the *Entwurf* field equations in the form[3]

$$\frac{\partial}{\partial x^\alpha}(\sqrt{-g}g^{\alpha\beta}\Gamma_{\sigma\beta}^\nu) = -\chi(\mathcal{T}_\sigma{}^\nu + t_\sigma{}^\nu), \tag{4}$$

where the stress-energy tensor[4] $t_\sigma{}^\nu$ for the gravitational field is defined as

$$t_\sigma{}^\nu \stackrel{\text{def}}{=} \frac{\sqrt{-g}}{\chi}\left(g^{\nu\tau}\Gamma_{\mu\sigma}^\rho \Gamma_{\rho\tau}^\mu - \tfrac{1}{2}\delta_\sigma^\nu g^{\tau\alpha}\Gamma_{\mu\tau}^\rho \Gamma_{\rho\alpha}^\mu\right), \tag{5}$$

$\Gamma_{\mu\sigma}^\rho$ being the Christoffel symbols. Differentiating equation (4) with respect to x^ν, Einstein obtained the conservation law for matter plus gravitational field in the form

$$\frac{\partial}{\partial x^\nu}(\mathcal{T}_\sigma{}^\nu + t_\sigma{}^\nu) = 0. \tag{6}$$

It must be stressed, however, that, already in 1914, Einstein noticed that

> $t_\sigma{}^\nu$ does not transform as a tensor under arbitrary justified transformations, but only under linear transformations; nevertheless, we will call $t_\sigma{}^\nu$ the [stress-]energy tensor[5] of the gravitational field. Something analogous holds for the components $\Gamma^\nu_{\sigma\beta}$ of the gravitational field strength. (Einstein 1914, p. 1077)

In the spring of 1915, in private correspondence with Einstein, Levi-Civita sharply attacked Einstein's proofs of the covariance of certain fundamental quantities of his *Entwurf* theory (Cattani and De Maria 1989b); however, he did not explicitly criticize the pseudotensor character of $t_\sigma{}^\nu$.

3. Lorentz's Variational Approach (1915)

In 1915, Lorentz published a paper (Lorentz 1915) in which he criticized both the *Entwurf* theory and the variational formulation Einstein had given to it in 1914. In the second part of his paper, Lorentz proposed a more general variational derivation of gravitational field equations. Lorentz did not specify the form of the Lagrangian; he just assumed it to be a function of the metric tensor and its first-order derivatives. Requiring that the action integral be stationary not only for arbitrary infinitesimal variations of the coordinates, as Einstein had required, but also for arbitrary infinitesimal variations of the components of the metric tensor, Lorentz obtained the gravitational field equations in the form

$$\frac{\partial \mathcal{R}^*}{\partial g_{\mu\nu}} - \frac{\partial}{\partial x^\sigma}\left(\frac{\partial \mathcal{R}^*}{\partial g^{\mu\nu}_\sigma}\right) = -\chi \frac{\partial \mathcal{M}}{\partial g_{\mu\nu}}, \tag{7}$$

where $\mathcal{R}^*$ and $\mathcal{M}$ are the Lagrangians for the gravitational field and matter, respectively. Furthermore, Lorentz showed that equations (7) turn into the *Entwurf* field equations when the function H chosen by Einstein is substituted for $\mathcal{R}^*$. As is well known, Einstein himself later realized that his choice of a Lagrangian was, in fact, quite arbitrary (Cattani and De Maria 1989b). Unlike Levi-Civita, Lorentz at this point was unaware of the mathematical mistakes Einstein made in his early variational approach, and praised him for "his ingenious mode of reasoning" (Lorentz 1915, p. 1089).

4. Hilbert's Variational Approach (1915)

On November 20, 1915, Hilbert presented a paper, entitled "The Foundations of Physics" (Hilbert 1915), in which he discussed a variational principle for general relativity. Hilbert cited both Einstein (1914, 1915a, 1915b,

1915c) and Mie (1912), the former for his gravitational field equations, the latter for his work on nonlinear electrodynamics and his electromagnetic theory of matter. Like Mie, Hilbert restricted his investigation to the situation of an electromagnetic field in the presence of a gravitational field.

Hilbert was critical of Einstein's 1914 variational approach as the following quotation from his paper illustrates:

> Einstein gave the fundamental original idea of general invariance a simple expression; however, for Einstein the Hamilton principle only plays a subordinate role and his function H is not at all generally invariant. Moreover, the electrical potentials are not included [in his theory]. (Hilbert 1915, I, p. 396, footnote)

Hilbert proceeded as follows. He assumed that the quantities characterizing the fields are the ten gravitational potentials $g_{\mu\nu}$ and the four electromagnetic potentials q_μ. He defined a unique invariant world function according to the following axioms:

> Axiom 1 (of Mie about the world function). The law of physical events is determined through a world function [Lagrangian] $\mathcal{H} = \sqrt{-g}H$ that contains the following arguments:
>
> $$g_{\mu\nu}, \quad \frac{\partial g_{\mu\nu}}{\partial x^\alpha}, \quad \frac{\partial^2 g_{\mu\nu}}{\partial x^\alpha \partial x^\beta}; \qquad q_\sigma, \quad \frac{\partial q_\sigma}{\partial x^\alpha}$$
>
> and specifically the variation of the action integral must vanish for [changes in] every one of the 14 potentials $g_{\mu\nu}, q_\sigma$.
>
> Axiom 2 (of general invariance). The world function $\mathcal{H}$ is invariant with respect to arbitrary transformations of the world parameters [coordinates] x^α. (Hilbert 1915, I, p. 396)

He then defined two Lagrangian functions, one for the gravitational field and one for matter. For the gravitational field he used the Riemann curvature scalar $\mathcal{R}$. For the matter part he introduced a function $\mathcal{M}$. As long as the gravitational field equations contain no derivatives of $g_{\mu\nu}$ higher than of second order, the total Lagrangian $\mathcal{H}$ must be the direct sum of these two functions:

$$\mathcal{H} = \mathcal{R} + \mathcal{M}. \tag{8}$$

By evaluating the "Lagrangian derivatives" (Hilbert 1915, I, p. 397) of $\mathcal{H}$ with respect to the various variables, Hilbert obtained the evolution equations for both gravitational and electromagnetic potentials. His next step was to show that Axiom 2 allows one to give an explicit proof of the covariance of these evolution equations. Splitting the Lagrangian into two parts, the scalar curvature invariant for the gravitational field and a Lagrangian for the electromagnetic field, Hilbert arrived at the correct gravitational field equations:

$$G_{\mu\nu} = -\chi \frac{1}{\sqrt{-g}} \mathcal{T}_{\mu\nu}, \tag{9}$$

where

$$G_{\mu\nu} \stackrel{\text{def}}{=} R_{\mu\nu} - \tfrac{1}{2} R g_{\mu\nu}. \tag{10}$$

Finally, Hilbert obtained the evolution equations for electrodynamics in a curved space-time by generalizing Mie's derivation for flat Minkowski space-time.

In conclusion, we want to stress the limits of Hilbert's method:

(1) Hilbert derived the field equations in the context of Mie's electromagnetic theory of matter. As a consequence, his variational method could not readily be generalized to other matter. To accomplish that, one would have to specify how the matter Lagrangian depends on the gravitational potentials $g_{\mu\nu}$.

(2) Although Hilbert obtained generally covariant field equations, he made use of Lagrangian derivatives that were not generally covariant.

(3) Hilbert was unaware of the contracted Bianchi identities, so that he arrived at the explicit form of the gravitational tensor in a rather clumsy way.

5. Lorentz's Variational Approach (1916)

In 1916, Lorentz published a long paper in four parts on general relativity (Lorentz 1916, I–IV). In part III, he derived the correct gravitational field equations and an expression for the "stress energy complex" for the gravitational field. In part IV, he discussed the conservation law for the gravitational field.

As opposed to the unspecified Lagrangian of his 1915 article, Lorentz now chose the Riemann curvature scalar $\mathcal{R}$ as the Lagrangian for the gravitational field. He had come to realize that the Lagrangian has to be a generally covariant scalar (Lorentz 1916, I, p. 248, p. 251; see also Janssen 1992).

Lorentz split the variation of the action $\mathcal{R}$ into two parts. The first part, which is no longer a scalar quantity, leads to gravitational field equations; the second part vanishes identically on account of the boundary conditions. Moreover, he showed that the form of his gravitational tensor coincided with Einstein's "only *for one special* choice of coordinates" (Lorentz 1916,

p. 281, italics in the original). Lorentz obtained the correct gravitational field equations (Lorentz 1916, III, p. 285). We want to stress, however, that Lorentz made some mathematically unwarranted assumptions in deriving his results. He assumed that the infinitesimal variations of the components of the metric tensor have tensor character. Moreover, he had to make a special choice of coordinates.

Lorentz also discussed the conservation of energy-momentum of matter plus gravitational field, and arrived at the equations (6) obtained by Einstein in 1914 (Lorentz 1916, III, p. 292). Lorentz too was aware of the fact that the complex ${t_\sigma}^\nu$ is not a tensor (Lorentz 1916, III, p. 294). Whereas this was perfectly acceptable to Einstein, Lorentz wrote that

> [e]vidently it would be more satisfactory if we could ascribe a stress-energy-*tensor* to the gravitation field. Now this can really be done. (Lorentz 1916, III, p. 295, italics in the original)

A "natural" candidate for this tensor, according to Lorentz, was the gravitational tensor $G_{\mu\nu}$ of Einstein's generally covariant field equations. Therefore he suggested one interpret these equations as conservation laws. In Lorentz's opinion this interpretation of the field equations

> and the conception to which they have led, may look somewhat startling. According to it we should have to imagine that behind the directly observable world with its stresses, energy etc. the gravitation field is hidden with stresses, energy etc. that are everywhere equal and opposite to the former; evidently this is in agreement with the interchange of momentum and energy which accompanies the action of gravitation. On the way of a lightbeam, e.g., there would be everywhere in the gravitation field an energy current equal and opposite to the one existing in the beam. If we remember that this hidden energy-current can be fully described mathematically by the quantities g_{ab} and that only the interchange just mentioned makes it perceptible to us, this mode of viewing the phenomena does not seem unacceptable. At all events we are forcibly led to it if we want to preserve the advantage of a stress-energy-*tensor* also for the gravitation field. (Lorentz 1916, III, p. 296, italics in the original)

In part IV of his paper, Lorentz compared his definition of the stress-energy components of the gravitational field with the definition given by Einstein. While his expression contained first and second order derivatives of the metric, "Einstein on the contrary has given values for the stress-energy components which contain the first derivatives only and which therefore are in many respects much more fit for application" (Lorentz 1916, IV, p. 297). Thus Lorentz defined a stress-energy complex with components ${t_\sigma}^{\nu'}$ that are homogeneous and quadratic functions of the first-order derivatives of the metric and do not contain any higher-order derivatives. The divergence of

Lorentz's complex coincides with the divergence of Einstein's $t_\sigma{}^\nu$. Lorentz showed that when $\sqrt{-g} = 1$ and $g_{\alpha\beta} = \delta_{\alpha\beta}$ his complex is the same as Einstein's. He added that "it seems very probable that the agreement will exist in general" (Lorentz 1916, IV, p. 299).

In conclusion, we want to stress that Lorentz showed, for the first time, that the quantity representing gravitational stress-energy was not uniquely defined.

6. Einstein's Variational Approach (1916)

In 1916, Einstein returned to a variational approach to derive his gravitational field equations. He remarked that both Lorentz and Hilbert had succeeded in giving general relativity a clear form by deriving the field equations from a single variational principle. His aim now was to present the basic relations of the theory as clearly as possible and in a more general way. In fact, he considered his new approach more general and "in contrast especially with Hilbert's treatment" (Einstein 1916b, p. 1111), since he rejected some of Hilbert's restrictive hypotheses on the nature of matter.

His starting point was the universal function $\mathcal{H} \stackrel{\text{def}}{=} H\sqrt{-g}$, assumed to be a function of the metric tensor and its first-order derivatives and a linear function of its second-order derivatives. Furthermore, he generalized the variational principle to any physical phenomenon by assuming $\mathcal{H}$ to be dependent on matter variables q_ρ (not necessarily of electromagnetic origin) and their first-order derivatives. Thus, he replaced his 1914 Lagrangian by

$$\mathcal{H} = \mathcal{H}\left(g^{\mu\nu}, \frac{\partial g^{\mu\nu}}{\partial x^\sigma}, \frac{\partial^2 g^{\mu\nu}}{\partial x^\rho \partial x^\sigma}; \; q_\rho, \frac{\partial q_\rho}{\partial x^\alpha}\right). \tag{11}$$

Integrating a Lagrangian of this form with the usual boundary conditions, one arrives at the variational principle

$$\delta \int \mathcal{H}^* \, \mathrm{d}\tau = 0, \tag{12}$$

where $\mathcal{H}^*$ no longer depends on the second-order derivatives of the metric. Einstein had to start from a function of the form of (11) because, according to his principle of general relativity, the Lagrangian $\mathcal{H}$ must be invariant under arbitrary coordinate transformations. However, the reduction of $\mathcal{H}$ to $\mathcal{H}^*$ (i.e., the reduction to a quadratic function of the metric's first-order derivatives) enabled Einstein to make use of the mathematical machinery developed in his 1914 paper. Meanwhile, the problems he had struggled with in 1914 had been overcome: the theory was now generally covariant and his choice of a Lagrangian was no longer arbitrary (Norton 1984; Cattani and De Maria 1989b).

Einstein's next step was to split, like Hilbert, the Lagrangian into a gravitational and a matter part (see equation (8) above). Einstein concluded that in order to satisfy his principle of general relativity, the gravitational part of the Lagrangian "(up to a constant factor) must be the scalar of the Riemann curvature tensor; since there is no other invariant with the required properties" (Einstein 1916b, p. 1113). Closely following his 1914 variational approach, Einstein showed, using an infinitesimal coordinate transformation $x^{\mu\prime} = x^{\mu} + \Delta x^{\mu}$, that the condition $\mathcal{B}_{\mu} = 0$ (see equation (3) above) still holds. In fact, Einstein proved that this condition could be obtained by showing that $\Delta \int \mathcal{R}\, d\tau = \Delta \int \mathcal{R}^{*}\, d\tau$ where

$$\mathcal{R}^{*} = \sqrt{-g}\, g^{\mu\nu}\Big(\Gamma^{\beta}_{\mu\alpha}\,\Gamma^{\alpha}_{\nu\beta} + \Gamma^{\alpha}_{\mu\nu}\,\Gamma^{\beta}_{\alpha\beta}\Big).$$

Therefore, the relation $\mathcal{B}_{\mu} = 0$ now holds in every coordinate system, due to the invariance of $\mathcal{R}$ and to the principle of general relativity. $\mathcal{B}_{\mu}$ played a fundamental role in Einstein's new derivation of the conservation laws. In fact, according to Einstein, the gravitational equations could be explicitly written as equations (7). These equations allowed him to obtain, in a very straightforward way, the conservation laws. By multiplying equations (7) by $g^{\mu\nu}$ he obtained

$$\frac{\partial}{\partial x^{\alpha}}\left(\frac{\partial \mathcal{R}^{*}}{\partial g^{\sigma\mu}_{\alpha}}\, g^{\nu\mu}\right) = \chi\, (\mathcal{T}_{\sigma}{}^{\nu} + t_{\sigma}{}^{\nu}), \tag{13}$$

where

$$\mathcal{T}_{\sigma}{}^{\nu} = g^{\mu\nu}\frac{\partial \mathcal{M}}{\partial g^{\mu\sigma}} \tag{14}$$

and

$$t_{\sigma}{}^{\nu} \stackrel{\text{def}}{=} \frac{1}{\chi}\left(\frac{\partial \mathcal{R}^{*}}{\partial g^{\mu\sigma}_{\alpha}}\, g^{\mu\nu}_{\alpha} + \frac{\partial \mathcal{R}^{*}}{\partial g^{\mu\sigma}}\, g^{\mu\nu}\right).$$

When conditions (2)–(3) are imposed, it follows that

$$t_{\sigma}{}^{\nu} = \tfrac{1}{2}\left(\mathcal{R}^{*}\delta^{\nu}_{\sigma} - \frac{\partial \mathcal{R}^{*}}{\partial g^{\mu\nu}_{\alpha}}\, g^{\sigma}_{\mu\alpha}\right). \tag{15}$$

When equation (13) is differentiated with respect to x^{ν}, the left-hand side turns into $\mathcal{B}_{\mu}$. Since $\mathcal{B}_{\mu}$ vanishes, the relation obtained in this way is just equation (6), expressing conservation of total energy-momentum.

As in his previous theory, Einstein identified $\mathcal{T}_\sigma{}^\nu$ as representing the stress-energy density for matter and $t_\sigma{}^\nu$ as representing the stress-energy density of the gravitational field (Einstein 1916b, p. 1116). He concluded that although $t_\sigma{}^\nu$ was not a tensor, the equations expressing the conservation of total energy-momentum are generally covariant, since they were obtained directly from the principle of general relativity (Einstein 1916b, p. 1116). As we shall see, this claim led Levi-Civita, in 1917, to dispute not only the tensor character of $t_\sigma{}^\nu$ but also the equations Einstein used as his conservation laws for matter plus gravitational field (Cattani and De Maria 1989a).

7. Einstein's First Paper on Gravitational Waves (1916)

In another paper from 1916, Einstein tried to compute the components of $t_\sigma{}^\nu$ for the special case of a weak field, and in doing so discovered the existence of gravitational waves. The metric for the weak field is written, as usual, in the form

$$g_{\mu\nu} = \eta_{\mu\nu} + \gamma_{\mu\nu}, \tag{16}$$

where $\eta_{\mu\nu}$ is the Minkowski metric and $\gamma_{\mu\nu}$ (and its first-order derivatives) are infinitesimal quantities. In the weak-field approximation the field equations reduce to

$$\sum_{\alpha=1}^{4} \frac{\partial^2 \gamma'_{\mu\nu}}{\partial x^{\alpha 2}} = 2\chi \mathcal{T}_{\mu\nu}, \tag{17}$$

where

$$\gamma'_{\mu\nu} = \gamma_{\mu\nu} - \tfrac{1}{2}\gamma\delta_{\mu\nu}, \qquad \gamma \stackrel{\text{def}}{=} \gamma^\mu_\mu. \tag{18}$$

The quantities $\gamma'_{\mu\nu}$ are defined only up to a gauge transformation. Einstein therefore imposed the gauge condition

$$\sum_{\nu=1}^{4} \frac{\partial \gamma'_{\mu\nu}}{\partial x^\nu} = 0.$$

In this way, he found solutions of the weak-field equations, vanishing at infinity, that are the analogs of retarded potentials in electrodynamics. Therefore, according to Einstein, "gravitational fields propagate as waves with the speed of light" (Einstein 1916a, p. 692). Multiplying equation (17) by $\partial\gamma'_{\mu\nu}/\partial x^\sigma$, Einstein obtained the conservation law for the total energy-momentum in the usual form (6), where

$$t_{\mu\nu} = \frac{1}{4\chi}\left[\sum_{\alpha\beta} \frac{\partial \gamma'_{\alpha\beta}}{\partial x^\mu}\frac{\partial \gamma'_{\alpha\beta}}{\partial x^\nu} - \tfrac{1}{2}\delta_{\mu\nu}\sum_{\alpha\beta\tau}\left(\frac{\partial \gamma'_{\alpha\beta}}{\partial x^\tau}\right)^2\right]. \tag{19}$$

In deriving the conservation law, however, Einstein made a trivial mathematical error (he used $\gamma'^{\alpha\beta}$ instead of $\gamma^{\alpha\beta}$ in the conservation law for matter). As we shall see, two years elapsed before Einstein discovered this "regrettable error in computation" (Einstein 1918b, p. 154). The error caused some "strange results" (Einstein 1916a, p. 696). Einstein obtained three different types of gravitational waves compatible with equation (17): not just longitudinal and transversal ones but also a "new type" of wave (Einstein 1916a, p. 693). Using equation (19) to compute the energy carried by these waves, he found the paradoxical result that no energy transport was associated with either the longitudinal or the transversal waves. He tried to explain this absurdity by treating these waves as fictitious:

> The strange result that there should exist gravitational waves without energy transport... can easily be explained. They are not "real" waves, but "apparent" ones, because we have chosen as the coordinate system the one vibrating as the waves. (Einstein 1916a, p. 696)

Einstein found that only the third kind of waves transport energy. He concluded, however, that the mean value of the energy radiated by this new type of waves was very small, because of a damping factor $1/c^4$ and because of the small value of the gravitational constant χ $(= 1.87 \cdot 10^{-27})$ that entered into its expression. Still, the possibility of gravitational radiation was bothersome. As Einstein stated in his paper:

> Nevertheless, due to the motion of the electrons in the atom, the atoms should radiate not only electromagnetic energy, but also gravitational energy, though in a little quantity. Since, this does not happen in nature, it seems that the quantum theory should modify not only the electrodynamics of Maxwell, but also the new theory of gravitation. (Einstein 1916a, p. 696)

8. Levi-Civita's 1917 Article

Einstein's choice of a noncovariant stress-energy complex (Einstein 1916b) and his strange results on gravitational waves (Einstein 1916a) motivated Levi-Civita to try and find a satisfactory definition of a gravitational stress-energy tensor in Einstein's theory (Levi-Civita 1917). In Levi-Civita's opinion, it was Einstein's use of pseudotensor quantities that led to his physically unacceptable results on gravitational waves. He wrote:

> The idea of a gravitational [stress-energy] tensor belongs to the majestic construction of Einstein. But the definition proposed by the author is unsatisfactory. First of all, from the mathematical point of view, it lacks the invariant character it should have in the spirit of general relativity.

> More serious is the fact, noticed also by Einstein, that it leads to a clearly unacceptable physical result regarding gravitational waves. He thought that the way out of this last problem was through the quantum theory. . . . Indeed, the explanation is closer at hand: everything depends on the correct form of the gravitational [stress-energy] tensor. (Levi-Civita 1917, p. 381)

In Levi-Civita's opinion, general relativity called for a generally covariant gravitational stress-energy tensor. Since no differential invariants of the first order exist, one cannot have a stress-energy tensor containing only first-order derivatives of the metric; and, since the definition of $t_\sigma{}^\nu$ in (Einstein 1916b) only contains first-order derivatives, Levi-Civita concluded that "Einstein's choice of the gravitational tensor is not justified" (Levi-Civita 1917, p. 391). Levi-Civita, in fact, showed that Einstein's stress-energy complex was covariant under linear transformations only. He proposed a new candidate for the gravitational stress-energy tensor, and, consequently, a new candidate for the conservation law.

Starting from the Ricci tensor $R_{\mu\nu}$, Levi-Civita, like Hilbert in 1915, defined $G_{\mu\nu} = R_{\mu\nu} - \frac{1}{2}\, g_{\mu\nu}\, R$ and wrote the gravitational field equations in the form of (9). Using, for the first time, the contracted Bianchi identities, Levi-Civita showed that the covariant divergence of $G_\mu{}^\nu$ vanishes: $\nabla_\nu G_\mu{}^\nu = 0$. Consequently, $\nabla_\nu \mathcal{T}_\mu{}^\nu = 0$. This conservation law for matter will hold, Levi-Civita pointed out, since "$\mathcal{T}_\mu{}^\nu$ includes the complete contribution of all phenomena (but gravitation) which take place at the point in time under consideration" (Levi-Civita 1917, p. 389).

Levi-Civita now made a move similar to the one we saw Lorentz make earlier: he proposed to interpret equation (9) both as field equations and as conservation laws. Defining the stress-energy tensor for the gravitational field as

$$A_{\mu\nu} \stackrel{\text{def}}{=} \frac{1}{\chi}\mathcal{G}_{\mu\nu} = -\mathcal{T}_{\mu\nu} \quad \Rightarrow \quad A_{\mu\nu} + \mathcal{T}_{\mu\nu} = 0, \tag{20}$$

he identified

> $A_{\mu\nu}$ as the components of a [stress-]energy tensor of the space–time domain, i.e., depending only on the coefficients of ds^2. Such a tensor can be called both gravitational and inertial, since gravity and inertia simultaneously depend on ds^2. (Levi-Civita 1917, p. 389)

According to Levi-Civita, $A_{\mu\nu}$ completely characterizes the contribution of gravity to the local mechanical behavior. With this interpretation, it follows from equation (20) that no net flux of energy can exist. This equilibrium is guaranteed by the "real" existence of both quantities which, being tensors, are independent of the choice of coordinates. Hence,

> [n]ot only the total force applied to every single element vanishes, but also (taking into account the inertia of the $A_{\mu\nu}$) the total stress, the flux, and the energy density. (Levi-Civita 1917, p. 389)

So, for Levi-Civita, the gravitational stress-energy is characterized by the only element independent of the coordinates, the Riemann tensor.

In Levi-Civita's approach, the problems that Einstein ran into are avoided. Einstein had to admit the possibility that gravitational waves transporting energy are generated in the absence of sources. Einstein's weak-field equations have solutions for $\mathcal{T}_{\mu\nu} = 0$ representing such spontaneous gravitational waves. Moreover, the energy flux, computed on the basis of equation (17), could be zero in one coordinate system and nonzero in another. Einstein invoked the help of quantum theory to solve these problems. Levi-Civita claimed that it was enough to define the gravitational stress-energy tensor the way he suggested and to reinterpret the field equations accordingly. This precludes all counterintuitive situations of the sort Einstein encountered, for, according to (20), the gravitational stress-energy tensor $A_{\mu\nu}$ vanishes whenever the stress-energy tensor $\mathcal{T}_{\mu\nu}$ for matter vanishes.

9. Einstein's Response to Levi-Civita

In the summer of 1917, the Great War still raging on, Einstein went on a vacation trip to his home country, neutral Switzerland. While there, the mathematician Adolf Hurwitz gave him a copy of Levi-Civita's paper (Levi-Civita 1917), which had just been published in *Rendiconti dell'Accademia dei Lincei*. From Lucerne, on August 2, 1917, Einstein wrote a long letter to Levi-Civita,[6] still in Padua (which was very close to the war front), in order to rebut the latter's criticism of his theory, especially his use of a pseudotensor to represent gravitational stress-energy. Einstein gave some physical considerations to show that the stress-energy of the gravitational field cannot be represented by a generally covariant tensor.

Einstein began his letter expressing his admiration for Levi-Civita's "beautiful new work":

> I admire the elegance of your method of calculation. It must be nice to ride through these fields upon the horse of true mathematics, while people like me have to make their way laboriously on foot.... I still don't understand your objections to my view of the gravitational field. I would like to tell you again what causes me to persist in my view. (Einstein to Levi-Civita, August 2, 1917, p. 1)

He then proceeded to discuss the example of a counterweight pendulum clock to show that Levi-Civita's choice of a tensor to represent the stress-energy of the gravitational field is problematic from a physical point of view:

> I start with a Galilean space, i.e., one with constant $g_{\mu\nu}$. Merely by changing the reference system [i.e., by introducing an accelerated reference system], I obtain a gravitational field. If in K' a pendulum clock driven by a weight is set up in a state in which it is not working, gravitational energy is transformed into heat, while relative to the original system K, certainly no gravitational field and thereby no energy of this field is present.[7] Since, in K, all components of the energy "tensor" in question vanish identically, all components would also have to vanish in K', if the energy of gravitation could actually be expressed by a tensor. (Einstein to Levi-Civita, August 2, 1917, p. 1)

If gravitational stress-energy could be expressed by a tensor, no gravitational process could occur in K', in which case, contrary to experience, gravitational energy could not be transformed into heat. In short, the pendulum clock example shows that it should be possible for the components of gravitational stress-energy to be zero in one reference frame and nonzero in another. Therefore, gravitational stress-energy cannot be represented by a generally covariant tensor. Notice how Einstein's reasoning here is deeply rooted in his conception of the equivalence principle.

To the physical argument of the pendulum clock, Einstein adds an argument against the tensor character of gravitational stress-energy of a more mathematical nature:

> In general, it seems to me that the energy components of the gravitational field should only depend upon the first-order derivatives of $g_{\mu\nu}$, because this is also valid for the *forces* exerted by the fields.[8] *Tensors* of the first order (depending only on $\partial g_{\mu\nu}/\partial x^{\sigma} = g_{\sigma}^{\mu\nu}$), however, do not exist. (Einstein to Levi-Civita, August 2, 1917, pp. 1–2)

In his letter, Einstein went on to criticize Levi-Civita's interpretation of the gravitational field equations (20) as conservation laws. Einstein gave some examples showing that such conservation laws would have strange and undesired consequences. He wrote to Levi-Civita,

> You think that the field equations ... should be conceived of as energy equations, so that $\left[\mathcal{G}_{\mu}^{\sigma}\right]$ would be the [stress-]energy components of the gravitational field. However, with this conception it is quite incomprehensible how something like the energy law could hold in spaces where gravity can be disregarded. Why, for example, should it not be possible on your view for a body to cool off without giving off heat to the outside? (Einstein to Levi-Civita, August 2, 1917, p. 2)

On Levi-Civita's proposed definition of the conservation laws, the only way for matter to lose energy, it seems, is to transfer it locally to the gravitational field. It does not seem to allow for the possibility of energy transfer from one place to another.

At the same time, Levi-Civita's proposal did seem to allow for processes one would like to rule out. Einstein wrote:

> The equation
> $$\mathcal{G}_4^4 + \mathcal{T}_4^4 = 0 \tag{21}$$
> allows $\mathcal{T}_4^4$ to decrease everywhere, in which case this change is compensated for by a decrease of the, physically not perceived, absolute value of the quantity $\mathcal{G}_4^4$.... I maintain, therefore, that what you [Levi-Civita] call the energy law has nothing to do with what is otherwise so designated in physics. (Einstein to Levi-Civita, August 2, 1917, p. 2)

On these grounds, Einstein rejected Levi-Civita's interpretation of the field equations as conservation laws, and held on to his earlier formulation of the conservation laws (6). He argued that this formulation was perfectly sensible from a physical point of view, even though it involved a pseudotensor representing gravitational stress-energy:

> [My] conclusions are correct, whether or not one admits that the t_ν^σ are "really" the components of the gravitational [stress-]energy. That is to say, the relation
> $$\frac{\mathrm{d}}{\mathrm{d}x^4}\left\{\int \left(\mathcal{T}_4{}^4 + t_4{}^4\right) \mathrm{d}V\right\} = 0$$
> holds true with the vanishing of $\mathcal{T}_\sigma{}^\nu$ and $t_\sigma{}^\nu$ at [spatial] infinity, where the integral is extended over the whole three-dimensional space. For my conclusions, it is only necessary that $\mathcal{T}_4{}^4$ be the energy density of matter, which neither one of us doubts. (Einstein to Levi-Civita, August 2, 1917, p. 2)

Finally, Einstein pointed out that, in his definition, the gravitational stress-energy exhibits the desired behavior at spatial infinity:

> ... (in the static case) the field at infinity must be completely determined by the energy of matter and of the gravitational field (taken together). This is the case with my interpretation.... (Einstein to Levi-Civita, August 2, 1917, p. 2)

10. Levi-Civita's Response to Einstein

At the end of August 1917, Einstein received Levi-Civita's answer,[9] full of flattery as well as criticism:

> I am very grateful that you kindly appreciate the mathematics of my last articles but the credit of having discovered these new fields of research goes to you. (Levi-Civita to Einstein, August 1917, draft, p. 1)

In his letter, Levi-Civita criticized Einstein's definition of the gravitational field energy, wondering why a function of first-order derivatives of the metric tensor should be taken as stress-energy (pseudo)tensor, and asking for a more convincing motivation of this choice.

On the other hand, Levi-Civita granted Einstein that his interpretation of the field equations as conservation laws was not very *fecund*:

> I recognize the importance of your objection that, in doing so, the energy principle would lose all its heuristic value, because no physical process (or almost none) could be excluded a priori. In fact, [in order to get any physical process] one only has to associate with it a suitable change of the ds^2. (Levi-Civita to Einstein, August 1917, draft, p. 1)

Levi-Civita seems to be referring to Einstein's example of a stress-energy tensor for matter whose energy component decreases everywhere. Einstein's conservation laws (4) rule out such a stress-energy tensor. It looks as if Levi-Civita's conservation laws, i.e., the gravitational field equations, do not. It looks as if it would be possible for almost any matter stress-energy tensor to find a metric field such that the field equations are satisfied. The conservation laws thus seem to lose their "heuristic value" of restricting the range of acceptable matter stress-energy tensors. Of course, through the contracted Bianchi identities, the field equations do, in fact, restrict the range of acceptable matter stress-energy tensors.

In his letter, Levi-Civita stressed having no prejudice against a definition of gravitational stress-energy dependent on the choice of coordinates, or, as he put it,

> dependent on the expression of ds^2, in analogy with what happens for the notion of force of the field.... In the case of the equations of motion, written in the form
>
> $$\frac{d^2x^\nu}{ds^2} = -\begin{Bmatrix} \nu \\ \sigma\ \mu \end{Bmatrix} \frac{dx^\sigma}{ds}\frac{dx^\nu}{ds},$$
>
> one can explicitly connect the right-hand side (which does *not* define either a covariant or a contravariant system) with the ordinary notion of force. According to you, the same should happen for your $t_\sigma{}^\nu$ (which do not constitute a tensor). I am not in principle opposed to your point of view. On the contrary, I am inclined to presume that it is right as are all intuitions of geniuses. But I would like to see each conceptual step [canceled: logical element] to be clearly explained and described, as is done (or, at least, as is known can be done) in the case of the equation above, where we know how to recover the ordinary notion of force. (Levi-Civita to Einstein, August 1917, draft, pp. 1–2)

At the same time, Levi-Civita insisted that, at least from a logical point of view, there was nothing wrong with his own choice of a generally covariant tensor to represent gravitational stress-energy:

> [canceled: Let me add some opinions for a *logical* defense]. While I maintain an attitude of prudent reserve and wait, I still want to defend the *logical flawlessness* of my tensor $\mathcal{G}_{\mu\nu}$. (Levi-Civita to Einstein, August 1917, draft, p. 2)

Next, Levi-Civita attacked the counterweight pendulum-clock example:

> I want to stress that, contrary to what you claim, there is no contradiction between the accounts of the pendulum-clock in the two systems K and K', the first one fixed (in the Newtonian sense), the second one moving with constant acceleration. You say that:
>
> (a) in K, the energy tensor is zero because the $g_{\mu\nu}$ *are constant*;
> (b) in K', this is not the case; instead, there is a physical phenomenon with an observable transformation of energy into heat;
> (c) due to the invariant character of a null tensor, the simultaneous validity of (a) and (b) implies that there is something wrong with the premises.
>
> I contest (a), since we can assume that $g_{\mu\nu}$ are constant outside of the ponderable bodies, but [not] in the space taken up by your pendulum-clock. (Levi-Civita to Einstein, August 1917, draft, p. 2)

In other words, Levi-Civita denied that Einstein's pendulum clock example is incompatible with the tensor character of $A_{\mu\nu}$, observing that since the pendulum is not massless, strictly Euclidean coordinates cannot be assumed in K. Therefore, the energy tensor for gravitational field is different from zero both in K and in K'.

Finally, Levi-Civita responded to Einstein's comment on the behavior of the gravitational field at infinity:

> With regard to the last consideration of your letter (point 4), if I am not wrong, it [the behavior of the gravitational field at infinity] is not a consequence of the special form of your $t_\sigma{}^\nu$, but is equally valid for my $A_{\mu\nu}$. It seems to me that the behavior at infinity can be obtained from [our equation (20)] by using the circumstance that the divergence of the tensor $A_{\mu\nu}$ is identically zero; therefore, the divergence of $\mathcal{T}_{\mu\nu}$ also vanishes, and it reduces asymptotically to $\frac{\partial}{\partial x^\nu}\mathcal{T}_{\sigma\nu} = 0$, because the $g_{\mu\nu}$ tend to the values $\epsilon_{\mu\nu}$ [i.e., the constant Minkowski values of the metric tensor]. (Levi-Civita to Einstein, August 1917, draft, p. 2)

So, Levi-Civita invoked the contracted Bianchi identities to show that his conservation laws, like Einstein's, exhibit the desired behavior at spatial infinity.

In an addendum, Levi-Civita finally remarked:

> An indication in favor [of our equation (20)] is the negative value of the energy density of the gravitational field A_{00} (assuming $\mathcal{T}_{00} > 0$). This is in agreement with the old attempts to localize the potential energy of a Newtonian body, and explains the minus sign as due to the exceptional role of gravity compared to all other physical phenomena. (Levi-Civita to Einstein, August 1917, draft, p. 2)

11. Einstein's Second Paper on Gravitational Waves (1918)

After Levi-Civita's August 1917 letter, the polemic between the two scientists stopped until Einstein in 1918 published a new paper on gravitational waves (Einstein 1918b). In the introduction, he recognized that his earlier approach to gravitational waves (in Einstein 1916a)

> was not transparent enough, and it was marred by a regrettable error in computation. Therefore, I have to turn back to the same argument. (Einstein 1918b, p. 154)

Because of this error, he had obtained the wrong expression for his stress-energy complex. Correcting the error, Einstein could easily derive the correct expression for the stress-energy complex. As a consequence, he obtained only two kinds of waves, thereby resolving all the physical paradoxes of his previous results. Einstein could now assert with confidence that

> [a] mechanical system which always maintains its spherical symmetry cannot radiate, contrary to the result of my previous paper, which was obtained on the basis of an erroneous calculation. (Einstein 1918b, p. 164)

In the last section of (Einstein 1918b), entitled "Answer to an objection advanced by Mr. Levi-Civita,"[10] Einstein publicly gave his final reply to Levi-Civita's old objections. Einstein gave improved versions of some of the arguments already given in his August 1917 letter to Levi-Civita. He stressed that at least the time component of equation (6) must be looked upon as the energy equation, even if the $t^{\nu}{}_{\sigma}$ cannot be considered components of a tensor.

In this section of his paper, Einstein gave ample credit to Levi-Civita for his contributions to general relativity:

> In a recent series of highly interesting studies, Levi-Civita has contributed significantly to the clarification of some problems in general relativity. In one of these papers [Levi-Civita 1917], he defends a point

> of view regarding the conservation laws different from mine, and disputes my conclusions about the radiation of energy through gravitational waves. Although we have already settled the issue to the satisfaction of both of us in private correspondence, I think it is fitting, because of the importance of the problem, to add some further considerations concerning conservation laws.... There are different opinions on the question whether or not $t^{\nu}{}_{\sigma}$ should be considered as the components of the [stress-]energy of the gravitational field. I consider this disagreement to be irrelevant and merely a matter of words. But I have to stress that [our equation (6)], about which there are no doubts, implies a simplification of views that is important for the significance of the conservation laws. This has to be underscored for the fourth equation ($\sigma = 4$), which I want to define as the energy equation. (Einstein 1918b, p. 166)

Without entering into the mathematical details of $t_{\sigma}{}^{\nu}$, Einstein defended his energy equation with the following argument:

> Let us consider a spatially bounded material system, whose matter density and electromagnetic field vanish outside some region. Let S be the boundary surface, at rest, which encloses the entire material system. Then, by integration of the fourth equation over the domain inside S, we get

$$-\frac{\mathrm{d}}{\mathrm{d}x^4}\int_{\mathcal{V}}\left(\mathcal{T}_4{}^4 + t_4{}^4\right)\mathrm{d}V = \int_S\left(t_4{}^1\cos(nx_1) + t_4{}^2\cos(nx_2) + t_4{}^3\cos(nx_3)\right)\mathrm{d}\sigma.$$

> One is not entitled to define $t_4{}^4$ as the energy density of the gravitational field and $(t_4{}^1,\ t_4{}^2,\ t_4{}^3)$ as the components of the flux of gravitational energy. But one can certainly maintain, in cases where the integral of $t_4{}^4$ is small compared to the integral of the matter energy density $\mathcal{T}_4{}^4$, that the right-hand side represents the material energy loss of the system. It was only this result that was used in this paper and in my first article on gravitational waves. (Einstein 1918b, pp. 166–167)

Einstein then considered Levi-Civita's main objection against his choice of conservation laws:

> Levi-Civita (and prior to him, although less sharply, H.A. Lorentz) proposed a different formulation... of the conservation laws. He (as well as other specialists) is against emphasizing [equations (6)] and against the above interpretation because $t_{\sigma}{}^{\nu}$ is not a tensor. (Einstein 1918b, p. 166)

Although Einstein obviously had to admit that $t_{\sigma}{}^{\nu}$ is not a tensor, he concluded:

> I have to agree with this last criticism, but I do not see why only those quantities with the transformation properties of the components of a tensor should have a physical meaning. (Einstein 1918b, p. 167)

Finally, Einstein stressed that, even though there is no "logical objection" (Einstein 1918b, p. 167) against Levi-Civita's proposal, it has to be dismissed on physical grounds.

> I find, on the basis of [equation (20)], that the components of the total energy vanish everywhere. [Equation (20)], (contrary to [equation (6)]), does not exclude the possibility that a material system disappears completely, leaving no trace of its existence. In fact, the total energy in [equation (20)] (but not in [equation (6)]) is zero from the beginning; the conservation of this value of the energy does not guarantee the persistence of the system in any form. (Einstein 1918b, p. 167)

In fact, this result is due to the algebraic form of Levi-Civita's "conservation law" (according to which the total stress-energy is equal to zero everywhere). In Levi-Civita's opinion, the local vanishing of the matter stress-energy does not allow any energy flux. From a mathematical point of view, Levi-Civita's approach, with a generally covariant gravitational stress-energy tensor, was certainly more general than Einstein's, and apparently more in line with the spirit of general relativity. Einstein's choice, on the other hand, was more convincing on the basis of physical arguments, as Levi-Civita himself admitted. At the time, Einstein stood alone in his defense of a noncovariant definition of gravitational energy. Modern general relativists, however, follow Einstein's rather than Levi-Civita's approach to conservation laws.

12. Schrödinger's Example against Einstein's Stress-Energy Complex and Einstein's Reply

Lorentz and Levi-Civita were not the only two scientists to criticize Einstein's definition of gravitational stress-energy. In November 1917, Erwin Schrödinger showed, in a straightforward calculation, that, given a symmetrical distribution of matter, Einstein's gravitational stress-energy complex $t_\sigma{}^\nu$ can be zero in a suitable coordinate system. Schrödinger evaluated the stress-energy complex, starting from the Schwarzschild metric for the case of an incompressible fluid sphere of matter, and noticed that

> to determine $t_\sigma{}^\nu$, we must always specify the coordinate system, since their values do not have tensor character and do not vanish in every system, but only in some of them. The result we get in this particular case, i.e. the possibility of reducing $t_\sigma{}^\nu$ to be identically zero, is so surprising that I think it will need a deeper analysis.... Our calculation shows that there are some real gravitational fields whose [stress-]energy components vanish; in these fields not only the momentum and the energy flow but also the energy density and the analogs of the Maxwell

> stresses can vanish, in some finite region, as a consequence of a suitable choice of the coordinate system. (Schrödinger 1918, p. 4)

Thus, Schrödinger concluded,

> This result seems to have, in this case, some consequences for our ideas about the physical nature of the gravitational field. Since we have to renounce the interpretation of $t_\sigma{}^\nu$ ... as the [stress-]energy components of the gravitational field, the conservation law is lost, and it will be our duty to somehow replace this essential part in the foundation [of the theory]. (Schrödinger 1918, pp. 6–7)

About two and a half months later (on February 5, 1918), Einstein replied to Schrödinger in the same journal (Einstein 1918a). Oddly enough, Einstein started by raising further doubts about his choice of the quantities $t_\sigma{}^\nu$ to represent gravitational stress-energy:

> Schrödinger's calculations have shown that in a suitably chosen coordinate system all [stress-]energy components $t_\alpha{}^\sigma$ of the gravitational field [generated by a] sphere vanish outside of this sphere. Understandably, he was puzzled by this result, and so was I at first; in particular, he wondered whether $t_\alpha{}^\sigma$ should really be interpreted as [stress-]energy components.... To these doubts I can add two more:
>
> (1) the [stress-]energy components of matter $\mathcal{T}_\sigma{}^\nu$ represent a tensor, while this is not true for the "[stress-]energy components" of the gravitational field $t^\sigma{}_\nu$;
>
> (2) the quantities $\mathcal{T}_{\sigma\tau} = \sum_\nu \mathcal{T}_\sigma{}^\nu g_{\nu\tau}$ are symmetric in the indices σ and τ, while this not true for $t_{\sigma\tau} = \sum_\nu t_\sigma{}^\nu g_{\nu\tau}$.
>
> For the same reason as mentioned in point (1), Lorentz and Levi-Civita also raised doubts about interpreting $t_\alpha{}^\sigma$ as the [stress-]energy components of the gravitational field. Even though I can share their doubts, I am still convinced that it is helpful to give a more convenient expression for the energy components of the gravitational field. (Einstein 1918a, p. 115)

Einstein then offered the following explanation for Schrödinger's apparently strange result. He pointed out that a gravitational field generated by only one body, as in Schrödinger's example, is different from physical gravitational fields that always involve more than one body: "in gravitational fields mediating exchange effects between different bodies the quantities $t^\sigma{}_\nu$ cannot vanish identically" (Einstein 1918a, p. 115). As an example, Einstein considered two material bodies, M_1 and M_2, connected by a rigid rod. Using his conservation law, he found that since the stresses for matter are nonzero, the gravitational energy flux is nonzero as well. Therefore,

> [t]hese considerations hold *mutatis mutandis* in all those cases where the field transmits exchange effects between different bodies. But this is not the case for the field considered by Schrödinger. (Einstein 1918a, p. 116)

He concluded peremptorily:

> Hence, the formal doubts (1) and (2) cannot lead to a rejection of my proposal for the expression of the energy-momentum. It does not seem justified to put any further formal demands [on the properties of a quantity representing gravitational stress-energy]. (Einstein 1918a, p. 116)

13. Bauer's Example against Einstein's Stress-Energy Complex and Einstein's Final Reply

About one month after Einstein's reply to Schrödinger, Hans Bauer attacked Einstein's choice of $t^{\sigma}{}_{\nu}$ (Bauer 1918). He discussed an example complementary to Schrödinger's. Schrödinger had shown that Einstein's gravitational stress-energy sometimes vanishes despite the presence of a gravitational field. Bauer now showed that it does not always vanish in the absence of a gravitational field. He stressed that

> the partial nonvanishing of the [stress-]energy components has nothing to do with the presence of a gravitational field, but it is due only to the choice of a coordinate system.... This behavior is not surprising, since $t^{\sigma}{}_{\nu}$ is not a tensor. (Bauer 1918, p. 165)

So, Bauer thought he had thrown another stone at the physical plausibility of Einstein's proposal:

> we have to conclude that the "[stress-]energy components" $t^{\sigma}{}_{\nu}$ are not related to the presence of a gravitational field as they depend only on the choice of coordinates. They can vanish in presence of a field, as shown by Schrödinger, and do not always vanish in absence of a field, as shown below. Hence, their physical significance seems to be very dubious. (Bauer 1918, p. 165)

Einstein replied to Bauer's criticism without delay. In May 1918, he published a new reply to Schrödinger and Bauer (Einstein 1918c). He once again justified his choice with physical arguments. In his opinion,

> the theory of general relativity has been accepted by most theoretical physicists and mathematicians, even though almost all colleagues stand against my formulation of the energy-momentum law. Since I am convinced that I am right, I will in the following present my point of view on these matters in more detail. (Einstein 1918c, p. 448)

Einstein reminded his readers how special relativity combines the ordinary conservation laws of energy and momentum into one differential equation (i.e., the vanishing of the four-divergence of the stress-energy tensor) which is equivalent to the integral form of these conservation laws verified in experience. The generalization of this conservation law to general relativity, he explained, was particularly delicate. Einstein showed how, with his choice, "the classical concepts of energy and momentum are established as concisely as we are accustomed to expect in classical mechanics" (Einstein 1918c, p. 449). Then he demonstrated that the energy and momentum of a closed system are uniquely determined only when the motion of the system (considered as a whole) is expressed "with respect to a given coordinate system" (Einstein 1918c, pp. 449–450). In particular, he showed that the stress-energy of such closed systems can only be expected to transform as a tensor under certain coordinate transformations, viz. those coordinate transformations that reduce to the identity transformation at infinity. The transformations used in Schrödinger and Bauer's examples do not meet this requirement, so they do not count as counterexamples.

After this article by Einstein, the debate on the correct formulation of conservation laws in general relativity apparently came to the end.

14. Conclusions

In this chapter, we have described the polemic between Einstein and Levi-Civita on the correct formulation of conservation laws in general relativity during the years 1917–1918. Prompted by a mistake Einstein made in his first paper on gravitational waves, Levi-Civita criticized the use of non-covariant quantities in a generally covariant theory. This, in turn, stimulated Einstein to give a new and correct description of gravitational waves. Meanwhile, Lorentz had shown that there is no unique definition of the stress-energy of the gravitational field in general relativity. Following up on this insight, Lorentz proposed to interpret the field equations as conservation laws. Levi-Civita independently made the same proposal in a mathematically more satisfactory way, using the contracted Bianchi identities. Einstein held on to his old formulation of the conservation laws involving the pseudotensor $t_\sigma{}^\nu$ to represent the gravitational stress-energy. Schrödinger and Bauer showed that, in certain cases, Einstein's choice of $t_\sigma{}^\nu$ led to paradoxical results.

This episode makes for an interesting case study in the history of general relativity for at least two reasons: (1) it clarifies the connections between variational methods and conservation laws in general relativity and their cross-fertilization; (2) it shows the extent of Einstein's scientific isolation in his efforts to complete the edifice of general relativity during 1916–1918. Some of the most celebrated mathematical physicists, such as Lorentz and Levi-Civita, attacked his choice of a pseudotensor to represent gravitational stress-energy on the basis of formal mathematical arguments very much in the spirit of general relativity. Moreover, two young theoretical physicists, Schrödinger and Bauer, came up with some apparently damning counterexamples against Einstein's choice. Yet Einstein, masterfully exploiting the equivalence principle as a heuristic tool, stubbornly defended his choice and justified it with strong physical arguments. By today's standards, he was right.

ACKNOWLEDGMENTS. The authors wish to thank J. Stachel for his critical reading of a preliminary version of the manuscript and M. Janssen for many useful suggestions and his thorough editing of this article.

NOTES

[1] See also Cattani's chapter "Levi-Civita's Influence on Palatini's Contribution to General Relativity" in this volume.

[2] With his 1914 choice of H, $\mathcal{B}_\mu$ explicitly is

$$\mathcal{B}_\mu = \frac{\partial^2}{\partial x^\nu \partial x^\alpha}\left((-g)^{1/2} g^{\alpha\beta} g_{\sigma\mu} \frac{\partial g^{\mu\nu}}{\partial x^\beta}\right).$$

[3] For a more extensive discussion of these calculations, see Norton (1984).

[4] Einstein defined the pseudotensor t_σ^ν as (Einstein 1914, p. 1077)

$$t_\sigma{}^\nu \overset{\text{def}}{=} \frac{1}{\chi}\left(-g^{\nu\tau}\frac{\partial H(-g)^{1/2}}{\partial g^{\sigma\tau}} - g_\alpha^{\nu\tau}\frac{\partial H(-g)^{1/2}}{\partial g_\alpha^{\sigma\tau}}\right),$$

in order to show explicitly its dependence on H.

[5] In this period physicists meant stress-energy tensor when they said energy-tensor.

[6] Einstein to Levi-Civita, August 2, 1917, Einstein Archive, Boston (EA 16-253). English translation by J. Goldstein and E.G. Straus with some modifications.

[7] Let us examine Einstein's pendulum clock example a little more closely. In K, the reference frame in which there is no gravitational field, the clock is not working since the counterweight that should drive it is not subjected to a gravitational field. Let us take a concrete example. Suppose our clock is in a spacecraft far from any masses with its engines turned off (frame K). In this case, the clock is in a situation of "absence of weight," and consequently cannot work. When the engines are turned on, the spacecraft accelerates (frame K'). Consequently, all objects inside the spacecraft experience an apparent gravitational field. Our clock will want to start working under the influence of this field. If, in K', we want to prevent this, the clock's gravitational energy will be transformed into heat.

[8] Here Einstein presumably alludes to the fact that in general relativity gravitational forces are expressed in terms of the Christoffel symbols, which contain first-order derivatives of the metric only.

[9] Levi-Civita to Einstein, August 1917. Only a draft of this letter survives (Levi-Civita Papers, Accademia dei Lincei, Rome). It seems reasonable, though, to assume that the actual letter was not all that different from the draft.

[10] "Antwort auf einen von Hrn. Levi-Civita herrührenden Einwand," Einstein 1918b, pp. 166–167.

REFERENCES

Bauer, Hans (1918). "Über die Energiekomponenten des Gravitationsfeldes." *Physikalische Zeitschrift* XIX: 163–166.

Cattani, Carlo and De Maria, Michelangelo (1989a). "Gravitational Waves and Conservation Laws in General Relativity: A. Einstein and T. Levi-Civita, 1917 Correspondence." In *Proceedings of the Fifth M. Grossmann Meeting on General Relativity*, D.G. Blair and M.J. Buckingham, eds. Singapore: World Scientific, pp. 1335–1342.

—— (1989b). "The 1915 Epistolary Controversy between A. Einstein and T. Levi-Civita." In *Einstein and the History of General Relativity*, D. Howard and J. Stachel, eds. Boston: Birkhäuser, pp. 175–200.

Einstein, Albert (1914). "Die formale Grundlage der allgemeinen Relativitätstheorie." *Königlich Preussische Akademie der Wissenschaften* (Berlin). *Sitzungsberichte*: 1030–1085.

—— (1915a). "Zur allgemeinen Relativitätstheorie." *Königlich Preussische Akademie der Wissenschaften* (Berlin). *Sitzungsberichte*: (I) November 4, 778–786; (II) November 11, 799–801.

—— (1915b). "Erklärung der Perihelbewegung des Merkur aus der allgemeinen Relativitätstheorie." *Königlich Preussische Akademie der Wissenschaften* (Berlin). *Sitzungsberichte*: November 18, 831–839.

—— (1915c). "Feldgleichungen der Gravitation." *Königlich Preussische Akademie der Wissenschaften* (Berlin). *Sitzungsberichte*: November 25, 844–847.

—— (1916a). "Näherungsweise Integration der Feldgleichungen der Gravitation." *Königlich Preussische Akademie der Wissenschaften* (Berlin). *Sitzungsberichte*: 688–696.

—— (1916b). "Hamiltonsches Prinzip und allgemeine Relativitätstheorie." *Königlich Preussischen Akademie der Wissenschaften* (Berlin). *Sitzungsberichte*: 1111–1116.

—— (1918a). "Notiz zu E. Schrödingers Arbeit: Die Energiekomponenten des Gravitationsfeldes." *Physikalische Zeitschrift* XIX: 115–116.

—— (1918b). "Über Gravitationswellen." *Königlich Preussische Akademie der Wissenschaften* (Berlin). *Sitzungsberichte*: 154–167.

—— (1918c). "Der Energiesatz in der allgemeinen Relativitätstheorie." *Königlich Preussische Akademie der Wissenschaften* (Berlin). *Sitzungsberichte*: 448–459.

Einstein, Albert and Grossmann, Marcel (1913). *Entwurf einer verallgemeinerten Relativitätstheorie und einer Theorie der Gravitation. I. Physikalischer Teil von Albert Einstein. II. Mathematischer Teil von Marcel Grossmann.* Leipzig and Berlin: B.G. Teubner. Reprinted, with added "Bemerkungen," in *Zeitschrift für Mathematik und Physik* 62 (1914): 225–261.

—— (1914). "Kovarianzeigenschaften der Feldgleichungen der auf die verallgemeinerte Relativitätstheorie gegründeten Gravitationstheorie." *Zeitschrift für Mathematik und Physik* 63: 215–225.

Hilbert, David (1915). "Die Grundlagen der Physik." *Königliche Gesellschaft der Wissenschaften zu Göttingen, Mathematisch-physikalische Klasse, Nachrichten:* (I) (1915): 395–407; (II) (1916): 53–76.

Janssen, Michel (1992). "H.A. Lorentz's Attempt to Give a Coordinate-Free Formulation of the General Theory of Relativity." In *Studies in the History of General Relativity*, Jean Eisenstaedt and A.J. Kox, eds., Boston: Birkhäuser, pp. 344–363.

Levi-Civita, Tullio (1917). "Sulla espressione analitica spettante al tensore gravitazionale nella teoria di Einstein." *Rendiconti Accademia dei Lincei* ser. 5, vol. XXVI: 381–391.

Lorentz, Hendrik Antoon (1915). "Het beginsel van Hamilton in Einstein's theorie der Zwaartekracht." *Koninklijke Akademie van Wetenschappen te Amsterdam. Verslagen van de Gewone Vergaderingen der Wis- en Natuurkundige Afdeeling* 23: 1073–1089; English translation: "On Hamilton's Principle in Einstein's Theory of Gravitation." *Koninklijke Akademie van Wetenschappen te Amsterdam. Proceedings of the Section of Sciences* 19: 751–767.

—— (1916). "Over Einstein's theorie der Zwaartekracht." *Koninklijke Akademie van Wetenschappen te Amsterdam. Verslagen van de Gewone Vergaderingen der Wis- en Natuurkundige Afdeeling* (I) 24, (1916): 1389–1402; (II) 24, (1916): 1759–1774; (III) 25, (1916): 468–486; (IV) 25, (1917): 1380–1396. English translation: "On Einstein's Theory of Gravitation," in H.A. Lorentz, *Collected Papers*. Vol. 5. P. Zeeman and A.D. Fokker, eds. The Hague: Martinus Nijhoff, 1937, pp. 246–313.

Mehra, Jagdish (1974). *Einstein, Hilbert and the Theory of Gravitation.* Dordrecht: D. Reidel.

Mie, Gustav (1912). "Grundlagen einer Theorie der Materie." *Annalen der Physik* (I) 37, (1912): 511–534; (II) 39, (1912): 1–40; (III) 40, (1913): 1–66.

Norton, John (1984). "How Einstein Found His Field Equations: 1912–15." *Historical Studies in the Physical Sciences*, 14: 253–316. Also printed in *Einstein and the History of General Relativity*, D. Howard and J. Stachel, eds. Boston: Birkhäuser, 1989, pp. 101–160.

Schrödinger, Erwin (1918). "Die Energiekomponenten des Gravitationsfeldes." *Physikalische Zeitschrift* XIX: 4–7.

The General-Relativistic Two-Body Problem and the Einstein–Silberstein Controversy

Peter Havas

1. Introduction: The "Problem of Motion" in the General Theory of Relativity

In 1933, Ludwik Silberstein, a Polish physicist, wrote Einstein that he had found an exact solution of the field equations of the general theory of relativity for the problem of two masses at rest. A lengthy correspondence ensued, which became more and more acrimonious and finally spilled over into the newspapers. To be able to understand the details of this controversy, it is necessary to outline earlier work on this problem, both by Einstein himself and by other scientists.

The two-body problem is an important part of the "Problem of Motion" in the general theory. I gave a talk on the early history of this problem at our 1985 conference, of which a slightly extended version is being published in the *Proceedings* (Havas 1989). To understand the problem under consideration and to put it in its proper historical perspective, it will be necessary, however, to repeat some of the earlier discussion as well as to elaborate on part of it and to provide some technical details.

In his initial formulation of the general theory, Einstein had assumed that—just as in Newtonian mechanics—the laws of motion are independent of the force laws or field equations responsible for the interactions between bodies, and he had postulated that a single mass point would move along a geodesic of the metric $g_{\mu\nu}$ describing the field. For a single body at rest, this assumption poses no difficulties, and the exact solution for such a body, obtained very early on (Schwarzschild 1916; Droste 1916a), remains untouched by the subsequent investigations of the problem of motion.

The first attack on the two-body problem is also due to Droste, a student of H.A. Lorentz who, in accordance with Einstein's ideas, assumed that it was possible to solve the field equations under the assumption that the bodies were permanently at rest and thus their field was static. He obtained an approximate solution (Droste 1915); he also obtained an approximate solution for *n* slowly moving bodies (Droste 1916b), but did not proceed far enough to realize that his method would lead to inconsistencies. Both Droste 1916a and 1916b were based on his University of Leiden thesis, which he defended in December (Droste 1916c). He briefly continued working with Lorentz; their important joint paper (Lorentz and Droste 1917) is discussed in Havas 1989. But then he moved into mathematics and did not publish anything further in relativity.

The general theory was developed and the investigations mentioned were carried out while the First World War was raging in Europe. Schwarzschild died shortly after finding his solution. Of the other early investigators in general relativity, Lorentz and his school were working in neutral Holland, Einstein in Berlin, Eddington and others in England. Although they were not completely isolated from each other, communication was difficult, and it is not possible to establish when (or sometimes if) they became aware of each other's results. De Donder, on the other hand, was working in complete isolation in German-occupied Belgium; although he seems to have obtained some important results before anybody else, he was not able to communicate them even to Lorentz in neighboring Holland without delays of many months.

Eddington was able to complete a report on the general theory of relativity for the Physical Society of London by June 1918 (Eddington 1918). From general considerations he came to the conclusion (p. 65) for particles of matter considered as singularities of the field that "the laws of motion of the singularities must be contained in the field equations." He later published a popular discussion of the theory of relativity in *Space, Time and Gravitation*; the French edition of this book (Eddington 1921) contained a 149-page mathematical supplement (apparently completed in October 1920) in whose section IV a much more detailed derivation of the law of motion is given.

This book was used in the preparation of an excellent introduction to relativity by Jean Becquerel, based on a course given by him for several years, whose section 87 is entitled "The law of motion of the free mass point is contained in the law of gravitation" (Becquerel 1922)[1] and essentially repeats the derivation given in Eddington 1921.

Thus, through both a French textbook and a French edition of an English book, French scientists had had access to this important result of the general theory of relativity for a year by the time Einstein visited Paris in 1922. However, there is no indication that Einstein himself was aware of it then or that it had been pointed out to him during his visit, although his host, Paul Langevin, had written the introduction to Eddington 1921. It is also doubtful that he ever studied Eddington's contributions to his theory in any detail.

Eddington was by no means the only scientist who had realized the connection between the field equations and the laws of motion by 1921. It was clearly recognized by De Donder in Belgium, whose derivation of the geodesic law from the field equations (De Donder 1919) is also presented in chapter III of his exposé of Einstein's theory (De Donder 1921). A derivation from the variational principle underlying the field equations was given by a Swiss physicist working in Göttingen (Humm 1918). The most important contributions, however, are due to Hermann Weyl, a German mathematician who was a professor at the ETH in Zurich from 1913 to 1930. He was therefore a colleague of Einstein before he left for Berlin, and since Switzerland was neutral during the war, they also had no difficulty communicating later.

2. Static Solutions in General Relativity

Initially, Weyl was concerned with static axially symmetric exact solutions of Einstein's field equations (Weyl 1917, 1919b; Bach 1922) (as was the Italian mathematician Levi-Civita [1917–1919]). In the course of this work he came to realize that two bodies interacting only gravitationally cannot be in equilibrium. More precisely, this is always the case for two extended bodies that can be separated by an open surface; if this is not possible, i.e., if one body encloses the other, equilibrium may be possible [the latter case was discussed much later in Marder (1959)]. This is exactly analogous to the situation in Newtonian mechanics. Weyl, however, was mainly concerned with bodies considered as singularities of the field (which of course can always be separated by a plane) and the remainder of this paper will be restricted to this case,[2] as well as to purely gravitational interactions; in the presence of other interactions, again just as in Newtonian mechanics, equilibrium may be possible.

In his first paper discussing axially symmetric static solutions, Weyl (1917) assumed that the bodies were held at rest by stresses counteracting the gravitational forces, without going into any detail. After the paper was criticized by Levi-Civita, he elaborated on this and indicated how the stresses can be calculated (Weyl 1919b). It is implicit in these papers that in Einstein's theory bodies cannot be in equilibrium under the influence of gravitational forces alone, but somewhat surprisingly it was nowhere stated that the fact that this follows from the field equations alone is an important new result of this theory. This was only done explicitly in Weyl's "Addendum" to R. Bach's paper on new solutions of Einstein's equations (Bach 1922),[3] which discussed "The Static Two-Body Problem" in full generality. After showing that Bach's calculations imply that two mass points are attracted by a force that, for masses whose gravitational radii are small compared to their separation, reduces to that given by Newton's law, Weyl concluded that

> The physical importance of this result should not be exaggerated; for the solution of the real two-body problem, the determination of the motion of two gravitationally attracting bodies, nothing is gained by it.

Nevertheless, the importance of his proof that there is no static solution for two masses that are free to move was widely, though not universally, recognized.

Within the next few years, a number of scientists attacked the static two-body problem, not always realizing the need for stresses to maintain equilibrium. (This requirement is now frequently stated as the need for a "strut" or "rod" between the bodies.) At about the time of the publication of Bach 1922, but clearly not aware of it and of earlier results by Weyl, a German mathematician published a paper claiming an exact solution for the static field of two mass points (Trefftz 1922). This claim was immediately disputed by Einstein himself (Einstein 1922) who showed that if one attempted to interpret Trefftz's solution as the field of two massive spheres, this would require the presence of a true singularity of the field outside the two masses and that

> therefore it is not permitted to continue the solution up to that spot. In reality it presupposes the existence of other extended masses distributed with spherical symmetry, as already shown by H. Weyl.

No reference to Weyl is given, and the papers by Weyl referred to earlier do not put his results into this form. Nevertheless, this passage shows that in late 1922 Einstein was aware of some of Weyl's work, although he did not realize that (just like Eddington's results) it implied that the field equations contained the equations of motion. Weyl had attacked the problem of finding the explicit form of these equations earlier within the context of his own generalization of Einstein's theory (Weyl 1919a) and elaborated on it in Weyl 1921a and in the third, more clearly the fourth, and especially in the fifth addition of his book *Raum-Zeit-Materie* (Weyl 1919c, 1921b, 1923). In the third and fourth editions, this elaboration was still done in the context of his own theory, which attempted to geometrize the electromagnetic field in addition to the gravitational one; however, a careful reading of his presentation leaves no doubt that all the mathematical and physical arguments remain valid in the absence of the electromagnetic field, in which case Weyl's theory reduces to Einstein's. In the fifth edition, however, Weyl considered the problem of motion purely in the context of Einstein's theory. Nevertheless, Einstein, having raised various objections to Weyl's theory earlier, apparently did not recognize the validity of Weyl's considerations on the problem of motion within his own theory and did not realize the connection between his field equations and the laws of motion until 1927.

In that year, he and his assistant Jakob Grommer published a paper containing a derivation of the geodesic law (Einstein and Grommer 1927) which until recently has been widely credited with being the first to recognize the connection between the field equations and the equations of motion. It also contained a discussion of Einstein's reasons for not having recognized this connection earlier. No discussion of earlier work was given, showing that he had not been aware that this connection had been discussed even in several standard presentations of his theory. This paper and the various arguments presented by Einstein are discussed in some detail in Havas 1989, and I shall not repeat this discussion here. I shall only note that Weyl reacted to the paper as soon as he had seen the galleys (shown to him by Herglotz) and wrote Einstein that "I must confess that I did not understand what in it goes beyond my earlier developments" (letter by Weyl, February 3, 1927, EA 24-086, in German). He then referred to his "Addendum" and to Weyl 1923, and to make sure that he would not be misunderstood he outlined the arguments given there in some detail. Einstein responded almost three months later, raising objections which only refer to electrically charged particles (EA 24-088), although Weyl's derivation was only concerned with neutral ones, apparently still under the impression that Weyl's treatment was restricted to his extension of Einstein's theory. Weyl's answer appears lost. In any case, in his later work Einstein never acknowledged the priority of Weyl's or any other author's contributions to recognizing that the field equations imply the laws of motion.

As noted before, much work on the relativistic two-body problem continued after Weyl's fundamental work. In 1922, the American mathematician Horace Levinson obtained his Ph.D. at the Department of Astronomy of the University of Chicago with a thesis on the gravitational field of masses at rest (Levinson 1922); the next year he received a *Doctorat d'Université* from the University of Paris with a thesis on the field of two mass points at rest (Levinson 1923a). Both theses derive only approximate solutions and show no recognition of the problem of motion, nor do his publications on the subject (Levinson 1923b, 1928). Although at least one of the examiners, Elie Cartan, was quite familiar with general relativity, clearly no objections were raised.[4]

Levinson continued investigations in general relativity as a sideline while working in business; his most significant contribution was a letter to Einstein (August 25, 1948, EA 16-300) criticizing the mathematical methods used in the famous EIH paper (Einstein, Infeld, and Hoffmann 1938) and its sequel (Einstein and Infeld 1940) to derive the approximate equations of motion of n bodies from the field equations. This criticism, discussed further in a lengthy correspondence, prompted Einstein to take up the problem again and, together with Infeld, to devise an alternate derivation (Einstein and Infeld 1949).

The problem of determining the field of two bodies at rest was also attacked in Palatini 1923 and Chazy 1923a, 1923b, 1924, apparently without any knowledge of Bach and Weyl's work or recognition of the need for stresses. Both authors gave exact solutions, but it was pointed out by Chazy that Palatini's solution did not reduce to Schwarzschild's if the two masses coalesced, while Chazy's did. The need for stresses was explicitly realized by Straneo (1924a, 1924b, 1924c). An excellent discussion of the early work on the two-body problem was given a few years later in a slender French monograph on general relativity (Darmois 1927). Some mathematical problems of the n-body problem were discussed in a thesis at the University of Paris (Racine 1934), which apparently has been universally overlooked, although the examination committee consisted of the most knowledgeable French physicists—Cartan, Chazy, and Darmois; this may be considered as divine retribution for the fact that it did not contain a single reference to non-French papers, except for Levi-Civita's, not even Weyl's.

The n-body problem was also treated by the British mathematician Harry Curzon (1880–1935), who had been "Recognized Teacher of Mathematics" at Goldsmiths' College of the University of London since 1906. His papers (Curzon 1924a, 1924b), his only contribution to physics, do not contain any references, and it seems that he was not aware of any previous work on the problem. However, he used the same method as Weyl and Levi-Civita to obtain static axially symmetric solutions, which leads to a two-dimensional Laplace equation in cylindrical coordinates. But while Weyl and Levi-Civita had recognized that the solution corresponding to that of Schwarzschild and Droste required a line singularity on the axis in the particular coordinate system employed, Curzon, without any comment, used point singularities instead, which, transformed to spherical coordinates, do not describe mass points but what later became known as Curzon singularities.[5] He also treated the case of such singularities carrying electric charges for which, as noted before, equilibrium is possible without the need for stresses. However, this problem was not discussed there. Curzon also does not seem to have recognized that his solutions did not represent mass points.

Curzon's one- and two-body solutions were rediscovered later by Silberstein, and will therefore be discussed in the context of the latter's controversy with Einstein.[6] It appears that Curzon's paper was totally ignored for a decade and not referred to in the literature before 1936.

3. The Einstein–Silberstein Controversy: A Tragicomedy of Errors in Two Acts

3.1 Dramatis Personae

Ludwik Silberstein was born in Warsaw in 1872. After initially studying in Cracow, he continued on to Heidelberg and Berlin, where he obtained his Ph.D. in 1894. He was *Assistent* in physics at the University of Lemberg (now Lvov, Ukraine) from 1895 to 1897, but was apparently unable to obtain a permanent position in Poland. He was *Libero Docente* (lecturer) in mathematical physics at the University of Bologna from 1899 to 1903, and from 1903 until 1912 he was at the University of Rome in the same capacity. While in Italy, he wrote a number of excellent texts in mathematics and physics (in Polish). In 1912 he moved to London, lectured on relativity at University College, and wrote one of the first treatments of the special theory of relativity (Silberstein 1914). It should be noted that he was one of a very small number of physicists working in relativity who was older than Einstein. He lectured on relativity and gravitation at Cornell University in 1920 and at the Universities of Toronto and of Chicago in 1921. Based on these lectures he wrote *Theory of General Relativity and Gravitation* (Silberstein 1922). In all of his writings, he showed great originality and revealed an independent and critical mind, occasionally more critical than the facts warranted.

At that time Silberstein was certainly not antagonistic either toward the theory of relativity or toward its creator. On the contrary, he wrote in the introduction of his book on general relativity:

> Some of my readers will miss, perhaps, the enthusiastic tone which usually permeates the books and pamphlets that have been written on the subject (with the notable exception of Einstein's own writings). Yet the author is the last man to be blind to the admirable boldness and the severe architectonic beauty of Einstein's theory. But it has seemed

that beauties of such a kind are rather enhanced than obscured by the adoption of a sober tone and an apparently cold form of presentation.

Nevertheless, Silberstein remained skeptical and ambiguous in his attitude toward Einstein's theory. Quite early he attempted to formulate a theory of gravitation that was generally covariant, but did not contain the principle of equivalence, which he considered to be the weak point of Einstein's theory on both theoretical and observational grounds (Silberstein 1918), no red shift having yet been observed. Here and on other occasions he was ready to accept experimental results uncritically if they seemed to contradict either the special or the general theory's predictions.

Silberstein stayed in London until 1920 and became a British subject. During his stay he continued working on relativity and earned his living as "Scientific Advisor" for Adam Hilger Ltd., a leading optical instrument maker, from 1915 until 1920; his expertise in optics dated to a period (1898–1899) as scientific codirector of an optical firm in Warsaw. (Some of the biographical information is taken from an undated—1921?—letter by Silberstein, University of Toronto Archives A67-0007/65 Falconer Papers.) In 1920 he was invited to join the research laboratory of Eastman Kodak as their leading scientific advisor. He moved to the United States in June and stayed with Eastman Kodak until his death in 1948—seven years before Einstein, but at the same age. It is not clear whether he went into industry by choice or, more likely, because he was unable to obtain a permanent academic position either in Britain or in the United States, possibly due his age and to the prevailing anti-Semitism at British and American universities between the two world wars. At Eastman Kodak he worked mostly in optics, but he maintained his interest in relativity.

Einstein and Silberstein had corresponded at length since 1918, mostly on inquiries by Silberstein concerning the theory of relativity, but also on various other matters, and appear to have become quite close. After Silberstein's move to the United States they met during Einstein's trip to this country in the cause of Zionism in the spring of 1921, at Princeton and possibly in Chicago. Although it is not directly related to our topic, I would like to discuss one exchange of letters just after Silberstein's stay at the University of Chicago, as it shows both the close relationship of the two men and their attitude toward the situation in Germany at the time, and reveals a little-known offer to Einstein. Having just returned to Rochester, Silberstein wrote on September 4,1921 (EA 21-046, in German; underlinings, here and in all subsequent quotations, in the original; signatures omitted):

On September 1st, Dr. Gale (full professor, coordinated with Millikan at the Ryerson Lab, and Dean of the Science Faculty, Univ. of Chicago)

> has asked me and urgently requested to feel you out "informally" whether you would be inclined to accept a position as professor in the Physics Department (seat of the Ryerson Lab), as "head" (leader) of studies and investigations in theoretical physics (not necessarily lectures, as long as it does not suit you) and, more or less, what your conditions would be.... You would receive all conceivable support enabling you to devote yourself freely to your research, in completely free cooperation with the experimental physicists in the Ryerson Lab. You would have to devote only as much (or as little) time to lectures as is convenient for you—especially as the faculty intends to engage an Assistant Professor of Theoretical Physics in America to help you*

(Added in a footnote: "* Dr. Gale offered me the prospect of this position; I told him I would be only too happy to work with you as my superior.")

> whose duties would include systematic lecturing in agreement with you. In short, you would have ideal conditions for your investigations.
>
> For my part, I would like to urge you to say "yes," the more so [the last three words in English] since I have recognized in Chicago in the past three months that the intellectual and also the social atmosphere there is really excellent. Instead of envy and hostile demonstrations you would find in Chicago the best sympathy, veneration, and friendship—and these are important factors for such an ideal (and affectionate) and sensitive man as you are.
>
> Although Frau Einstein had told me (in Princeton) that you had a moral "duty" (a perfectly mystical concept in the present case) [the phrase in parentheses in English] "not to leave the Germans who have, after all, lost almost everything" just now. But I am deeply convinced that Germany is not the right place for you.

(Added as a footnote: "By this I mean the atmosphere of the German professors, the Geheimräthe, the Hofräthe, etc.—since the working class in Germany is free of Junkerdom and other dirt."

> The Lenards, the Gehrkes, etc.—their name is legion—(possibly with the exception of Planck and the late Rudolph Virchow) are petty and simultaneously brutal individuals, Junkers and simultaneously miserable slaves of the Kaiser regime.[7]

The letter continued in the same vein, expressing sentiments exactly like those expressed by Einstein about Germany after the next world war—and about the German academic atmosphere since his early youth. Nevertheless, Einstein answered almost immediately, on October 4 (EA 21-048, in German):

> I was very touched that colleagues Gale and Michelson [note that Silberstein had mentioned Millikan, not Michelson] are ready to offer me this

> wonderful position. The prospect of working with these men and especially with you in close cooperation is extraordinarily attractive to me. I am also convinced that such a cooperation would be very satisfactory and fruitful. But still I cannot accept this beautiful call. While it is true that I have experienced some evil by my colleagues and students here, still I am rooted here so firmly by family and friendship ties that in the absence of a real emergency I could not make the decision to move to a totally new, even if very tempting, environment. If one has lived so long and has acquired human relationships, one would leave behind a large piece of oneself, and at my age I am not able to regenerate sufficiently to change my environment so completely without significant damage. Please transmit my heartfelt thanks to the colleagues; they will certainly be able to appreciate the inner conflict which does not permit me to make such a radical decision.

The attitude expressed here was not uncommon among assimilated Jews in Central Europe before Hitler came to power, especially within intellectual circles and among individuals active in the trade unions and in the various political parties of the left. But it is noteworthy because Einstein's letter was written precisely at the time he had embraced Zionism, completed a propaganda tour for it, and elsewhere—but nowhere in this letter—put more and more stress on his Jewishness.[8]

Chicago's offer and Einstein's refusal are not mentioned in any of his biographies, as far as I am aware, nor in Millikan's autobiography (Millikan 1950) or in Michelson's biography by his daughter (Livingston 1973). The prospective offer of a position for Silberstein was never mentioned by him again and seems to have been entirely contingent on Einstein's acceptance.

Silberstein answered Einstein's letter on December 11 (EA 21-051, in German), writing that he had passed on the letter to Dean Gale and had only received an answer two days earlier, from which he concluded that Einstein's "words had reached him and he had liked them very much in spite of the result which is sad for all of us."

3.2 Prologue

During his stay in Chicago, Silberstein had suggested to Michelson that he undertake a new test of the hypothesis that the ether is carried along by the earth, essentially a repetition of Sagnac's experiment with more powerful methods, and even promised to pay for it (Livingston 1973), an offer possibly made on behalf of Eastman Kodak. Michelson wrote later (Michelson 1925):

> ... at the urgent instance of Dr. Silberstein the writer was convinced of the importance of the work, notwithstanding serious difficulties which

> were anticipated in the way of raising the necessary funds.... Funds for this experiment, amounting to about $17,000, were furnished by the University of Chicago, with an additional contribution of $491.55 made through the efforts of Dr. Silberstein.

After unsuccessful open-air experiments had been performed at Mt. Wilson in the summer of 1923, the funds provided allowed the construction of a pipeline one mile long and a foot in diameter that could be evacuated. It was installed in Clearing, Illinois, and Silberstein wrote to Einstein on the progress of the experiment on April 15 (EA 21-052). It was carried out in late 1924 in his presence (Michelson and Gale 1925).[9]

The results of this experiment, like those of all of Michelson's previous ones, were in full agreement with those expected from the special theory of relativity. However, other results obtained by a former collaborator seemed to contradict the theory (Miller 1925). At the request of Science Service, a Washington-based organization that published a science news bulletin, Silberstein wrote a brief analysis of the as yet unpublished results, which appeared in the bulletin under the headline "NEW EXPERIMENTS MEAN DOWNFALL OF RELATIVITY" provided by the editor (EA 21-053), stating that those results could be explained "by means of the Stokes ether concept, as modified by Planck and Lorentz." A similar comment appeared in a letter to *Nature* (Silberstein 1925a), which was contradicted in Eddington 1925 prompting a brief rejoinder by Silberstein (1925b), and caution in any interpretation was advised in Giorgi 1925. Giorgi had simultaneously written to Einstein about this, asking his opinion (letter of July 14, 1925, EA 21-054, in Italian). Miller's results were reanalyzed much later and finally discounted (Shankland et al. 1955).

Ironically, Silberstein had written to Einstein as early as March 10, 1920 (EA 21-041), about the Stokes–Planck–Lorentz ether theory, sending him a reprint on the subject (Silberstein 1920) and asking his opinion. No answer has been preserved. The Miller controversy does not seem to have affected the tone of Silberstein's letters to Einstein, which had always been friendly and frequently quite deferential, even when he informed him that he had submitted a paper with the "impertinent" title "SPECIAL RELATIVITY OVERTHROWN BY DOUBLE STARS" (EA 21-044, 21-045) to a journal (he withdrew it before publication). An example of his deference is the beginning of the letter quoted earlier (EA 21-051):

> Dear Herr Kollege!
>
> If I address you like this, copying your own letter, this is only for the sake of the sacred principle of equality and comradeship, even though I had really "Most revered master" in mind.

There was no indication anywhere in their exchanges over two decades of any latent hostility or veiled irritation. They continued their correspondence, though sporadically, for the next decade.

Eleven years after refusing Chicago's offer, Einstein found himself in the presence of "a real emergency," which forced him to abandon Germany even without the inducement of an offer and made him renounce his former pacifism and adopt a permanent hostility toward Germany. Being eleven years older, he—like thousands of other refugees—was even less able "to regenerate without significant damage" than he had been at the age of 48, but he had no choice anymore.

3.3 Act I

Shortly after arriving in the United States, Einstein received a letter from Silberstein (December 3, 1933, EA 21-059). It started out in German:

> Dear Professor Einstein,
> First of all, I would like to greet you most heartily on the occasion of your arrival and settlement in America. Everybody here reveres and loves you, so that you without any doubt will feel very happy in your new home country. Furthermore I would like to beg you for your kind instruction in a question of relativity which has haunted me for some time and which seems to me to be fundamental. But since little by little I have lost fluency in the German language, I take the liberty of writing in English, the more so since you yourself probably use this language more and more.

This last assumption can only induce a smile in anybody who met Einstein in this country; he never became comfortable with the English language, and wrote all his letters in German, having them translated if necessary. Nevertheless, from this point on Silberstein always wrote in English and Einstein always answered in German; therefore no further reference will be made to the language of the various quotations. As to Silberstein's assertion, it should be kept in mind that his German was flawless, at least in writing, although it was not his native language; his English, on the other hand, was not, and all the awkward turns of phrase, occasional wrong choices of words, as well as the British spelling, in subsequent quotations are his own.[10]

Silberstein continued:

> A "free particle" placed in a metrical field $g_{\iota\kappa}$ describes a geodesic in that field. Outside of matter, and rejecting the λ-term, the field is determined by
>
> $$G_{\iota\kappa} = 0 \qquad (1)$$
>
> These are two main assumptions of your theory.

> Now suppose we have found a solution of (1), $g_{\iota\kappa} = g_{\iota\kappa}(x_1, \ldots, x_4)$, which has one or more singular "points" (or rather, four-dimensionally, singular lines), such e.g. as $r = 0$ for all x_4 in the case of the familiar Schwarzschild solution. Such a singular point can be interpreted as a mass-centre or particle. Are we entitled to consider it as a "free particle"? If so, then it should describe a geodesic of the field. In other words, the singular lines of the solution $g_{\iota\kappa}$ ought to be geodesics of the field $g_{\iota\kappa}$. [All underlinings are Silberstein's.] In fact, in the simplest, radially symmetric case corresponding to a unique mass-centre the singular line ($r = 0$, any x_4) is a geodesic, i.e., satisfies $\ddot{x}_\iota + \left\{ {\alpha\beta \atop \iota} \right\} \dot{x}_\alpha \dot{x}_\beta = 0$. But cases of two or more mass-centres have not been analyzed from this point of view (quite apart from the difficulty of producing such solutions).
>
> I would greatly appreciate your opinion on this matter. Such considerations may perhaps be helpful also for establishing your law of motion of free particles not by an independent act, but in intimate connection with the field equations themselves. I believe that you have yourself expressed the desirability of some such unification, though on different lines.
>
> I have in mind non-stationary fields corresponding to at least two mass-centres. I am fully aware, of course, of the insuperable mathematical difficulties in constructing such solutions. But it should be possible to read off the properties of such singular lines from the differential eqs. $G_{\iota\kappa} = 0$ without ever solving them.... The problem, restated concisely, is: It being assumed that a field $g_{\iota\kappa}$ satisfying $G_{\iota\kappa} = 0$ has singularities distributed along lines, to find the differential properties, of 2nd order, of these lines.#

(Added in a footnote: "# Without introducing, of course, a tensor of matter. The vanishing of the divergence of such a tensor for a pressureless medium readily gives (under certain conditions about ρ) for each element of the medium the equations $\ddot{x}_\iota + \left\{ {\alpha\beta \atop \iota} \right\} \dot{x}_\alpha \dot{x}_\beta = 0$.")

> If this yielded the geodesics, it would be an elegant result. But the problem is much beyond my power, & I would greatly appreciate to have your views on the whole question.

Before continuing with Silberstein's letter, two comments are in order. First, Silberstein was clearly aware of the possibility of deriving the geodesic law in the presence of a matter tensor, an approach taken, e.g., in Eddington 1918, whether or not he knew of this or similar derivations by others. Second, he appears to have shared Einstein's view that one should work with the vacuum field equations alone, and probably also shared his failure to see that such equations with a singularity actually correspond to singular energy-momentum tensors, a point which is discussed in detail in Havas 1989.

Now Silberstein came to the crucial problem:

In connection with this subject, I should like to ask a somewhat different question, namely, about the physical admissibility of solutions of $G_{\iota\kappa} = 0$. Consider a stationary axially symmetric field corresponding to two mass-centres. Levi-Civita's general ax. symmetr. solution is

$$ds^2 = e^{2\nu}\, dx_4^2 - e^{-2\nu}\left[e^{2\mu}(dx_1^2 + dx_2^2) + x_1^2\, dx_3^2\right]$$

where ν is any solution of the ordinary cylindrical Laplacian equation $\nabla^2\nu = \frac{1}{x_1^2}\frac{\partial}{\partial x_1}\left(x_1\frac{\partial\nu}{\partial x_1}\right) + \frac{\partial^2\nu}{\partial x_2^2} = 0$ and

$$\begin{aligned} d\mu &= \frac{\partial\mu}{\partial x_1}\, dx_1 + \frac{\partial\mu}{\partial x_2}\, dx_2 \\ &= x_1\left[\left(\frac{\partial\nu}{\partial x_1}\right)^2 - \left(\frac{\partial\nu}{\partial x_2}\right)^2\right] dx_1 + 2x_1\frac{\partial\nu}{\partial x_1}\frac{\partial\nu}{\partial x_2}\, dx_2 \end{aligned} \tag{2}$$

(this being a total differential in virtue of $\nabla^2\nu = 0$). The solution corresponding to a single mass-centre is immediate. Passing to two mass-centres, i.e. putting

$$\nu = -\frac{M_1}{r_1} - \frac{M_2}{r_2},$$

I find by some simple artifices, as a solution of (2),

$$\mu = -\frac{x_1^2}{2}\left(\frac{M_1^2}{r_1^4} + \frac{M_2^2}{r_2^4}\right) + \frac{2M_1M_2}{a^2}\left[\sqrt{1 - \frac{a^2x_1^2}{r_1^2r_2^2}} - 1\right].$$

This is accompanied by a sketch showing that Silberstein was using bipolar coordinates and that a is the separation of the two centers.

In the above, Silberstein had rediscovered the solution given in Curzon 1924a and 1924b and had fallen into the same trap. He then continued:

This field ν, μ is, then, a rigorous solution of $G_{\iota\kappa} = 0$, and it has only the two singular points $r_1 = 0$ and $r_2 = 0$, in fine, the mass-centres themselves. The field being stationary, the mass-centres will remain at rest, at an x_2-distance a, instead of falling towards each other, as we know, unofficially, from Newtonian physics. Now, it does not seem satisfactory to imagine that M_1 & M_2 are forced to remain at relative rest by a stress-system (as does Dr. Weyl; "stuetzende Spannungen"; R.Z.M., 5th ed., p. 257) [Weyl 1923] or say by a stiff rod placed between them. For this would mean the existence of a material tensor $T_{\iota\kappa}$, i.e. $G_{\iota\kappa} \neq 0$ within the rod, and even if the rod is made ideally thin, it would mean that the field has singularities all along the segment M_1M_2 of the axis, whereas such is not the case; the solution becomes at any point of the axis $\nu = -\frac{M_1}{r_1} - \frac{M_2}{r_2}$, $\mu = 0$, and this is singular only at M_1, M_2, and perfectly regular along the included segment.

This passage is crucial for the following discussion, and therefore it had to be quoted in full. It shows that Silberstein was fully aware of Weyl's results, but thought that he had found a counterexample.

He continued:

Such being the case, this field ν, μ seems to be entirely inadmissible and yet it is a rigorous solution of the eqs. $G_{\iota\kappa} = 0$. Whence the moral: not every solution of the field-eqs. is admissible. Now, in this flagrant case we happen to know (not from Relativity) that the solution is inadmissible. But there might be other, more subtle, cases in which no such extraneous knowledge would warn us.

It would, therefore, seem necessary to set up some more general criterion of admissibility or non-admissibility of a solution of the field equations,—always supposing that the Theory of Relativity is to be a self-contained doctrine not borrowing special information from other sources.

You would greatly oblige me, dear Professor Einstein, by giving me your views on these two points, and especially on the first one.

Apologizing for my prolixity in stating these subjects,

with kind regards, yours sincerely,

The tone of this letter is that of a disciple asking his master for advice, the same tone as had been adopted by Silberstein in all previous correspondence. This was soon to change, however.

Einstein responded two weeks later, on December 17 (EA 21-061):

At first I was taken aback by your static example with two masses, since I believed you that the space outside the mass points is regular. I was even more astonished since I myself had shown earlier that singularities will appear already in calculating the second approximation.

Actually, however, the solution given by you is singular, as shown by the following consideration. Your spatial line element is given by

$$e^{2(\mu-\nu)}(\mathrm{d}x_1^2 + \mathrm{d}x_2^2) + e^{2\mu}x_1^2\,\mathrm{d}x_1^2 = \mathrm{d}\sigma^2.$$

Einstein then proceeded to calculate the ratio of the circumference to the radius of a circle perpendicular to the axis and surrounding it. The details will not be given here, since he had already made a trivial error in the equation quoted above, as noted in Silberstein's response. Einstein then continued, having obtained the value $e^{-2\nu} \cdot 2\pi$ for the ratio:

But this ratio would have to be 2π for an infinitely small circle in the limit, which is not the case here for the x_2-axis. The field calculated therefore is singular everywhere on the x-axis.

From this, first of all, it follows that your example is not valid. It would be more interesting to prove the nonexistence of a static solution (whose singularities have the character of simple poles). I have shown this earlier at least for the second approximation (and also that for a "correctly" accelerated mass the singularity disappears). It can thus hardly be doubted that the field equations contain the law of motion, so that the geodesic hypothesis is unnecessary.

However, a really complete theory would exist only if the "matter" could be represented in it by fields and without singularities.

With thanks also for your friendly personal words and with friendly regards

Your

Einstein clearly had put his finger on one crucial error in Silberstein's argument; however, his own "consideration" was wrong, as immediately noticed by Silberstein. It should also be noted that Einstein did not recognize that the proof he had asked for as being more interesting had been provided by Weyl more than a decade earlier—in spite of the fact that Silberstein had mentioned Weyl's work in his letter.

Silberstein answered by return mail on December 20 (EA 21-062):

I wish to thank you for your kind letter of December 17th. Your verdict, however, I am sorry to say, is quite wrong. You have inadvertently misplaced the two exponents ν and μ.

As in my first letter

$$ds^2 = e^{2\nu}\,dx_4^2 - e^{2\nu}\{e^{2\mu}(dx_1^2 + dx_2^2) + x_1^2\,dx_3^2\}. \tag{1}$$

Thus the circumference of the circle you are contemplating is

$$C = 2\pi R\, e^{-\nu},$$

and its radius,

$$\rho = Re^{\mu-\nu},$$

whence,

$$\underline{C/\rho = 2\pi\, e^{-\mu}} \quad (\underline{\text{not}}\ e^{-\nu}2\pi).$$

Now,

$$\mu = -\frac{x_1^2}{2}\left(\frac{M_1^2}{r_1^4} + \frac{M_2^2}{r_2^4}\right) + \frac{2M_1M_2}{a^2}\left[\sqrt{1 - \frac{a^2x_1^2}{r_1^2r_2^2}} - 1\right] \tag{2}$$

vanishes rigorously for $x_1 = R \to 0$, so that

$$\lim \frac{C}{\rho} = 2\pi.$$

Thus the solution (1), with (2) and $\nu = -M_1/r_1 - M_2/r_2$, satisfies also your own requirement of regularity (elementally Euclidean behaviour).[11] The statements made in my first letter remain, therefore, in full rigour. Against your expectations, a statical solution with two (and, similarly, 3 or more) "singularities of simple pole character" does exist and, in view of its physical implications, it is imperative to deal with it in a fundamental way in order to uphold your gravitational theory.

I shall expect, with much interest, your views on this matter.

Einstein scribbled some calculations on this letter about the metric components and the Christoffel symbols as "First approximation" to check Silberstein's assertions; having found that, in this approximation, indeed

$$\left\{\begin{matrix}\alpha\\ \mu\,\nu\end{matrix}\right\}_{,\alpha} - \left\{\begin{matrix}\alpha\\ \mu\,\alpha\end{matrix}\right\}_{,\nu} = 0,$$

he wrote *stimmt* (correct) at the bottom. He then immediately, on Christmas eve, wrote to Silberstein (EA 21-063):

> I beg you to excuse my mistake. So it is true that there exists a static solution with only two pointlike singularities. What does this signify for the general theory?
>
> First of all it is clear that the general basis of the theory implies the correct law of motion.

He then proceeded to insist that

> Singularities must be excluded in principle in a field theory.... In any case, your investigation shows clearly how carefully one has to handle singularities and how empty is a field theory which allows singularities without precisely stipulating their character.

As mentioned in the discussion of Curzon's papers before, the singularities introduced by Curzon and Silberstein are *not* simple poles of the field. It is surprising that Silberstein still considered them to be such poles since he was familiar with at least some of Weyl's work on the two-body problem, as well as with Levi-Civita's. Einstein, who was not, seems to have fully accepted in the first paragraph of his letter Silberstein's characterization of his results, and then to have hedged on this issue, but did not question directly their significance as interpreted by Silberstein. Neither of them knew of Curzon's papers.[12]

On December 30, Silberstein replied (EA 21-064):

> Many thanks for your excellent letter of the 24th. I fully agree with you. It seems that, for the present, the best plan is to make the complete field-eqs (i.e. with $T_{\iota\kappa} \neq 0$) the master equations of the theory, and if somebody finds solutions of $R_{\iota\kappa} = 0$ with singularities, he has to test them by considering these singularities as small regions (slender world tubes), seats of $T_{\iota\kappa}$.... This settles, for the present, the subject proposed in my first letter, and I wish once more to thank you most cordially for the patience and kindness with which you have discussed it with me.

The curtain falls on a scene of mutual kindness and reconciliation built upon a shared error.

3.4 Entr'acte

In the same letter (EA 21-064) Silberstein discussed at some length "a certain result which I have found a few days ago and which seems to me very remarkable (so far as I know, it is new)." This result was, as he showed in a two-page calculation, that the most general spherically symmetric solution of $R_{\iota\kappa} = 0$ is "a statical field (the familiar Schwarzschild field) around a centre of necessarily constant 'mass' m." He then quoted from Einstein's previous letter a suggested requirement that the singularities should have temporally constant and spatially central symmetric character and stated that "in view of my result it is enough to make them rad. symmetrical; for then they will, eo ipso, also be constant in time." Einstein's answer of February 13, 1934, is lost, but Silberstein's belated response of September 16 shows that he had suggested that Silberstein should "correspond with Levicivita [*sic*] for the possibly existing literature on the subject & then, perhaps, publish my proof if it differs from the others." Obviously neither of the two had heard of Birkhoff's theorem, which was already known when Silberstein 1922 was published (Jebsen 1921; Alexandrow 1921; Birkhoff 1927). For whatever reason, Silberstein only published his result four years later; it will be discussed in Section 3.6.

These subjects were put aside at that point; the next few exchanges were mainly concerned with the problem of helping Hitler's victims. The situation of the Jews in Germany in 1933, still eight years away from the Holocaust, appears to have affected Silberstein psychologically more than Einstein, as seems evident from a 13-page rambling letter Silberstein himself called "passionate," written on September 23, 1934 (EA 21-070), in a haphazard mixture of German and English and in a handwriting differing from that of all other letters.

Although not returning to it in their correspondence, Einstein clearly was deeply disturbed by Silberstein's results and felt that the entire problem of interacting masses had to be treated in a different manner. This was done jointly with one of his current assistants, Nathan Rosen. In early May 1935, they submitted the manuscript of the famous "bridge" paper to *The Physical Review*, which appeared in the July 1 issue (Einstein and Rosen 1935). Silberstein's results were fully accepted and given as the prime motivation for the investigation. In the Abstract they described their method and results as having been

> led to modify slightly the gravitational equations which then admit regular solutions for the static spherically symmetric case. These solutions involve the mathematical representation of physical space by a space of two identical sheets, a particle being represented by a "bridge" connecting these sheets.

In spite of this different approach, to which they did not return in later years, Einstein and Rosen also continued to work on the problem of motion with particles treated as singularities. On vacation, Einstein wrote to Rosen in Princeton (September 8, 1935, EA 20-209) that he had found a "better form for the calculation of the many-body problem in first approximation. I believe that Lanczos[13] has once published something similar, but don't know it anymore exactly." Rosen commented on this method on the 20th (EA 20-210, in German), and added:

> Now I am trying the following: One can easily generalize the (isotropic) Schwarzschild solution for the case of a uniformly moving particle.... With this solution as model I am now looking for a solution of the equations for two particles which move along a line (of course not uniformly). Probably nothing will come of it.

Nothing did.

But he added, returning to Einstein's calculation:

> I am of the opinion that we have to start from the ordinary Schwarzschild solution, even though it is not singularity-free, because it is necessary to have the functions appearing in the equations as simple as possible to be able to find solutions.

Nothing came of these calculations either.

3.5 Act II

In 1935–1936, Silberstein again spent some time at the University of Toronto. He invited Einstein for a visit during the meeting of the American Astronomical Society, but nothing came of it. On September 23, 1935, he wrote him again (EA 21-074), requesting a reprint of Einstein and Rosen 1935, which he had seen in manuscript form, and mentioning that

> Paul Epstein (Pasadena) asked me to disclose to him how I got the complete solution of your field-eqs for two mass-points. This I sent him.... In reply he wrote me...saying that this is a "very important contribution" and urging me to publish it in detail.... I shall write out the whole investigation and send it as a paper to Phil. Mag. but before doing so I would like to hear your opinion: Is this solution (with two singularities, point singularities, which necessitated a revision of your whole theory and gave rise to your new attempts, is it in itself important enough to be worth a publication—in toto? Or should I merely publish the result, i.e. the final ds^2, axially symmetrical, with two point singularities?

Apparently not waiting for an answer, he sent off the paper containing all the details to *The Physical Review* in November, where it was received on the 25th, and informed Einstein of its submission.

Maybe Einstein was stung by the suggestion that he had revised his whole theory; in any case, this letter induced him to take another look at Silberstein's calculations. He wrote him on December 21, 1935 (EA 21-076), that "I also have to inform you that your example of the two mass-points at rest (calculated by the method of Weyl and Levi-Civita) has a critical flaw." Then he proceeded to repeat precisely the same mistake he had made when he had first looked at Silberstein's calculations and repeated the same objection as in his letter of December 17, 1933 (EA 21-061). He added that "If it would be possible for you to withdraw your publication on this matter, it would be better."

Not surprisingly, Silberstein hit the ceiling; and, having realized that the master was not infallible, from then on he changed the tone of his letters from that of a disciple to that of a rival. On December 28, he responded (EA 21-077):

> I am greatly puzzled by your statement.... Is it possible that you have quite forgotten that you have made the very same "objections" in December 1933 and that I have then shewn to you that you have just made a "clerical" error (misquoting my formula), nay, that you have then (Jan. 1934) written me a long letter apologizing heartily for your mistake?! And now you repeat exactly the same thing....

He then repeated his calculation of EA 21-062.

At last, Einstein really took a close look at the problem. On December 30, he replied (EA 21-079):

> Dear Mr. Silberstein:
>
> Now I remember very well that you already informed me of your argument concerning the two-body problem after I had claimed the appearance of a singularity along the axis. However, I let myself be convinced incorrectly, since this proof was wrong.
>
> You claim that
>
> $$l = -\frac{x_1^2}{2}\left(\frac{M_1^2}{r_1^4} + \frac{M_2^2}{r_2^4}\right) + \frac{2M_1M_2}{\overline{AB}^2}\left\{\sqrt{1 - \frac{(AB)^2 x_1^2}{r_1^2 r_2^2}} - 1\right\}$$
>
> vanishes everywhere outside the singularities on the axis $x_1 = 0$. But this presupposes that (without violating continuity) one can take the square root as positive everywhere.
>
> That this, however, is not the case, one can recognize thus: Calling α the angle between r_1 and r_2 and Δ the triangle [showing the sketch of a triangle with one side AB opposite the angle α, and clearly meaning that Δ is the area], then
>
> $$2\Delta = r_1 r_2 \sin\alpha = \overline{AB} x_1,$$
>
> thus
>
> $$\{\ \} = \sqrt{1 - \sin^2\alpha} - 1 = \mp\cos\alpha - 1.$$
>
> The sign of $\cos\alpha$ can be freely chosen, but one has to take it as the same in all of space, if one does not want to introduce a discontinuity in the first derivative. However one chooses the sign, one can not achieve that l vanishes everywhere on the axis.

He then elaborated on this argument, but Silberstein did not accept it. He replied on January 3, 1936 (EA 21-080):

> Dear Professor Einstein,
>
> Many thanks for your prompt answer to my letter. I am sorry to say that you are again wrong.... Of course I assume $\sqrt{\ }$ to have a fixed sign, namely +1, once for all, i.e. between A & B and outside the segment AB.
>
> Now, such being the case, we have not, as you put it,
>
> $$\left\{\ \right\} = \cos\alpha - 1,$$
>
> but
>
> $$\left\{\ \right\} = |\cos\alpha| - 1,$$
>
> and therefore for $\alpha = 0$, as well as for $\alpha = \pi$, $\alpha = 0, \alpha = \pi, \alpha = 0$,
>
> $$\left\{\ \right\} = 0;$$
>
> for $x_1 = 0$, $\lambda \to 0$, and $\frac{\text{Perimeter}}{\text{Diameter}}$ of circle equal π. This, I hope, will settle the matter.

Einstein, however, did not accept this. He responded on the 8th (EA 21-081):

> Dear Mr. Silberstein,
>
> I am not yet giving up the hope of convincing you of your error. You think that you can put in your l-expression
>
> $$\left\{\ \right\} = |\cos\alpha| - 1.$$
>
> I already mentioned that the first differential quotient of this function is discontinuous (in $\alpha = (2n+1)\frac{\pi}{2}$). One must consider $\alpha = \frac{\pi}{2}$ [accompanied by a sketch]. In this surface the differential equations are violated by your solution.
>
> If you still don't admit your error, I will write nothing about it anymore. I only beg you not to conclude from such silence that I assent.
>
> With friendly greetings

Silberstein answered on January 15 (EA 21-082) that "University lectures & some social pastimes have delayed a reply to your letter of Jan. 8, all these days. I am now ready to answer it." He then proceeded with a lengthy discussion, concluding:

> In fine, the $g_{\iota\kappa}$'s become infinite only at A, B and their derivatives are discontinuous at a certain surface passing through them. What of that? Why don't you consider this as an admissible gravitational field surrounding two mass-centres? The Schwarzschild solution for one

centre... (S) has a much more formidable singularity at the sphere $r = 2M$, namely

$$g_{44} = 0, \quad g_{11} = \infty,$$

yet neither you nor any other relativist has ever hesitated to use the line-element (S) as representing the field around a mass-centre.... If we apply to $M_1 + M_2$ the same leniency as to a single M (Schwarzschild solution), we must admit that your field-equations, $R_{\iota\kappa} = 0$, misrepresent fact and experience—giving two stars placed opposite each other.

Einstein, as promised, did not answer. This is unfortunate, since it would have been important to stress the difference between a coordinate singularity and a real one, e.g., by providing an invariant characterization. This difference had been recognized for the Schwarzschild solution at least since Eddington 1923.[14]

Silberstein wrote him again on February 10 (EA 21-083):

I pointed out to you that your invocation of these little singularities is but a "futile exercise," and a quite hair-splitting one.... Now, it greatly surprises me that instead of answering my letter of Jan. 15... you have told some reporters at Princeton (Feb. 9) that my conclusion "was based on an error," etc.... I am sorry to say that, while our correspondence in the past has been just & unimpeachable, your behaviour now in relation to my last letter and to your Princeton reporters strikes me as quite unfair. And I say this with much regret because I have always had the highest opinion of your objectivity and fairness in scientific polemics.... It is quite possible that the reporters have distorted your (Feb. 9) statements, as they certainly have distorted or exaggerated of late some of my statements in this matter.

Einstein again did not answer, and Silberstein grew frantic. On March 6 he wrote him again (EA 21-084):

I desire to remind you that my letters... have been left unanswered by you. And as they were preceded by some unfair and, in part, nonsensical remarks which you have given out to some Princeton reporters (published by the press in Feb. 9), I feel justified in assuming that you do not desire to continue any direct correspondence with me and that you prefer to drop your previous principle of fair scientific discussion and to embark on a non-geodesical (in plain English, crooked) way in dealing with your previous friend, and with the radical defect of your "great" gravitation theory.

I shall thoroughly conform my further actions to this assumption which (in view of your silence) I consider to be true to actual facts.

Yours faithfully,

The paper Silberstein had submitted in November appeared in the February 1 issue of *The Physical Review* (Silberstein 1935). It carried the provocative title "Two-Centers Solution of the Gravitational Field Equations, and the Need for a Reformed Theory of Matter," and stated that the solution "has singularities at A and B only, and not (as in R. Bach's and H. Weyl's physically trivial solution) along the straight segment joining these two points" and that it had been communicated

> to Einstein, pointing out, rather emphatically, that this is a case of a perfectly rigorous solution of his field equations and yet utterly inadmissible physically, so that one cannot henceforth treat "matter particles" as singularities of the field. This has, in fact, induced Einstein to attempt, in collaboration with N. Rosen, a new theory of matter.

It was the publication of this article which had brought the reporters to Einstein's door.

The article prompted Einstein and Rosen to submit a letter to the editor on February 17, which appeared in the March 1 issue (Einstein and Rosen 1936). After repeating the arguments of Einstein's letter to Silberstein of December 30, 1935 (EA 21-079), it stated that

> a closer investigation shows that the calculation can be carried through without the introduction of the square root and the resultant ambiguity of sign. One then finds that in the correct solution
>
> $$[\,] = \cos\alpha - 1.$$
>
> This, however, also fails to satisfy the regularity conditions....
>
> We should like to remark that, as shown in a letter to one of us, Professor C. Lanczos of Purdue University has independently recognized the error in Silberstein's paper.

Lanczos had written to Einstein on February 15 (EA 15-256, in German): "The last issue of Phys. Rev. contains an article by Mr. Silberstein, which is in complete opposition to the general expectations." After briefly outlining his own earlier approach to the problem of motion, concluding that

> While the indeterminacy of the field due to the omission of the matter tensor is quite large, one can nevertheless derive center-of-mass theorems which are largely analogous to the usual mechanical theorems for rigid bodies,

he continued:

> That the singularity concept is insufficient and would cause a large indeterminacy in the field, we all know, of course. But one can not forge a weapon against this conception out of the gravitational field alone, as this yields approximately the same as one would expect on the basis of classical physics.
>
> The fallacy in Mr. S.'s paper is contained in the square-root term in formula (10), p. 270.

He then provided a proof that this term leads to a discontinuity and that "one obtains again the usual mass line, which prevents the two masses from falling on each other." The proof is substantially the same as that given by Einstein in EA 21-079.

Neither Lanczos' letter nor the Einstein–Rosen letter to the editor mentions that the mass centers do not correspond to simple poles, nor is there any discussion of the difference between the line singularity of Silberstein's solution and the coordinate singularity of Schwarzschild's. None of them mention Curzon; on the other hand, the *Science Abstracts* summary of Silberstein's paper, written by McVittie, states at the outset that it had "rediscovered" the Curzon solution (McVittie 1936).

On March 7, 1936, a note appeared in *The Evening Telegram* of Toronto with the headline "Fatal Blow to Relativity Issued Here" and the subtitle "Told by Einstein That He's Wrong, Toronto Savant Makes New Attack on Theory." It started with

> Relatively speaking, the battle between Professor Einstein and Dr. Ludwik Silberstein, visiting lecturer at the University of Toronto, over a theory is warming up to frizzling point.

It then summarized Silberstein's article as showing that

> Einstein's gravitational theory was invalid and that the general theory of relativity hadn't a leg to stand on. Professor Einstein agreed that his gravitational theory required revision, but, answering the criticism in the current issue of the *Physical Review*, he charged Dr. Silberstein with conjuring "mathematical spooks" which had nothing to do with relativity.

(Actually, no such "spooks" are mentioned in the Einstein–Rosen letter.) Then it mentioned that Silberstein had sent another paper "to the *Physical Review* yesterday.... It's a follow-up which Dr. Silberstein contends gives the *coup-de-grace* to Einstein's gravitational theory."

Silberstein enclosed this note in a letter (EA 21-085) addressed to "Messrs. Einstein, Rosen, & Lanczos" and mailed to Lanczos. He wrote:

> Gentlemen,
>
> The new paper by the undersigned (mentioned in the attached clipping from *Evening Telegram*) fully disposes of Dr. Einstein's & Dr. Lanczos' rash & foolish objection to my solution of Feb. 1 (*Phys. Rev.*).

After adding a few calculations and again concluding that there is "No 'matter' between the centres," he ended by "Einstein and Rosen's idea of calling my solution 'mathematical spooks,' etc., is as foolish as it is unfair. E. & R.'s attitude strikes me as also vulgar. They will soon repent it."

This letter was transmitted to Einstein by Lanczos as requested on March 12, together with a lengthy letter of his own (EA 15-257, in German), starting:

> Dear Mr. Einstein! Dutifully, I am sending you the enclosed letter, as it is also addressed to you and Mr. Rosen. The situation with Mr. Silberstein is very regrettable, since he is obviously more and more doggedly stuck with these fixed ideas. In a further letter to me he informs me of his additional results regarding axially symmetric solutions (especially, that for certain *Ansätze* of the line element there exist only static solutions [but setting the $g_{i4} = 0$!]), but which, it seems to me, are all well known through the papers of the Italian school. I have tried as gently as possible to point out his error concerning the problem of motion, but given his high-strung state all this will not help much. Given the rigorous criticism common for American journals possibly his paper for *Phys. Rev.* will not even be accepted, which I am almost afraid of, since the rejection might cause the total collapse of his mental vigor. It is sad that in such cases one cannot do anything sensible, but after all one can not demand that one should swallow obvious errors for humane reasons.

(The Lanczos–Silberstein correspondence is not available, nor are Einstein's answers to the two letters from Lanczos.)

Before Einstein had received Lanczos' letter containing Silberstein's, however, he had already dashed off an irate response to Silberstein's letter of March 6 (EA 21-084) on March 10 (EA 21-087):

> Dear Mr. Silberstein,
>
> I have alerted you in two letters in detail to your mistake, and advised you to withdraw publication. In addition, the newspaper contained the idiotic claim that I had revised the general theory of relativity because of an earlier letter by you. By this you made it necessary for me to correct your errors publicly. Pauli told me, e.g., that I should absolutely do this, since the error was not so obvious that it could be noticed by any knowledgeable reader. Whether I will answer later publications by you on this subject will depend on whether I consider it necessary.
>
> With friendly greetings

Silberstein responded on the 17th (EA 21-088):

> Sweet Mr. Einstein,
>
> Your letter of Mar. 10 to hand. Its tone, and the "airs" you give yourself therein, greatly surprises me. But let us adhere to the principle of "sense of humour," especially cultivated by the Anglo-Saxons. And so, instead of barking at you, I send you herewith a refutation of your objections which is yet simpler than my proof (given in a paper sent a week ago to *Phys. Review*) that all $R_{\iota\kappa} = 0$ on the axis AB, & all $T_{\iota\kappa} = 0$

(no "matter" between the centres A, B).... This, I trust, fixes you up (U.S.A. slang) once & for all.

With friendly greetings,
Yours sincerely

"The rest is silence" (Shakespeare 1602). Silberstein's solution was never mentioned again, and no letters were exchanged for five years.

3.6 Epilogue

During those five years Nazi Germany had remilitarized the Rhineland, the civil war in Spain took its course, Austria suffered the "Anschluss," Munich produced the annexation of the Sudetenland and subsequently the occupation of all of Czechoslovakia, the attack on Poland started World War II, and, within less than a year, half of Europe had been occupied by Nazi Germany and the Soviet Union. Nuclear fission had been discovered, and Einstein had been induced to write to President Roosevelt about it.

Einstein, and probably also Silberstein, spent more and more time trying to help the victims of world events. But while they previously had on occasion collaborated in these efforts, they did not communicate even on these subjects now.

The paper mentioned in Silberstein's last letter (EA 21-088) and in the Toronto *Telegram* article was never published; whether it was withdrawn or rejected by *The Physical Review* is not known. Rosen left Princeton to accept an appointment at the University of Kiev and did not return to the problem of motion for more than a decade. The scientific public accepted the Einstein–Rosen letter as the final word on Silberstein's claims. As many Jewish and anti-Nazi scientists had to worry more about survival than about their research, and some of those who escaped as well as their former or new colleagues on both sides devoted their energies to war work, very little effort went into investigations of fundamental problems.

Silberstein had mentioned results on spherically symmetric solutions to Einstein in his letter of December 30, 1934, and similar results on axially symmetric solution in a (lost) letter referred to by Lanczos on March 12, 1936. He submitted an extended version to *The Philosophical Magazine* in November 1937, which was published shortly thereafter (Silberstein 1937). It had two parts: Part I (whose results, he noted, had been "communicated to Dr. Einstein in a private letter of December 1933"), consisted of the proof that any spherically symmetric solution of $R_{\iota\kappa} = 0$ can be put in a static form, without giving any references to previous work; this indicates that Silberstein had not gotten in touch with Levi-Civita (and that the referee also was not aware of Birkhoff's theorem, which by then had become quite well known and was, e.g., accorded a section in Tolman's famous monograph (Tolman 1934)).

Part II first gave a proof that if one assumes the form of the axially symmetric line element to be that which had been proved by Levi-Civita to be the most general static one, this form allows *only* static solutions of $R_{\iota\kappa} = 0$. This seems to be the same proof Silberstein had mentioned to Lanczos. But then he went on to give a "proof" that *all* axially symmetric solutions are static, and thus, in particular, there does not exist a solution for two mass centers moving along a line. This result is nonsense, on the face of it. In spite of this, it not only got past the referee, but it apparently has never been challenged directly in the literature. It probably escaped attention for a few years because of world events, and then was not noticed by the next generation of relativists—or by Weyl, who had also ignored Silberstein 1936. Although Weyl later joined Einstein in Princeton, nobody seems to have drawn his attention to Silberstein's claims. Of course, numerous examples of explicitly time-dependent solutions have been exhibited by several authors in the last fifty years, indirectly disproving Silberstein's result.

This was Silberstein's last publication in the area of relativity, although he published dozens of papers on other subjects, mostly in optics, between 1937 and his death eleven years later.

In the meantime, Einstein, in collaboration with Banesh Hoffmann and Leopold Infeld, had attacked the problem of motion from a different angle. The result, which they called the "new approximation method" for obtaining the equations of motion of n slowly moving particles, was published in 1938 (Einstein, Infeld, and Hoffmann 1938). Three years later, on February 8, 1941, Silberstein wrote to "Drs. A. Einstein, L. Infeld, and B. Hoffmann" (EA 21-089):

> Gentlemen,
>
> On Jan. 26th I have pointed out to Prof. Veblen certain fundamental objections to the method of attacking the "Problem of Motion" adopted in your paper of Jan. 1938 . . . namely the non-existence of spherically symmetrical point singularities, i.e. their incompatibility with the very structure of Einstein's gravitational field equations,—singularities which, nonetheless, you assume throughout your investigation.
>
> My impression is that this essential objection still holds and it has seemed worthwhile of bringing it directly to your notice, as the authors of that otherwise very interesting (although, by necessity, extremely laborious) method.

He then stated that he had written Veblen again on January 27 telling him the result of his computation of the perihelion motion, which differed from Einstein's both in sign and in magnitude, but then had found an error and now agreed with the old result as well as the one obtained in Robertson 1938 on the basis of the EIH equations, but "Since my own derivation of it seems to be more lucid than Robertson's, I give it here, in toto, in the belief that it may interest you. . . ." He ended the letter with "Nonetheless, the objection as to the rigorous non-existence of spherically symmetrical point singularities persists." Then he added a P.S. containing a calculation claiming the existence of a secular acceleration of the center of mass for two comparable masses.

Einstein replied on February 18 (EA 21-090), apologizing for the delay. He then wrote:

> Your computation of the two-body problem is pretty. But in the present case it is important to give the solution for nonvanishing mass ratio. Now as to your objection concerning the spherical symmetry of the singularity. Here it should be noted that the main interest of the entire consideration is that only that part of the space matters in which the field is regular (surface conditions). . . . Of course, in a complete field theory the positing of singularities is altogether forbidden. In the present case the introduction of singularities is justified because it allows treatment on the basis of the gravitational field alone of a problem of which "matter" is a part, without having to use a theory of the latter.
>
> Infeld and Hoffmann are no longer in Princeton, and I preferred not to bother them for the time being.
>
> With friendly greetings,
> Yours,

Silberstein was delighted. He replied immediately (February 21, EA 21-091):

> Dear Professor Einstein,
>
> Your letter of Feb. 18 has given more pleasure than I can say in words. The very fact that you have written to me at all after my discourteous letter of 1937 (or so), the outcome of a momentary passion, and thus have forgiven me, is a precious gift to me. For, having been since 1921, instinctively, your true friend, I have these last four years often reproached myself bitterly for that explosion of bad temper (originated in the two mass-centres problem). Well, I thank you most heartily for your spirit of goodwill and forgiveness. . . . I am naturally glad that you have found my treatment of planetary motion ($m_2 \gg m_1$) "pretty" . . . and that you have recognized the validity of my objection, viz. the non-existence of radially symmetrical point singularities in a field $R_{\iota\kappa} = 0$. I accept, at the same time, your views as to the (practical) necessity of working—with your method—with just such singularities. . . .

He then added more than four pages describing various results he had obtained using the EIH equations, raising various objections in that context.

Einstein replied on March 4 (EA 21-093):

> It has to be said again that I cannot see it as an objection that the gravitational interaction disturbs the central symmetry of the fields surrounding the particles.... Now I come to your new objections.... You conclude that for the motion along a line...the "mass center" is not in uniform motion. But you have to bear in mind that our coordinate system has no absolute significance.... Something similar occurs in your application of II to the motion of the "mass center" of the masses circling one another with constant separation. This result would show the absurdity of our formulae <u>if it would not rest on an error in the calculation</u>.... Thus the objections amount to nothing.

Silberstein responded on March 8 (EA 21-094) with an eight-page letter which started with "Many thanks for your interesting, and actually instructing, letter" and concluded with "Please, Prof. Einstein, have patience with me and teach me to conquer my ignorance in dealing with these intricate subtleties." It does not appear that he meant this ironically. In an undated reply (EA 21-095), Einstein wrote that he did not have the time to work through all the details of Silberstein's letter, but elaborated further on the significance of coordinates. The correspondence continued until the end of 1946, dealing with various topics, including Silberstein's questioning of the universal validity of $E = mc^2$; the tone was generally friendly.

Sometime in early 1941 Infeld received two letters by Silberstein, apparently containing objections to the method of EIH. He wrote to Einstein (undated, probably March 1941, EA 14-055): "Of course he is wrong. But I doubt that you will be able to convince him, because he is mentally unbalanced as I learned from people who know him well."

Given Infeld's style, this judgment should probably not be taken literally. What is surprising is that his letter gives the impression that he did not know Silberstein personally; both were of Polish origin, although one generation apart, and since 1938 Infeld had been at the University of Toronto, where Silberstein frequently had visited from Rochester, just across Lake Ontario. I have been unable to question anybody who actually knew Silberstein in his later years; however, Lanczos' "high-strung" comment seems to be justified on the basis of the tone of some of Silberstein's letters.

It is also quite clear from the correspondence between Einstein and Silberstein, however, that here were two proud and stubborn men—Silberstein even more so than Einstein, even less inclined to accept criticism of his work and, after the break in 1936, incapable of taking the initiative and apologizing to Einstein, although he knew that he was at fault; both frustrated, Silberstein by the lack of recognition of his by no means insignificant earlier work, attempting to show his mettle by using every opportunity to try to prove that Einstein's theory was inadequate (various ether-drift experiments, double star and red shift observations, the two-body metric and other axially symmetric solutions), Einstein by his failure to find *the* pure field theory; both—perhaps because of their strong belief in their own mental powers—persisting in some easily correctable errors, having stopped reading the relevant literature as well as having failed to consult with easily accessible scientists who had worked in the same area, such as Weyl and Levi-Civita.

"Finita la commedia" (Alighieri 1321).

4. Conclusion

Scientifically, it is clear that Silberstein was wrong on the main issue, his solution for the two-body problem; but Einstein was not completely right either. His dislike of singularities (and vain search for a "pure" field theory of matter) made him go off on tangents repeatedly, without realizing that even Silberstein's one-center solution was not what it was purported to be, since it did not describe the field of a spherically symmetric source. Thus, the extended correspondence and the associated publications (Einstein and Rosen 1936; Silberstein 1936, 1937), while shedding much light on the modes of thinking and the character of the men, fundamentally added nothing to clarifying the problem of two particles.

In 1927, Darmois had summarized the status of that problem as follows (Darmois 1927, p. 44), after lauding Weyl's contribution (given *en français*, because it sounds so much better than any translation could):

> *Mais le veritable problème du mouvement libre de deux masses, éxigeant par conséquent un* ds^2 *à deux tubes massiques, n'est nullement resolu. Même pour le problème de deux masses égales, tournant circulairement autour du centre de gravité, on ne sait encore rien.*[15]

And the conclusion in 1988? "Plus ça change, plus c'est la même chose" (Karr 1849).

ACKNOWLEDGMENTS. I am deeply indebted for permission to quote from unpublished correspondence of Albert Einstein and Ludwik Silberstein by courtesy of the Albert Einstein Archives, Hebrew University of Jerusalem, and to the Einstein Project at Boston University. This research was supported in its initial stages by the National Science Foundation.

NOTES

This chapter is a slightly extended version of a talk given at the Second International Conference on History of General Relativity held in September 1988 at Campus Universitaire de Marseille-Luminy.

[1] All translations are my own.

[2] We shall also restrict ourselves to singularities corresponding to positive masses. Negative masses were considered much later in Synge 1960; Hoffmann 1962; Israel and Khan 1964; the general case of equilibrium configurations with multipole singularities along an axis was treated in Szekeres 1968.

[3] "R. Bach" is really Rudolf Förster; I am indebted to A.J. Kox for drawing my attention to the Förster correspondence in the Einstein Archives. Förster obtained a doctorate in mathematics and physics at the University of Leipzig in 1908. For a time he was an assistant at the Technische Hochschule Danzig (now Gdansk) according to the eulogy quoted below; I am indebted to Prof. L. Kostro of the University of Gdansk for his help in trying to verify this, but unfortunately we were unable to do so. During World War I he was a research engineer at Krupp. He started working in general relativity in total isolation and corresponded with Einstein in 1917–1918. As he wrote on December 28, 1917 (EA 25-065), his contract prohibited any outside writing, "the fate of the 'industrial slave,'" and therefore he chose to publish under a pseudonym. After leaving Krupp, Förster worked at Zündapp and, from 1924 until his death in 1941, at Siemens-Schuckert in Nuremberg. Although his work as "R. Bach" was known to his last employer, and, in the eulogy by a Dr. Bohloff (November 2, 1941, EA 25-070, in German), this use of a pseudonym was ascribed to his modesty rather than to any outside pressure (and lauded—without mentioning that relativity was proscribed in Nazi Germany), he did not publish anything after 1922. However, according to his widow, who wrote to Einstein after the war (January 20, 1948, EA 25-068, in German), he did continue his scientific work until his death. She asked Einstein for permission to send him Förster's notes, but Einstein apparently never answered. None of Einstein's earlier letters to Förster are known to survive, but it is clear from Förster's that his various results as well as his questions were taken quite seriously. Furthermore, Förster's widow stated that Einstein once had written: "I see from your letter that I am dealing with a man of unusual theoretical talent. It would be regrettable if you would not have enough leisure to think about these beautiful problems." This was probably a quote from the lost letter of February 19, 1918, since, in his answer of March 19 (EA 25-067, in German), Förster wrote:

> Concerning my profession, I can only tell you that I am very satisfied with it and would not exchange it with that of a teacher, not even an academic one, quite apart from the strangely low pay. At most, I might be tempted by a position at a research institute. The results of my work here have only a very distant relation to the mass murder of the nations. I do not construct any cannons, but am occupied with electrical measurements, apparatus, electrical propulsion of mechanical apparatus, etc.

Since in the above-mentioned eulogy Förster's work at Krupp was described as involving "controls for artillery," the letter may well represent only an *apologia* for war work directed to a man known for his opposition to the war raging at the time. Förster's most important work (Bach 1922) unfortunately is reprinted in Weyl's collected papers (Weyl 1968) as if Weyl were the author, and "Bach" is not mentioned editorially at all.

[4] The other members of the examination committee were M. Brillouin and Emile Borel (chairman). According to Levinson's CV, he was working under Brillouin.

[5] For later discussions of the Curzon singularity, see Mysak and Szekeres 1966; Gautreau and Anderson 1967; Stachel 1968; Cooperstock and Junevicus 1974.

[6] A good discussion of the Weyl–Levi-Civita method and of Curzon's paper is given in Synge 1960, chapter VIII; there is no mention of Silberstein, however.

[7] Philipp Lenard (1862–1947) at the time was a professor at the University of Heidelberg and director of the Physics and Radiology Institutes there; he later wrote the infamous four-volume *Deutsche Physik.* Ernst Gehrcke (1878–1960) was director of the State Physical–Technical Institute in Berlin and a. o. Professor there; he had published *Die Relativitätstheorie, eine wissenschaftliche Massensuggestion* a year before Silberstein's letter. (That same year had seen a right-wing coup attempt, the Kapp Putsch, which was defeated mainly through a general strike.) Both were leaders in the campaign against Einstein, which is discussed in detail in this volume (Goenner 1993). Rudolph Virchow (1821–1902) is considered to be the father of modern pathology. In evaluating Silberstein's comments, it should be kept in mind that he had not lived in Germany for more than a quarter century. This may explain the inclusion of Virchow, who had been dead for almost two decades, as one of only two "good Germans." Possibly Silberstein had known him in his student days.

[8] This period in his life is discussed in detail in Stachel 1990. In contrast to several treatments of Einstein's relation to Judaism and Zionism which were written by religious Jews or Zionists, this paper provides an excellent balanced survey, although it does not give full weight to the degree of assimilation and the frequently total absence of religious feelings among many Central European Jews, who were often not even "Jews in name only." In the 1920s and early 1930s, after the Nazis had given the broadest and vaguest possible "racial" labeling of "Jews," an extensive discussion took place in the German and Austrian press, in books, and in meetings of the left and the right about "What Is a Jew?" A few years later, Einstein wrote an article (Einstein 1938) containing a section with the same title, which totally ignored that discussion and the arguments given there, actually echoing the Nazi line of "Once a Jew, always a Jew" without realizing it. This is briefly discussed in Havas 1980. After coming to power, the Nazis, in the Nuremberg laws, had to give up any attempt at a racial definition, and had to resort to using the religion of one's four grandparents as the only criterion.

[9] In Livingston 1973, p. 310, this is described as follows, based on an interview with Tom O'Donnell, a collaborator of Michelson:

> Dr. Silberstein arrived from Rochester at a time when Michelson was not well enough to meet him. Henry Gale took Silberstein out to Clearing and disliked him immediately.... Gale loved his liquor, but disliked Silberstein and would only drink with "friends."

This sounds as if Gale had met Silberstein only then, while he had known him for at least three years, and well enough to consider offering him a position. On the other hand, any dislike on Gale's part would explain why he did not maintain his offer to Silberstein once Einstein was out of the picture.

[10] The problem Silberstein experienced with the German language was, however, clearly stated by him in a later letter (September 23, 1934, EA 21-071, in German, with the two words in square brackets in English):

> This time I will try to write to you in German (although, through Hitlerian association, even the language itself [itself] sounds hateful [odious] inside my "soul." I shall only insert English words here and there.

[11] In Silberstein 1922, p. 13, the term "elementally flat" was used equivalently with locally Minkowskian. Apart from the unusual spelling (curiously, the index of the book refers to "elementary flatness"), this is the first use of the term in the literature that I am aware of. This is rather ironic, given that it is precisely the question of elementary flatness that would be at issue in the entire controversy.

[12] The most detailed discussion of the question whether a given space-time actually is spherically symmetric is given in Takeno 1952, where it is proved that the one-body solution of Weyl and Levi-Civita is indeed spherically symmetric, whereas that of Silberstein (and thus of Curzon, whose paper was apparently not known to Takeno) is not. The most recent discussion of the line singularities of Curzon's and Silberstein's two-body solution is contained in Schleifer 1985a, 1985b. There it is shown that, although all scalar invariants vanish everywhere outside the two centers, the region between them along the axis does not constitute a Lorentzian manifold.

[13] Cornelius Lanczos had been Einstein's assistant in Berlin in 1928–1929. At that time Einstein was working on a unified field theory, and Lanczos had vainly attempted to interest him in the problem of motion (Havas 1989). He left Germany in 1931 and became a professor at Purdue University.

[14] A detailed discussion of the difference between real and coordinate singularities is given in Szekeres 1960; cf. also Mysak and Szekeres 1966.

[15] "But the real problem of the free motion of two masses, which thus requires a ds^2 of two mass tubes, is not at all resolved. Even about the problem of two equal masses rotating around the mass center in a circle one knows as yet nothing."

References

Alexandrow, W. (1921). "Über den kugelsymmetrischen Vakuumvorgang in der Einsteinschen Gravitationstheorie." *Annalen der Physik* 72: 141–152.

Alighieri, Dante (1321). *La commedia.* Handwritten manuscript (lost); also in *Il codice Laudiano* (1336), *Il codice Trivulziano* (1337), and numerous editions since the invention of printing.

Bach, Rudolf (1922). "Neue Lösungen der Einsteinschen Gravitationsgleichungen" (with an Addendum by Hermann Weyl). *Mathematische Zeitschrift* 13: 134–145.

Becquerel, Jean (1922). *Le principe de la relativité et la théorie de la gravitation.* Paris: Gauthier-Villars.

Birkhoff, George D. (1927). *Relativity and Modern Physics.* Cambridge: Harvard University Press.

Chazy, Jean (1923a, 1923b, 1924). "Sur le champ de gravitation de deux masses fixes dans la théorie de la relativité." *Comptes Rendues de l'Académie des Sciences* 177: 303–305; 177: 939–941; *Bulletin de la Société mathématique de France* 48: 17–38.

Cooperstock, Fred I. and Junevicus, Gerald J. (1974). "Singularities in Weyl Gravitational Fields." *International Journal of Theoretical Physics* 9: 59–68.

Curzon, Harry E.J. (1924a). "Bipolar Solutions of Einstein's Gravitation Equations." *The London Mathematical Society. Proceedings* 23: XXIX.

——— (1924b). "Cylindrical Solutions of Einstein's Gravitation Equations." *The London Mathematical Society. Proceedings* 23: 477–480.

Darmois, Georges (1927). *Les équations de la gravitation einsteinienne*. (Mémorial des Sciences Mathématiques, Fascicule XXV). Paris: Gauthier-Villars.

De Donder, Théophile (1919). "La Gravifique. 2e communication." *Académie Royale de Belgique. Classe des Sciences. Bulletin* 317–325.

——— (1921). *La Gravifique einsteinienne*. Paris: Gauthier-Villars.

Droste, Johannes (1915): "Het veld van twee bolvormige rustende centra in Einstein's theorie der zwaartekracht." *Koninklijke Akademie van Wetenschappen te Amsterdam. Verslagen van de Gewone Vergaderingen der Wis- en Natuurkundige Afdeeling* 24: 749–757. English translation: "On the field of two spherical fixed centres in Einstein's theory of gravitation." *Koninklijke Akademie van Wetenschappen te Amsterdam. Proceedings of the Section of Sciences* 18: 760–768.

——— (1916a). "Het veld van een enkel centrum in Einstein's theorie der zwaartekracht, en de beweging van een stoffelijk pumt in dat veld." *Koninklijke Akademie van Wetenschappen te Amsterdam. Verslagen van de Gewone Vergaderingen der Wis- en Natuurkundige Afdeeling* 25: 163–180. English translation (1917): "The field of a single centre in Einstein's theory of gravitation, and the motion of a particle in that field." *Koninklijke Akademie van Wetenschappen te Amsterdam. Proceedings of the Section of Sciences* 19: 197–215.

——— (1916b). "Het veld van *n* bewegende centra in Einstein's theorie der zwaartekracht." *Koninklijke Akademie van Wetenschappen te Amsterdam. Verslagen van de Gewone Vergaderingen der Wis- en Natuurkundige Afdeeling* 25: 460–467. English translation (1917): "The field of *n* moving centres in Einstein's theory of gravitation." *Koninklijke Akademie van Wetenschappen te Amsterdam. Proceedings of the Section of Sciences* 19: 447–455.

——— (1916c). *Het zwaartekrachtsveld van een of meer lichamen volgens de theorie van Einstein*. Thesis, Rijksuniversiteit te Leiden: N.V. Boekhandel en Drukkerij E.J. Brill.

Eddington, Arthur S. (1918). *Report on the Relativity Theory of Gravitation*. London: Fleetway Press.

——— (1921). *Espace, temps et gravitation. La théorie de la relativité généralisée dans ces grandes lignes, exposé rationnel suivi d'une étude mathématique de la théorie*. Paris: J. Hermann. French translation of *Space, Time and Gravitation: An Outline of the General Relativity Theory*. Cambridge: Cambridge

University Press, 1920. Includes a mathematical supplement not contained in the English edition.

——— (1923). *The Mathematical Theory of Relativity*. Cambridge: Cambridge University Press.

——— (1925). "Ether-drift and the Relativity Theory." *Nature* 115: 870.

Einstein, Albert (1922). "Bemerkungen zu der Abhandlung von E. TREFFTZ: 'Das statische Gravitationsfeld zweier Massenpunkte der EINSTEINschen Theorie'." *Preussische Akademie der Wissenchaften* (Berlin). *Physikalisch-Mathematische Klasse. Sitzungsberichte*: 148–149.

——— (1938). "Why Do They Hate the Jews?" *Collier's Magazine*: November 26. Reprinted in *Out of My Later Years* (1950) and *Ideas and Opinions* (1954).

Einstein, Albert and Grommer, Jakob (1927). "Allgemeine Relativitätstheorie und Bewegungsgesetz." *Preussische Akademie der Wissenchaften* (Berlin). *Physikalisch-Mathematische Klasse. Sitzungsberichte*: 2–13.

Einstein, Albert, Infeld, Leopold, and Hoffmann, Banesh (1938). "Gravitational Equations and the Problem of Motion." *Annals of Mathematics* 39: 65–100.

Einstein, Albert and Infeld, Leopold (1940). "Gravitational Equations and the Problem of Motion II." *Annals of Mathematics* 41: 455–464.

——— (1949). "On the Motion of Particles in General Relativity Theory." *Canadian Journal of Mathematics* 3: 209–241.

Einstein, Albert and Rosen, Nathan (1935). "The Particle Problem in the General Theory of Relativity." *Physical Review* 48: 73–77.

——— (1936). "Two-Body Problem in General Relativity Theory." *Physical Review* 49: 73–77.

Gautreau, Ronald and Anderson, James L. (1967). "Directional Singularities in Weyl Gravitational Theory." *Physics Letters* 25A: 291–292.

Giorgi, Giovanni (1925). "Ether-Drift and Relativity." *Nature* 116: 132.

Goenner, Hubert (1993). "The Reaction to Relativity Theory in Germany III: A Hundred Authors against Einstein." This volume, pp. 248–273.

Havas, Peter (1980). "Einstein és kora." *Fizikai Szemle* 30: 66–69.

——— (1989). "The Early History of the 'Problem of Motion' in General Relativity." In: *Einstein and the History of General Relativity*. D. Howard and J. Stachel, eds. Boston: Birkhäuser, pp. 234–276.

Hoffmann, Banesh (1962). "Static, Axially Symmetric Gravitational Fields in General Relativity Involving Mass Singularities of Both Signs." In *Les Théories Relativistes de la Gravitation*. Paris: Éditions du C.N.R.S., pp. 237–247.

Humm, Rudolf J. (1918). "Über die Bewegungsgleichungen der Materie. Ein Beitrag zur Relativitätstheorie." *Annalen der Physik* 57: 68–80.

Israel, Werner and Khan, K.A. (1964). "Collinear Particles and Bondi Dipoles in General Relativity." *Il Nuovo Cimento* 33: 331–344.

Jebsen, J.T. (1921). "Über die allgemeinen kugelsymmetrischen Lösungen der Einsteinschen Gravitationsgleichungen im Vakuum." *Arkiv for Matematik, Astronomi och Fysik* 15: nr. 18.

Karr, Alphonse (1849). *Les Guêpes*, Paris. January: 305.

Levi-Civita, Tullio (1917–1919). "ds^2 einsteiniani in campi newtoniani." *Rendiconti Accademia dei Lincei* 26: 307; 27: 3, 183, 220, 240, 283, 343; 28: 3, 101.

Levinson, Horace C. (1922). *The Gravitational Field of Masses Relatively at Rest According to Einstein's Theory of Gravitation.* Dissertation, University of Chicago.

—— (1923a). *Le champ gravitational de deux points materiels fixes dans la théorie d'Einstein.* Thesis, Université de Paris. Paris: Gauthiers-Villars.

—— (1923b). "Sur le champ gravitationnel de *n* corps dans la théorie de la relativité." *Comptes Rendues de l'Academie des Sciences* 176: 981–983.

—— (1928). "The Gravitational Field of *n* Moving Particles in the Theory of Relativity." *Proceedings of the International Mathematical Congress, Toronto, Vol. II.* University of Toronto Press: 243–251.

Livingston, Dorothy M. (1973). *The Master of Light. A Biography of Albert A. Michelson.* New York: Charles Scribner's Sons.

Lorentz, Hendrik Antoon and Droste, Johannes (1917). "De beweging van een stelsel lichamen onder den invloed van hunne onderlinge aantrekking, behandeld volgens de theorie van Einstein." *Koninklijke Akademie van Wetenschappen te Amsterdam. Verslagen van de Gewone Vergaderingen der Wis- en Natuurkundige Afdeeling* 26: 392–403, 649–660. English translation (1937) in H.A. Lorentz, *Collected Papers*, Vol. 5. P. Zeeman and A.D. Fokker, eds. The Hague: Martinus Nijhoff, pp. 330–355.

McVittie, George C. (1936). "Two-Centers Solution of the Gravitational Field Equations, and the Need for a Reformed Theory of Matter. L. Silberstein." *Science Abstracts, Sec. A. Physics* 39: 238.

Marder, Leslie (1959). "Two Bodies at Rest in General Relativity." *Cambridge Philosophical Society. Proceedings*: 82–86.

Michelson, Albert A. (1925). "The Effect of the Earth's Rotation on the Velocity of Light. Part I." *Astrophysical Journal* 61: 137–139.

Michelson, Albert A. and Gale, Henry G. (1925). "The Effect of the Earth's Rotation on the Velocity of Light, Part II." *Astrophysical Journal* 61: 140–145.

Miller, Dayton C. (1925). "Ether-Drift Experiments at Mount Wilson." *National Academy of Sciences. Proceedings*: 306–314 and *Nature* 116: 49–50.

Millikan, Robert A. (1950). *The Autobiography of Robert A. Millikan.* New York: Prentice-Hall.

Mysak, Lawrence and Szekeres, George (1966). "Behavior of the Schwarzschild Solution in Superimposed Gravitational Fields." *Canadian Journal of Physics* 44: 617–627.

Palatini, Attilio (1923). "Sopra i potenziali simmetrici che conducono alle soluzioni longitudinali delle equazioni gravitazionali di Einstein." *Rendiconti Accademia dei Lincei* 32: 263–267 and *Il Nuovo Cimento*, Serie VII. 26: 5–24.

Racine, Charles (1934). *Le problème des N corps dans la théorie de la relativité.* Thesis, Université de Paris. Paris: Gauthier-Villars.

Robertson, Howard P. (1938). "Note on the Preceding Paper: The Two-Body Problem in General Relativity." *Annals of Mathematics* 39: 101–104.

Schleifer, Nathan (1985a). "Condition of Elementary Flatness and the Two-Particle Curzon Solution." *Physics Letters* 112A: 204–207.

—— (1985b). "The Nature of the Curzon Bipolar Solution." Unpublished manuscript.

Schwarzschild, Karl (1916). "Das Gravitationsfeld eines Massenpunktes nach der Einsteinschen Theorie." *Königlich Preussische Akademie der Wissenchaften* (Berlin). *Sitzungsberichte*: 189–196.

Shakespeare, William (1602). *Hamlet, Prince of Denmark*. London: *Quarto* (1603), *Folio* (1623).

Shankland, Robert S., McCuskey, Sidney W., Leone, Fred. C., and Kuerti, Gustav (1955). "New Analysis of the Interferometer Observations of Dayton C. Miller." *Reviews of Modern Physics* 27: 167–178.

Silberstein, Ludwik (1914). *The Theory of Relativity*. London: MacMillan.

—— (1918). "General Relativity without the Equivalence Hypothesis." *Philosophical Magazine* 36: 94–128.

—— (1920). "The Stokes–Planck Ether." *Philosophical Magazine* 39: 161–170.

—— (1922). *The Theory of General Relativity and Gravitation*. New York: D. Van Nostrand.

—— (1925a). "D.C. Miller's Recent Experiments and the Relativity Theory." *Nature* 115: 798–799.

—— (1925b). "Ether Drift and the Relativity Theory." *Nature* 116: 98.

—— (1936). "Two-Centers Solution of the Gravitational FIeld Equations, and the Need for a Reformed Theory of Matter." *Physical Review* 49: 268–270.

—— (1937). "On Einstein's Gravitational Field Equations." *Philosophical Magazine* 41: 814–822.

Stachel, John (1968). "Structure of the Curzon Metric." *Physics Letters* 27A: 60–61.

—— (1990). "Einstein's Jewish Identity." Paper prepared for the Symposium on "Einstein in Context," Jerusalem.

Straneo, P. (1924a). "Intorno alla teoria dei campi einsteiniani a simmetria assiale." *Atti R. Accad. Lincei Rend. Cl. Sci. Fis. Mat. Nat.* 33: 404–410.

—— (1924b). "Deduzione e interpretazione di qualche ds^2 einsteiniano simmetrico intorno ad un asse." *Atti R. Accad. Lincei Rend. Cl. Sci. Fis. Mat. Nat.* 33: 468–474.

—— (1924c). "Considerazioni generali sui campi einsteiniani a simmetria assiale." *Atti R. Accad. Lincei Rend. Cl. Sci. Fis. Mat. Nat.* 33: 547–552.

Synge, John L. (1960). *Relativity: The General Theory*. Amsterdam: North Holland.

Szekeres, George (1960). "On the Singularities of a Riemannian Manifold." *Publicationes Mathematicae, Debrecen* 7: 285–301.

Szekeres, Paul (1968). "Multipole Particles in Equilibrium in General Relativity." *Physical Review* 176: 1446–1450.

Takeno, Hyôitirô (1952). "On the Spherically Symmetric Space-Times in General Relativity." *Progress of Theoretical Physics* 8: 317–326.

Tolman, Richard C. (1934). *Relativity Thermodynamics and Cosmology*. Oxford: Clarendon Press.

Trefftz, Erich (1922). "Das statische Gravitationsfeld zweier Massenpunkte in der Einsteinschen Theorie." *Mathematische Annalen* 86: 317–326.

Weyl, Hermann (1917). "Zur Gravitationstheorie." *Annalen der Physik* 54: 117–145.

—— (1919a). "Eine neue Erweiterung der Relativitätstheorie." *Annalen der Physik* 59: 101–133.

—— (1919b). "Bemerkung über die axialsymmetrischen Lösungen der Einsteinschen Gravitationsgleichungen." *Annalen der Physik* 59: 185–188.

—— (1919c). *Raum-Zeit-Materie*. 3rd ed. Berlin: Julius Springer.

—— (1921a). "Feld und Materie." *Annalen der Physik* 65: 541–563.

—— (1921b). *Raum-Zeit-Materie*. 4th ed. Berlin: Julius Springer. English translation: *Space, Time, Matter*. Henry Brose, trans. London: Methuen, 1922; reprint New York: Dover, 1925.

—— (1923). *Raum-Zeit-Materie*. 5th ed. Berlin: Julius Springer.

—— (1968). *Gesammelte Abhandlungen*. K. Chandrasekharan, ed. Vol. II: 303–314. Berlin and New York: Springer-Verlag.

Part II

THE EMPIRICAL BASIS OF GENERAL RELATIVITY

Einstein's Explanation of the Motion of Mercury's Perihelion

John Earman and Michel Janssen

1. Introduction[1]

On November 18, 1915, Einstein presented to the Berlin Academy a paper containing three predictions of his still incomplete theory of gravitation (Einstein 1915c). The verification of the three predicted effects—the gravitational red shift, the bending of starlight passing near the sun, and the advance of the perihelion of Mercury—became known as the "classical tests" of the general theory of relativity (GTR). Einstein had previously predicted the first two effects using heuristic arguments based upon his principle of equivalence. The red shift prediction had already been confirmed, or so Einstein claimed, by Erwin Freundlich, the Babelsberg astronomer and confidant of Einstein. However, Freundlich's analysis of the spectral lines of stellar sources was criticized by Seeliger (1916),[2] while other experimentalists, such as St. John at the Mt. Wilson Observatory, were unable to confirm Einstein's red shift prediction for the sun.[3] Thus, this first classical test threatened to be a failure rather than a success. Einstein also stated that his theory yielded a deflection value of 1.7″ of arc for starlight grazing the sun, which was twice the previous 1911 prediction based on the principle of equivalence. Freundlich had set off into Russia in 1914, hoping to use a solar eclipse to test this earlier prediction; but perhaps fortunately for the nascent general theory, the hostilities of the First World War broke out, and Freundlich's equipment was seized and he was briefly interned by the Russians. An expedition from the Lick Observatory met with rainclouds rather than rifles. It was thus left to Arthur Stanley Eddington, whose England was bitterly at war with Einstein's Germany, to complete the second classical test.[4] In the November 18 communication Einstein noted that in contrast to the deflection of light, the value of the red shift was not changed by his new theory "because this result depends only on g_{44}" (Einstein 1915c, p. 834). This remark is a little misleading since, as we will see below, Einstein most likely used the red shift to help him guess the form g_{44} should take.

Einstein's third prediction was a new one, requiring deduction from a formal theory rather than the heuristic reasoning which led to predictions of the first two effects. Einstein found that the perihelion advance of a planet per orbit should be $6\pi GM/a(1-e^2)c^2$, where G is the gravitational constant, M is the solar mass, a is the semi-major axis of the planetary orbit, e is the eccentricity of the orbit, and c is the velocity of light. For the planet Mercury, the predicted advance is $43''$ per century, which is in striking agreement with the actual value of the anomalous advance that had exercised some of the most acute minds in astronomy for over half a century. The resolution of this anomaly was the first solid triumph for Einstein's GTR, and after nearly three-quarters of a century of careful and detailed scrutiny, the triumph remains untarnished.[5]

The manner in which this triumph was initially achieved is a story that has received surprisingly little attention. The details of Einstein's explanation of the perihelion advance form a fascinating web of mathematical analysis and physical intuition. They reveal a delicate, almost precarious path of reasoning. In these same circumstances, many able physicists would have lost their way, and some actually did.

2. Mercury's Perihelion: Observation

According to G.M. Clemence,

> Observations of Mercury are among the most difficult in positional astronomy. They have to be made in the daytime, near noon, under unfavorable conditions of the atmosphere; and they are subject to large systematic and accidental errors arising both from this cause and from the shape of the visible disk of the planet. (Clemence 1947, p. 361)

In addition, the observations "are affected by the precession of the equinoxes, and the determination of the precessional motion is one of the most difficult problems in positional astronomy, if not the most difficult" (ibid., p. 361). Finally, the anomalous advance of Mercury's perihelion is a theoretical value that is arrived at by subtracting estimates of the perturbations on Mercury's orbit caused by other planets from the observed advance of some $570''$ of arc per century.[6] Given all of these hazards and the smallness of the residual—some few dozens of seconds of arc per century—it is remarkable that astronomers were able to agree on a definite value for the anomaly.[7]

The first systematic investigation of the problem was undertaken in the 1850s by Le Verrier, who found an anomalous advance of 38.3″ per century.[8] Le Verrier was able to demonstrate to his satisfaction that fiddling with the values of the planetary masses would not produce agreement between observation and Newtonian theory. For example, increasing Venus' mass estimate by some 10% would explain away the anomaly, but such an increase would have, according to the Newtonian laws of gravitation, other consequences that contradicted observation. Since a knowledge of the main features of planetary orbits was deemed secure, the discrepancy had to be due to some as yet unknown source of perturbation or else the blame had to be placed at Newton's doorstep.

The next major advance in the analysis of the observations was due to the American astronomer Simon Newcomb. Whereas Le Verrier had based computations of secular perturbations on different mass values for the same planets in different parts of the computation, Newcomb sought a consistent set of planetary masses. He estimated masses independently of the problem of secular perturbations by means of observations of satellites of the planet, deflection of passing comets, and periodic perturbations on other planets. Einstein, though probably not familiar with the primary literature, was well aware of Newcomb's accomplishment. Writing to his friend Besso on December 10, 1915, Einstein explained that the knowledge of the value of the perihelion advance is "perfectly assured from the point of view of astronomy, because the determination of the masses of the inner planets has been made by Newcomb from the *periodic* perturbations (and not the secular)" (Speziali 1972, p. 60).

Using his mass estimates for the planets, Newcomb (1895) arrived at a value of 8.48″ ±.43″ for the product of Mercury's eccentricity and its anomalous perihelion advance per century, a figure repeatedly cited in the literature. What is more than a little puzzling are the different values for the advance bandied about in the secondary literature. The range of figures from the period immediately surrounding Einstein's perihelion paper is indicated by the following list: Jeffreys (1919), 41″; Silberstein (1917), 42.9″; Droste (1915), 44″; and Einstein (1915c), 45″ ± 5″. Jeffreys' value is understandable if the favored modern value of $e = .2056$ is used in conjunction with Newcomb's product figure, yielding a centennial advance of 41.24″ which Jeffreys presumably rounded off to 41″. The most plausible explanation of Silberstein's value is that he confused the observed value of the anomaly with the theoretical value predicted by GTR. We find Droste's value of 44″ inexplicable. And Einstein's value of 45″ ± 5″ is even more mysterious since he gives 43″ as the theoretical prediction, which he presumably arrived at by using $e = .2056$. Nor is the 45″ figure due to a slip of the pen, for Einstein repeats it in a letter to Sommerfeld written two weeks after his perihelion paper (see Hermann 1968, p. 32, and Section 4 below).

There are similar but less severe discrepancies in the quoted values of the theoretical prediction of GTR, the most popular value being 43.03″ (see, for example, Clemence 1947 and Weinberg 1972), while 42.9″(Møller 1972) and 42.95″(Will 1981) are also cited. It is somewhat disconcerting to find such divergences on both theory and observation in a topic that turns on a small handful of seconds of arc per century.

To set the record straight, the correct theoretical value is 42.98″.[9] The observational estimates have remained remarkably consistent, although there is a slight upward tendency from Newcomb (1895) 41.24″±2.09″, to Clemence (1947) 42.56″±.94″, to Shapiro et al. (1976) 43.11″±.21″. More to the point of the present work, when Einstein offered his explanation of the motion of Mercury's perihelion, Newcomb's work was generally regarded as reliable, and thus there was general agreement both that there was an anomaly and that the product of the anomalous centennial advance and the eccentricity of Mercury's orbit is given by Newcomb's figure of 8.48″±.43″.

3. Mercury's Perihelion: Theory

The anomaly in Mercury's perihelion documented by Newcomb left astronomers with two main choices: either continue to maintain Newton's laws of motion and his $1/r^2$ law of gravitational attraction and search for additional sources of secular perturbation on Mercury's orbit, or else modify Newton's second law or his law of gravitation or both. The various attempted implementations are described in Roseveare's splendid study, *Mercury's Perihelion from Le Verrier to Einstein* (1982); in this section we will simply summarize some of his findings, leaving it to the reader interested in further details to consult his work.

The most obvious candidates for additional sources of perturbation were solar oblateness and intra-Mercurial matter. The results of 19th century optical measurements of the photosphere of the sun were summarized in 1895 by Newcomb:

> The general result is that the mean of the equatorial measures are [sic] slightly less than the mean of the polar measures, the difference, however, being within the probable errors of the results. I conclude that there can be no such non-symmetrical distribution of matter in the interior of the Sun as would produce the observed effect. (Newcomb 1895, pp. 111–112)

As for intra-Mercurial matter, the simplest hypothesis would have been an extra planet in an orbit between the sun and Mercury. Various claims to have observed such a planet "Vulcan" were reported in the 19th century, and astronomers continued searching for the hypothetical Vulcan well into the 20th century. But by 1915 this hypothesis was not taken seriously, both because the probability was low that a Vulcan sufficient to account for the perihelion anomaly could have eluded observation and also because theorists had convinced themselves that this Vulcan would engender other anomalies. This left Seeliger's (1906) hypothesis of bands of diffuse intra-Mercurial matter, a hypothesis that received independent observational support from the existence of zodiacal light and that was flexible enough to hold out the promise of a consistent, anomaly-free account of planetary motions in Newtonian terms. Seeliger's account will receive more attention below, but we now turn to a summary of attempts to deal with the perihelion anomaly in non-Newtonian terms.

Proposed modifications of Newtonian theory can be conveniently divided into four classes.

Nonrelativistic theories. At least two modifications of Newton's $1/r^2$ law received serious attention. The first, initially proposed by Clairaut in 1745 in connection with the moon's perigee, would add to Newton's $1/r^2$ gravitational force law a term C/r^4. To understand this proposal, recall that one of Newton's demonstrations of his law of gravitational attraction combined the assertion that the apsides of the planets are quiescent with the proof that if the central force law differed from $1/r^2$ the apsides would rotate. This was a double-edged argument. The moon's perigee was known to rotate some 3° per orbit, the obvious cause being the attraction exerted by the sun. However, when Clairaut, d'Alembert, and Euler tried to calculate the influence of the sun, they could only account for half of the observed 3° advance.[10] Newton's demonstration would then seem to imply that a departure from the $1/r^2$ law was involved. The anomalous motion of Mercury's perihelion can be accounted for by inserting Clairaut's proposed force law into Newton's second law and adjusting the value of C. As Newcomb (1882) noted, however, at small distances the $1/r^4$ would dominate, producing effects that would contradict Cavendish type experiments. In a sense, Einstein's GTR revives Clairaut's law (see Section 6, Eq. (43)).

Another ad hoc modification, proposed by Hall (1894) and initially championed by Newcomb, replaced $1/r^2$ by $1/r^{(2+\delta)}$. Newcomb (1895) found that $\delta = .0000001574$ would account for an advance of $42.4''$ per century in Mercury's perihelion. But it was difficult to believe that such an ugly law could be true, and in any case De Sitter (1913) showed that such a law would lead to problems with the motion of the Moon's perigee.

More radical modifications of Newtonian gravitational theory were inspired by the work of Weber, Gauss, Riemann, and Clausius on action-at-a-distance electrodynamics. Applying their velocity dependent force laws to gravitation gives values for Mercury's perihelion advance ranging from 7″ to 14″ per century. Gerber (1898, 1902) also published a velocity-dependent force law and derived from it a formula for the perihelion advance of $24\pi^3 a^2/T^2 V^2(1-e^2)$, where T is the orbital period of the planet and V is the velocity of propagation of the gravitational potential. If V is identified with the velocity of light c, Gerber's formula is exactly the one that appears in Einstein's 1915 paper. No one, however, pretended to be able to find a coherent physical foundation for Gerber's theory. Ritz's theory of electromagnetism and gravitation (1909) also employed velocity dependent force laws. His gravitational force law contained a free parameter that could be adjusted so as to yield the correct advance for Mercury's perihelion and that, when so adjusted, gave reasonable values for the perihelia advances of Venus and Earth. But after the success of Einstein's special theory of relativity (STR) few could take seriously Ritz's emission theory of light and the electromagnetic basis of his law of gravitation.[11]

Transitional theories. If Newtonian mechanics is modified by adding Lorentz's mass transformation $m = m_0\sqrt{1 - v^2/c^2}$, the result is somewhat analogous to the velocity-dependent laws mentioned above. The upshot is an advance in Mercury's perihelion of 7″ per century.

Special-relativistic theories. Both Poincaré (1906) and Minkowski (1908) offered Lorentz-invariant forms of Newton's $1/r^2$ law using a retarded action-at-a-distance scheme. De Sitter (1911) found that Minkowski's version of this scheme gave no secular advance to Mercury's perihelion while Poincaré's gave 7.15″ per century. He also found that Poincaré's law could easily be generalized by multiplying by an integral power of a certain factor, in which case the secular advance would be $n \times 7.15''$. Choosing $n = 6$ gives 42.9″. However, De Sitter mentioned the generalization only in passing, and neither he nor anyone else offered $n = 6$ as an explanation of the anomaly—at least, that is, until Silberstein (1917) rediscovered a version of De Sitter's finding (see Section 8 below).

Post–special-relativistic theories. By 1907 Einstein was hard at work on a theory of gravitation, and he managed to convince himself, essentially on the basis of considerations of the principle of equivalence, that a successful theory of gravitation could not be constructed within the confines of STR.

His first formal theory of gravitation (Einstein 1912a, 1912b) relied upon the notion of a variable speed of light, a notion that from the modern perspective is either incoherent or else implicitly involves a bimetric approach, with a curved space-time metric on top of a flat background metric. On one construal, this theory gives a perihelion advance for Mercury of 28.7″ per century (Whitrow and Morduch 1960, 1965). In the same year, Max Abraham also concocted two theories (1912a, 1912b, 1912c) involving a variable speed of light; the first gives an advance of 14.52″(Pavanini 1912, 1913), and the second yields a retrogression of 3″(Roseveare 1982, p. 152). Nordström (1912, 1913) likewise offered a pair of theories, both of which yielded a retrogression for Mercury's perihelion (Roseveare 1982, p. 153; Whitrow and Morduch 1965). No perihelion prediction was worked out in Mie's theory (1913), but presumably it does not give a secular advance since it posits a Newtonian attraction. Finally, the Einstein–Grossmann theory (1913) predicts an advance of 18″per century (Droste 1915).

Though brief to the point of superficiality, this summary serves to establish that prior to Einstein's general theory, no modification of Newtonian theory that did not contemplate additional sources of perturbation offered a non-problematic resolution of the perihelion anomaly. In particular, it seemed that special-relativistic laws of gravitation would not suffice without resort to ad hoc trickery, and among the post–special-relativistic theories that endeavored to build gravitation upon new first principles, not one came within hailing distance of the observed anomaly.

4. Preliminary Remarks about Einstein's Perihelion Paper

Einstein's November 18 paper (1915c), to which we will refer as the "perihelion paper," marks the first occasion on which the perihelion problem appears in Einstein's published work. The first known mention occurs in 1907, the year in which he began to think seriously about a relativistic theory of gravitation, in a letter to Conrad Habicht: "Now I am busy on a relativistic theory of the gravitational law with which I hope to account for the still unexplained secular changes of the perihelion movement of Mercury. So far I have not managed to succeed" (Seelig 1956, p. 76). But as far as we are aware, there is no other mention of the perihelion problem in Einstein's correspondence prior to late 1915.

It is now known that this apparent neglect of the perihelion problem between 1907 and 1915 is only apparent. A recently discovered manuscript of some 50 pages, partly in Einstein's hand and partly in Michele Besso's, shows that, probably during a visit from Besso to Einstein in Zurich in late May or early June 1913, Einstein and Besso collaborated on calculations of the perihelion motion of Mercury.[12] Although these calculations were based on the *Entwurf* theory (Einstein and Grossmann 1913), a theory that Einstein abandoned in 1915, some of the techniques developed in the Einstein–Besso collaboration were put to good use in the November 18 paper (see Sections 5–7 below). This paper was apparently produced within the span of a week, an impressive feat even for someone of Einstein's ability. The fact that Einstein was not calculating *de nuovo* does not diminish the magnitude of his achievement, but it does make it more comprehensible.

The neglect of the problem in his published writings does call for some comment. The cynical explanation of his failure to mention the result of the Einstein–Besso calculation based on the *Entwurf* theory would be that the result was wrong. As mentioned earlier, the *Entwurf* theory predicts a perihelion advance of 18″ per century. The figure given in the Einstein–Besso manuscript is 1800″. The manuscript gives the correct formula for the perihelion advance, the erroneous factor of 100 is due to a mistake that occurs when actual numbers are inserted into this formula. There are several indications in the manuscript that Einstein and Besso discovered this mistake, but the correct figure is not explicitly stated. The so-called "scratch notebook" (EA 3–013), dated 1909–1914, also has Einstein inserting numbers into a formula for the perihelion advance equivalent to the one given in the Einstein–Besso manuscript except for the fact that the orbit's eccentricity is neglected. This time Einstein does not make any mistakes and he arrives at the figure of 17″.[13]

The neglect of the perihelion problem in published writings seems to have been fairly general in this period. Of the seven post–special-relativistic theories mentioned in the preceding section, a formula for the perihelion advance of a planet is worked out in only one case by the primary author—Nordström in the case of his second theory (Nordström 1914)—and in that case no numerical value for the predicted perihelion shift of Mercury is given. Once again, a cynical explanation is suggested by the fact that none of these theories gives anything like the correct value. While a touch of cynicism is perhaps healthy, a much more plausible explanation has been offered by Roseveare (1982, pp. 156ff.); namely, that while the advance of the perihelion of Mercury was an anomaly for Newtonian gravitational theory, it was not generally regarded as an effect that had to be explained on the basis of new principles since Seeliger's hypothesis was widely accepted as offering the means of a satisfactory resolution. Newcomb, who had initially adopted Hall's hypothesis, felt that he was "forced back upon the hypothesis of a mass of matter surrounding the Sun sufficient to cause the motion of the perihelion of Mercury" (Newcomb 1912, p. 227). In lectures given in 1906–1907, Poincaré (1953) endorsed a ring of matter as the probable cause of Mercury's perihelion shift. De Sitter, who, to judge by the number of publications and the respect with which they were received, was regarded as an authority on this matter, adopted Seeliger's explanation as a working hypothesis:

> Can we, then, consider the problem as finally solved [by Seeliger]? I think not. One more step remains to be done. The fate of Hall's hypothesis should be remembered. It is true that Seeliger's explanation differs from Hall's hypothesis in being vastly less hypothetical; in fact, it may be considered as nothing more nor less than a determination of mass of a material body whose existence is known beforehand. But, taking this point of view, we cannot consider that determination as final before it has been ascertained that it is not in contradiction with other possible determinations; in other words, before it has been verified that the attraction of the zodiacal masses does not give rise to other effects, which might be in contradiction with observations. (De Sitter 1913, pp. 302–303)

Across the English channel Harold Jeffreys was at work, also trying to show that the attraction of zodiacal masses sufficient to account for Mercury's perihelion shift would not give rise to other effects in contradiction with observations (Jeffreys 1916, 1918).

But if Seeliger's hypothesis was generally accepted as a working hypothesis, why did Einstein concern himself with the perihelion problem in November 1915? No doubt part of the answer lies in Einstein's healthy disrespect for generally accepted wisdom; but that part of the answer does not speak to the timing factor. To deal with that factor we need to remind ourselves that for the past two years Einstein had been struggling with his new gravitational theory. By means of a mistaken but ingenious and ultimately profound argument, Einstein had managed to convince himself that suitable gravitational field equations could not be generally covariant.[14] According to a letter to Sommerfeld, dated November 28, 1915, Einstein's dissatisfaction with the results of his experimentation with non-generally covariant equations was connected with the perihelion problem. One of the three reasons he gives for abandoning the Einstein–Grossmann theory is that "The motion of the perihelion of Mercury yields 18″ instead of 45″ [*sic*]."[15] One should perhaps not attach too much weight to this retrospective assessment, made in the flush of the success of the deduction of the missing 43″. Of course, once Einstein had abandoned his previous efforts, there would have been an understandable desire to convince himself that the new approach he was exploring toward the end of 1915 was not also heading up a blind alley. Meeting the challenge he had set for himself in 1907—accounting for the "still unexplained secular changes in the perihelion movement of mercury"—would restore confidence. Meeting this challenge would also serve to establish priority for the still developing theory, a concern one can reasonably attribute to Einstein since he knew that Hilbert was also at work on the same problem.[16]

A third possible factor relevant to the timing issue was the appearance of a paper by Freundlich attacking Seeliger's hypothesis (Freundlich 1915). Freundlich's paper is cited as a "noteworthy article" in a footnote in Einstein's perihelion paper.[17] However, in a letter to Sommerfeld, written in February 1916, Einstein gives the impression that he was not much influenced by Freundlich's critique, which he likened to "kicking in an open door" (*Einrennen einer offenen Thür*), a standard German idiom for making much to do about something that was completely obvious to begin with (Hermann 1968, p. 39). Apparently, Einstein was confident that a correct theory of gravitation would render the hypothesized zodiacal matter superfluous, at least insofar as it affected the perihelion problem. It is not clear, though, how Einstein could have been so confident that this was indeed an open door until he had the 43″ in hand.

By the same token, if the other two hypothetical motivations of restoring confidence and establishing priority were to be served, it was essential to predict an advance close to the observed value. In the next section we will see another equally compelling reason why Einstein had to come close to the accepted value of the advance.

In the end, however, it may be a mistake to look for any deep motivation behind Einstein's perihelion paper. In that paper he mentions that an application of Huygens' principle and a "simple calculation," which is omitted, show that a light ray passing a distance Δ from the sun will suffer an angular deflection of $4GM/\Delta c^2$, as opposed to his earlier prediction of $2GM/\Delta c^2$. It was only natural for him to wonder whether his new theory would also increase the previous prediction of the perihelion advance.

5. Finding an Approximate Solution for the Gravitational Field of the Sun

In the next three sections, we will closely examine the logic of Einstein's perihelion paper. The argument naturally breaks down into two parts: finding a solution of the field equations representing the gravitational field of the sun and finding the perihelion motion of planetary orbits in this field. Einstein tackles these tasks respectively in section 1 ("The Gravitational Field") and section 2 ("The Motion of the Planets") of his paper. Our discussion will likewise be divided into two parts. In this section, we will look at how Einstein calculated the field of the sun; in Sections 6 and 7, we will analyze his calculation of the perihelion motion in this field. In both cases we will pay special attention to the various approximations Einstein makes along the way.

When writing the perihelion paper, Einstein had not yet arrived at the final field equations, but he does refer in a footnote to a "forthcoming communication" on this matter so that we can infer that he was at work on the paper that was presented to the Berlin Academy on November 25 (Einstein 1915d) and that contains what are now called Einstein's gravitational field equations.[18] In his communication of November 11 he had taken as his field equations (Einstein 1915b, p. 800):

$$G_{\mu\nu} = -\kappa T_{\mu\nu}, \tag{1}$$

where $G_{\mu\nu}$ is the Ricci tensor, $T_{\mu\nu}$ is the energy-momentum tensor of the gravitational sources, and κ is a constant. For the exterior field of a massive body, such as the sun, Eq. (1) reduces to

$$G_{\mu\nu} = 0, \tag{2}$$

which coincides with the implications of the final Einstein field equations.

The Ricci tensor can be split into two parts (Einstein 1915a, p. 782, 1915b, p. 800; cf. Norton 1984, p. 269, pp. 304ff.):

$$G_{im} = R_{im} + S_{im}, \tag{3}$$

with

$$\begin{aligned} R_{im} &\equiv -\sum_l \frac{\partial}{\partial x_l}\begin{Bmatrix} im \\ l \end{Bmatrix} + \sum_{\rho l}\begin{Bmatrix} il \\ \rho \end{Bmatrix}\begin{Bmatrix} \rho m \\ l \end{Bmatrix}, \\ S_{im} &\equiv \sum_l \frac{\partial}{\partial x_m}\begin{Bmatrix} il \\ l \end{Bmatrix} + \sum_{\rho l}\begin{Bmatrix} im \\ \rho \end{Bmatrix}\begin{Bmatrix} \rho l \\ l \end{Bmatrix}, \end{aligned} \tag{4}$$

where both Latin and Greek indices take on the values 1, 2, 3, and 4. In modern notation, the Christoffel symbols $\begin{Bmatrix} \mu\nu \\ \alpha \end{Bmatrix}$ are defined as

$$\begin{Bmatrix} \mu\nu \\ \alpha \end{Bmatrix} \equiv \tfrac{1}{2} g^{\alpha\beta}\left(\partial_\nu g_{\mu\beta} + \partial_\mu g_{\nu\beta} - \partial_\beta g_{\mu\nu}\right). \tag{5}$$

In equation E.2[19] Einstein defines the components $\Gamma^\alpha_{\mu\nu}$ of the gravitational field as minus these Christoffel symbols (Einstein 1915c, p. 832, 1915a, p. 783):

$$\Gamma^\alpha_{\mu\nu} \equiv -\begin{Bmatrix} \mu\nu \\ \alpha \end{Bmatrix}. \tag{6}$$

In coordinate systems satisfying the coordinate condition E.3

$$\sqrt{-g} = 1, \tag{7}$$

where $g = \det g_{\mu\nu}$, the second part of the Ricci tensor, S_{ij}, vanishes, as follows from the relation $\left\{ {\mu\nu \atop \nu} \right\} = \partial_\mu \log \sqrt{-g}$. So, with this coordinate condition, the field equations (2) reduce to $R_{\mu\nu} = 0$, or, with the help of Eqs. (4) and (6), to equation E.1:

$$\sum_\alpha \frac{\partial \Gamma^\alpha_{\mu\nu}}{\partial x^\alpha} + \sum_{\alpha\beta} \Gamma^\alpha_{\mu\beta} \Gamma^\beta_{\nu\alpha} = 0. \tag{8}$$

Eqs. (7) and (8) are the relations Einstein used in the explanation of Mercury's perihelion motion.[20] The unphysical coordinate condition Eq. (7) not only simplified the field equations but was turned to other advantages that will become evident below.

Einstein sought a space-time metric $g_{\mu\nu}$ satisfying Eqs. (7) and (8) and the following conditions, which are obvious posits for the exterior field of the sun:

(C.1) The space-time metric is stationary;
(C.2) it is spherically symmetric;
(C.3) it is time orthogonal;
(C.4) it is asymptotically Minkowskian.[21]

Ideally, one would like to find first an exact solution to the field equations, or rather a family of exact solutions parameterized by the value of the central mass, and then to demonstrate that this is the unique family of solutions satisfying (C), as was, in effect, accomplished shortly afterward through the work of Schwarzschild and Droste. Einstein did not attempt to produce an exact solution. Instead he used a somewhat tricky variant on the iterative approximation procedure that he and Besso had used in their 1913 calculations on the perihelion problem. Moreover, he essentially simply bracketed the uniqueness problem.

After writing down equations E.1 and E.3—our Eqs. (8) and (7)—Einstein briefly addresses the uniqueness problem:

> However, one should keep in mind that the $g_{\mu\nu}$ for a given solar mass are not fully determined mathematically by the equations (1) and (3). This follows from the fact that these equations are covariant under arbitrary transformations with determinant 1. It might, however, be justified to assume that all solutions can be reduced to one another by such transformations and that they therefore (given the boundary conditions) differ only formally, not physically. Following this conviction, I am satisfied, for the time being, to derive one single solution, without entering into the issue of whether it is the only possible one. (Einstein 1915c, p. 832)

So, on the one hand, Einstein seems to have been clearly aware of the fact that there might be a uniqueness problem over and above the trivial non-uniqueness of the solution's coordinate representation; on the other hand, he felt perfectly comfortable ignoring any such complications.

This gives another reason, beyond the ones already given in Section 4, why it was important for Einstein to get very close to 43″. One can well imagine how his derivation of the perihelion motion could have been criticized for relying on a solution that not only is approximate rather than exact, but that might not be unique either. However, since Einstein found the magic 43″ in the end, all worries, it seems, were laid to rest. A stupendous stroke of ill fortune would seem to be needed to make some flawed approximation procedure yield exactly the right value; and in the brilliance of the moment the uniqueness problem recedes into the background where it can be left until a better opportunity arises to tackle it.

Einstein's pleasure over the perihelion result is evident from his letter to Sommerfeld of December 9, 1915: "The result for the motion of the perihelion of Mercury fills one with great satisfaction. How we are helped here by the pedantic precision of astronomy, which I often secretly poked fun at!" (Hermann 1968, p. 37; translation by H. and R. Stuewer). What does not come through in this letter is the psychological drama of the moment when Einstein inserted the numbers into his perihelion formula and the 43″ popped out. Something of the drama is conveyed by A.D. Fokker's report that Einstein suffered heart palpitations following the discovery.[22] Little wonder then that the psychological resolution of the above problems sufficed for the time being.

Einstein computed the field of the sun in Cartesian coordinates (x, y, z, ct). This means that to lowest order the metric field $g_{\mu\nu}$ is the usual diagonal Minkowski metric $\eta_{\mu\nu} \equiv \text{diag}(-1, -1, -1, 1)$ as specified in Einstein's equations E.4 and E.4a. Then follows a rather cryptic paragraph in which Einstein explains his iterative approximation procedure:

> In the following, we assume that the $g_{\mu\nu}$ differ from the values given in equation (4a) only by quantities that are small compared to unity. We will treat these deviations as small quantities of "first order," and functions of the *n*th degree in these deviations as "quantities of nth order." Equations (1) and (3) [our Eqs. (8) and (7)] enable us, starting with equation (4a), to calculate through successive approximations the gravitational field up to quantities of *n*th order. In this sense we will speak of the "*n*th approximation." The equations (4a) form the "zeroth approximation." (Einstein 1915c, p. 833)[23]

We need to look carefully at this paragraph. First, notice that it is ambiguous whether "gravitational field" refers to the quantities $\Gamma^{\alpha}_{\mu\nu}$ or to the metric field $g_{\mu\nu}$. Given Einstein's definition of the gravitational field earlier (see Eq. (6)) and given his usage of the term elsewhere in the paper,[24] one would be inclined to say that the reference is to $\Gamma^{\alpha}_{\mu\nu}$. On the other hand, the two references in this passage to equation E.4a, in which Einstein gives the Minkowski metric, suggest that by "gravitational field" he means $g_{\mu\nu}$ here.

Unfortunately, this is not the only ambiguity in this passage. Still, there seem to be enough clues—mainly coming from a careful analysis of the actual application of the approximation procedure outlined here—to make a solid case for the following interpretation. In presenting this interpretation it will be helpful to introduce some additional notation to distinguish the different orders of approximation.

First, we contend, Einstein assumes that the metric field can be written as a rapidly converging power series, where as his expansion parameter he takes the leading terms in the deviation of the metric field $g_{\mu\nu}$ from its Minkowski values. Using superscripts to distinguish terms of different order and suppressing all indices, we can write:

$$g = \overset{(0)}{g} + \overset{(1)}{g} + \overset{(2)}{g} + \cdots, \tag{9}$$

where $\overset{(0)}{g} \equiv \mathrm{diag}(-1, -1, -1, 1)$, where $\overset{(1)}{g}$ contains the leading terms in the deviations from $\overset{(0)}{g}$ produced by the gravitational field of the sun, and where $\overset{(2)}{g}$ (or more generally, $\overset{(n)}{g}$) contains terms of second (nth) order in $\overset{(1)}{g}$.

We realize that, when taken in isolation, the first two sentences of the passage quoted above are perhaps more naturally interpreted in a somewhat different way. Instead of a power series, Einstein could have had in mind something of the much simpler form $g_{\mu\nu} = \eta_{\mu\nu} + \delta_{\mu\nu}$, where the quantities $\delta_{\mu\nu}$ are used as expansion parameters for functions of $g_{\mu\nu}$, but not as the leadoff terms in a power series for $g_{\mu\nu}$ itself. Clearly, it is crucial for this interpretation that "gravitational field" in the passage quoted above refers to $\Gamma^{\alpha}_{\mu\nu}$ rather than to $g_{\mu\nu}$. As we just saw, it is not clear whether this is true. The interpretation also seems hard to reconcile with two other passages where Einstein talks about the metric to first order in a way that strongly suggests that there are higher-order terms.[25] Actually, we do not need to consider such subtleties to decide against this interpretation. It has some very serious difficulties, which will emerge below. For the time being, we will just list some independent support for our interpretation.

As we mentioned earlier, what Einstein in all likelihood is doing in his perihelion paper is applying the method he used in his 1913 perihelion calculations with Besso in the context of his new theory. Although this is not the place to substantiate this claim,[26] it is clear that in those earlier calculations Einstein and Besso were actually working with a power series expansion for $g_{\mu\nu}$. It is no coincidence that this is more easily seen in these 1913 calculations than in Einstein's 1915 perihelion paper. In 1913, the gravitational field was represented directly by $g_{\mu\nu}$, whereas in 1915 it had come to be represented by $\Gamma^{\alpha}_{\mu\nu}$ (see Eq. (6)). As a consequence, Einstein was now only interested in the quantities $\Gamma^{\alpha}_{\mu\nu}$ giving a perihelion motion, not in the $g_{\mu\nu}$ corresponding to $\Gamma^{\alpha}_{\mu\nu}$. In fact, the only metric field that is explicitly given in the paper (apart from $\eta_{\mu\nu}$) is what in our notation would be $\overset{(0)}{g} + \overset{(1)}{g}$. On the face of it, this would lend support to the reading that Einstein is not thinking in terms of a power series at all. Given his interest in $\Gamma^{\alpha}_{\mu\nu}$ rather than $g_{\mu\nu}$, however, it is perfectly understandable that he does not bother to compute the higher-order terms in the expansion of $g_{\mu\nu}$. On the basis of these considerations and others that will be brought out below, we feel strongly justified reading the passage quoted above as a somewhat cryptic explication of the approximation scheme we suggest.

We continue our exposition of this approximation scheme by looking at the relevant functions of the metric field. In our skeletal notation, the gravitational field $\Gamma^{\alpha}_{\mu\nu}$ to what Einstein calls first and second order can be written as[27]

$$\overset{(1)}{\Gamma} = \overset{(0)}{g}\, \partial \overset{(1)}{g}, \tag{10}$$

$$\overset{(2)}{\Gamma} = (\overset{(0)}{g} + \overset{(1)}{g})\, \partial \overset{(1)}{g} + \overset{(0)}{g}\, \partial \overset{(2)}{g}. \tag{11}$$

$\overset{(2)}{\Gamma}$ contains corrections to $\overset{(1)}{\Gamma}$ of the order of $\overset{(1)}{g}$. Notice that in the reading we just rejected, the last term in $\overset{(2)}{\Gamma}$ would not be present. The importance of this observation will become clear below.

The field equations can likewise be written to first and to second order respectively as

$$\partial \overset{(1)}{\Gamma} = 0, \tag{12}$$

$$\partial \overset{(2)}{\Gamma} + \overset{(1)}{\Gamma}\overset{(1)}{\Gamma} = 0. \tag{13}$$

From these equations it is clear that Einstein can compute $\overset{(2)}{\Gamma}$ without computing $\overset{(2)}{g}$. One simply substitutes the solution $\overset{(1)}{\Gamma}$ of Eq. (12) into Eq. (13) and solves for $\overset{(2)}{\Gamma}$. Since the geodesic equation likewise only depends on the metric field through $\Gamma^{\alpha}_{\mu\nu}$, there is no need to compute $\overset{(2)}{g}$ explicitly to find the perihelion motion.

We are now ready to tackle the remainder of section 1 of Einstein's paper, which comprises two subsections called "First Approximation" and "Second Approximation," respectively.

In E.4b, directly under the heading "First Approximation," we find an expression for—in our notation—$\overset{(0)}{g}_{\mu\nu} + \overset{(1)}{g}_{\mu\nu}$. We offer the following reconstruction of how Einstein arrived at this metric. In the paper on the *Entwurf* theory mentioned earlier, Droste goes over a simple argument to establish the form of a static spherically symmetric metric in Cartesian coordinates (Droste 1915, pp. 999–1000).[28] A sketchy version of the same argument can be found in the Einstein–Besso manuscript, on a page in Einstein's hand. The argument runs as follows.

Suppose we want to find the static spherically symmetric metric $g_{\mu\nu}$ at a point P with Cartesian coordinates (x, y, z). To this end we rotate the coordinate system in such a way that the x-axis now goes through P. In the new, rotated coordinate system, P will have coordinates $\left(x' = (x^2 + y^2 + z^2)^{1/2},\, y' = 0,\, z' = 0\right)$. In this special coordinate system, the metric at P will have a very simple form. From its static character and spherical symmetry, it follows that all off-diagonal components are zero and that $g'_{22} = g'_{33}$. Hence, $g'_{\mu\nu}$ at P can be written as

$$g'_{\mu\nu} = \mathrm{diag}(A, B, B, C), \tag{14}$$

where A, B, and C are as yet undetermined functions of the coordinates. Transforming back to the arbitrary Cartesian coordinates we started from, we find that $g_{\mu\nu}$ at P can be written as

$$g_{ij} = B\delta_{ij} + \frac{x^i x^j}{r^2}(A - B), \quad g_{i4} = g_{4i} = 0, \quad g_{44} = C, \tag{15}$$

where i and j take on the values 1, 2, and 3. Since $g_{\mu\nu}$ is asymptotically Minkowskian, it follows that $B = -1$. Moreover, Einstein knew that in order to recover both Newton's theory in the weak-field, slow-motion approximation and the desired result for the gravitational red shift, C had to be set equal to $1 - \alpha/r$, with $\alpha \equiv 2GM/c^2$ and $r \equiv (x^2 + y^2 + z^2)^{1/2}$. The third constant, A, is then fixed by the coordinate condition E.3. From $\sqrt{-g} = 1$ to first order in α/r, and

$$g \equiv \det g_{\mu\nu} = \det g'_{\mu\nu} = AB^2C = A\left(1 - \frac{\alpha}{r}\right), \tag{16}$$

it follows that

$$A = -\left(1 + \frac{\alpha}{r}\right). \tag{17}$$

Inserting these values for A, B, and C into Eq. (15), we arrive at the first-order metric in equation E.4b, corresponding to the line element:

$$\mathrm{d}s^2 = \left(1 - \frac{\alpha}{r}\right)c^2\,\mathrm{d}t^2 - \sum_{i,j}\left(\delta_{ij} + \alpha\frac{x^i x^j}{r^3}\right)\mathrm{d}x^i\,\mathrm{d}x^j. \tag{18}$$

Presumably, Einstein arrived at equation E.4b in a similar way. To make Eq. (18) more perspicuous, transform to spherical coordinates, in which the line element becomes

$$ds^2 = \left(1 - \frac{\alpha}{r}\right)c^2\,dt^2 - \left(1 + \frac{\alpha}{r}\right)dr^2 - r^2\left(d\theta^2 + \sin^2\theta\,d\phi^2\right), \quad (19)$$

which is the first-order in α/r approximation to the exact Schwarzschild line element[29]

$$ds^2 = \left(1 - \frac{\alpha}{r}\right)c^2\,dt^2 - \left(1 - \frac{\alpha}{r}\right)^{-1}dr^2 - r^2\left(d\theta^2 + \sin^2\theta\,d\phi^2\right). \quad (20)$$

Notice that Einstein cannot use spherical coordinates at this point, because they do not satisfy the coordinate condition $\sqrt{-g} = 1$.

Einstein still has to show, of course, that this metric field is a solution of the field equations to first order. He leaves it to the reader to verify this, giving only the field equations to first order (cf. Eq. (12)). One easily checks that the metric field in Eq. (18) is indeed a solution by computing $\overset{(1)}{\Gamma}{}^{\alpha}_{\mu\nu}$ and showing that $\partial_\alpha \overset{(1)}{\Gamma}{}^{\alpha}_{\mu\nu} = 0$. Expressions for the various components of $\overset{(1)}{\Gamma}{}^{\alpha}_{\mu\nu}$ are given at the end of the "first approximation" subsection (cf. equations E.6a[30] and E.6b):

$$\begin{aligned} \overset{(1)}{\Gamma}{}^{k}_{ij} &= -\frac{\alpha}{2}\frac{x^k}{r^3}\left(2\delta_{ij} - 3\frac{x^i x^j}{r^2}\right), \\ \overset{(1)}{\Gamma}{}^{i}_{44} = \overset{(1)}{\Gamma}{}^{4}_{4i} &= -\frac{\alpha}{2}\frac{x^i}{r^3}, \end{aligned} \quad (21)$$

where i, j, and k take on the values 1, 2, and 3. When either one index or all three indices are equal to 4, $\overset{(1)}{\Gamma}{}^{\alpha}_{\mu\nu}$ vanishes.

As John Norton has argued, the form of the weak static metric field in Eq. (18) freed Einstein from a prejudice: weak static fields need not be spatially flat to recover the correct Newtonian limit (Norton 1984, p. 257, p. 261, pp. 278–279, pp. 310–311). As Norton also pointed out, and as we just saw, the coordinate condition $\sqrt{-g} = 1$ played a central role in obtaining this result. Immediately after giving the first-order metric in E.4b, Einstein addresses the worry that a reader with his own old prejudice would have at this point:

> From our theory it follows that, in the case of masses at rest, the components g_{11} through g_{33} are already different from zero in quantities of the first order. We shall see later that through this no contradiction arises with Newton's law in the first approximation. (Einstein 1915c, p. 834; quoted in Norton 1984, p. 311)

The first sentence provides strong textual support for our interpretation of the approximation procedure. Translated into our notation, what Einstein seems to be saying is that the components $\overset{(1)}{g}_{ij}$ are already nonzero, not just the higher-order terms ($\overset{(2)}{g}_{ij}$, etc.). Einstein will make good on the promise in the second sentence at the beginning of section 2 of his paper, where he will show that for slow motion the geodesic equation to lowest order of approximation—in which only the components $\overset{(1)}{\Gamma}{}^{i}_{44}$ play a role—reduces to Newton's second law.

In the *Entwurf* theory (using Cartesian coordinates), only the 44 component of $\overset{(1)}{g}_{\mu\nu}$ differs from zero. As in the 1915 theory, the geodesic equation to lowest order of approximation reduces to Newton's second law. The perihelion motion of 18″ per century predicted by the *Entwurf* theory is completely due to second-order terms in the power series expansion of g_{44}. In this respect the situation in the 1915 theory is very different. As Einstein explained to Besso in a letter from January 3, 1916[31]:

> The strong increase of the effect compared to our calculation [on the basis of the *Entwurf* theory] stems from the fact that, according to the new theory, the g_{11}–g_{33} occur among the quantities of first order as well [i.e., along with g_{44}] and thus contribute to the perihelion motion. (Speziali 1972, p. 63)

It is important to keep straight which effects are found at which order of approximation in which theory (and, as we will see below, in which coordinates). In the 1915 theory, using Cartesian coordinates, the situation is as follows. The contribution of the first-order terms in the metric to the perihelion motion, to which Einstein is referring in his letter to Besso, comes from the $\overset{(1)}{\Gamma}{}^{i}_{kl}$ terms and from some $\overset{(1)}{g}\,\partial\overset{(1)}{g}$ terms in $\overset{(2)}{\Gamma}$ (cf. Eq. (35) in Section 6). It seems natural to assume that the magic 43″ are due partly to these terms and partly to the $\overset{(0)}{g}\,\partial\overset{(2)}{g}$ terms in $\overset{(2)}{\Gamma}$. From the phrasing of the letter to Besso—especially from the word "contribute" (*beitragen*)—and given Einstein's experience with the *Entwurf* theory, it seems safe to conclude that Einstein tacitly made this assumption. Notice that this would go against the alternative interpretation of the approximation procedure, according to which there are no $\overset{(0)}{g}\,\partial\overset{(2)}{g}$ terms in $\overset{(2)}{\Gamma}$. On the other hand, the assumption that both the $\overset{(1)}{g}\,\partial\overset{(1)}{g}$ and the $\overset{(0)}{g}\,\partial\overset{(2)}{g}$ terms contribute to the magic 43″ gives rise to a little puzzle. We have seen that Einstein's first-order metric is just a first-order approximation to the Schwarzschild metric in Cartesian coordinates (see Eqs. (18)–(20)); and in the usual derivation of the perihelion motion we do not have to go beyond first order. In particular, we do not need any

$\overset{(0)}{g}\,\partial\overset{(2)}{g}$ terms. The solution of this little puzzle will further strengthen the case for our interpretation of Einstein's approximation procedure.

In order to find the perihelion motion, Einstein needs the components Γ^i_{44} $(i = 1, 2, 3)$ to second order. These are computed in the short subsection "Second Approximation." Einstein writes down the relevant components of the second-order field equations $\partial\overset{(2)}{\Gamma} + \overset{(1)}{\Gamma}\overset{(1)}{\Gamma} = 0$ (see Eq. (13)). Using Eq. (21) for the relevant $\overset{(1)}{\Gamma}$ components, one obtains[32]

$$\partial_i \overset{(2)}{\Gamma}{}^i_{44} = -\frac{\alpha^2}{2r^4}. \tag{22}$$

One easily verifies that, up to some divergence-free term, $\overset{(2)}{\Gamma}{}^i_{44}$ has to be equal to $\alpha^2 x^i/2r^4$. The divergence-free term, of course, is just $\overset{(1)}{\Gamma}{}^i_{44}$. So we arrive at

$$\overset{(2)}{\Gamma}{}^i_{44} = \overset{(1)}{\Gamma}{}^i_{44} + \frac{\alpha^2}{2}\frac{x^i}{r^4} = \overset{(1)}{\Gamma}{}^i_{44}\left(1 - \frac{\alpha}{r}\right) = -\frac{\alpha}{2}\frac{x^i}{r^3}\left(1 - \frac{\alpha}{r}\right). \tag{23}$$

Notice that we indeed have a correction of the order of $\overset{(1)}{g}$ here to $\overset{(1)}{\Gamma}{}^i_{44}$ (cf. Eq. (11)).

There is another way of deriving this expression for $\overset{(2)}{\Gamma}{}^i_{44}$. This method is suggested by the alternative interpretation of Einstein's approximation procedure, according to which Einstein simply meant to write the metric field as $g_{\mu\nu} = \eta_{\mu\nu} + \delta_{\mu\nu}$. Recall that in this interpretation $\overset{(2)}{\Gamma}$ would not contain terms of the form $\overset{(0)}{g}\,\partial\overset{(2)}{g}$. Instead of $\overset{(2)}{\Gamma}$ in Eq. (11), we would have

$$\overset{(2)}{\Gamma}{}^* = \left(\overset{(0)}{g} + \overset{(1)}{g}\right)\partial\overset{(1)}{g}, \tag{24}$$

where we used Γ^* to indicate that we are dealing with an alternative interpretation of the approximation procedure. We can directly compute the components of $\overset{(2)}{\Gamma}{}^*$ from the first-order metric Einstein gives in E.4b (cf. Eq. (18)). For instance, we have

$$\overset{(2)}{\Gamma}{}^{i\;*}_{44} = \overset{(1)}{\Gamma}{}^i_{44} - \tfrac{1}{2}\overset{(1)}{g}{}^{i\mu}\left(2\partial_4\overset{(1)}{g}_{\mu 4} - \partial_\mu\overset{(1)}{g}_{44}\right), \tag{25}$$

where we used that $\Gamma^* = \Gamma$ to first order. Since the metric is static, $\partial_4\overset{(1)}{g}_{\mu 4} = 0$. Moreover, one easily verifies that $\partial_j\overset{(1)}{g}_{44} = \alpha x^j/r^3$ and $\overset{(1)}{g}{}^{ij} = \alpha x^i x^j/r^3$. Inserting these relations into Eq. (25), we obtain:

$$\overset{(2)}{\Gamma}{}^{i\;*}_{44} = \overset{(1)}{\Gamma}{}^i_{44} + \frac{\alpha^2}{2}\frac{x^i}{r^4}. \tag{26}$$

Comparing Eqs. (23) and (26), we see that $\overset{(2)}{\Gamma}{}^{i}_{44}{}^{*}$ and $\overset{(2)}{\Gamma}{}^{i}_{44}$ are equal to one another.

This result solves the little puzzle we drew attention to earlier, namely, that it is strange that we should consider a second-order approximation at all given that the line element in Eq. (18) is just the Schwarzschild line element to first order—albeit in Cartesian rather than spherical coordinates—from which we usually derive the perihelion motion without ever having to worry about second-order terms. Contrary to what Einstein presumably thought, the magic 43″ are due solely to the $\overset{(1)}{g}$ terms, the $\overset{(2)}{g}$ terms only contribute in higher and completely negligible order, not via the $\overset{(0)}{g}\,\partial\overset{(2)}{g}$ terms in $\overset{(2)}{\Gamma}$.

The result ironically brings out a serious difficulty for the interpretation of Einstein's approximation procedure that inspired its derivation. The problem is that the equality of $\overset{(2)}{\Gamma}{}^{i}_{44}{}^{*}$ and $\overset{(2)}{\Gamma}{}^{i}_{44}$ only holds in certain coordinate systems. So, in this interpretation we have to ascribe a considerable amount of good fortune to Einstein. He just happened to pick a coordinate system in which the assumption that the metric field can be written as $g_{\mu\nu} = \eta_{\mu\nu} + \delta_{\mu\nu}$ is compatible with the relevant components of the "second order" field equations. In many coordinate systems this would not be the case and the approximation procedure, in this reading, would be inconsistent. In our interpretation, on the other hand, Einstein's approximation procedure works, at least in principle, in arbitrary coordinates. So, on top of the considerations we already gave in favor of our interpretation, the alternative makes for a very uncharitable reading of Einstein's text.

To conclude this section, we will show that the equality $\overset{(2)}{\Gamma}{}^{i}_{44}{}^{*} = \overset{(2)}{\Gamma}{}^{i}_{44}$ that we found in Cartesian coordinates does not hold in isotropic coordinates. These coordinates will be used again in the next section to show that it is a coordinate dependent matter whether the perihelion motion comes out as an effect of the first- or the second-order terms in some power series expansion of the metric field.

In isotropic coordinates (r, θ, ϕ, ct), the Schwarzschild line element has the form (see, e.g., Adler et al. 1975, p. 198, Eq. 6.69)[33]

$$\mathrm{d}s^2 = \left(\frac{1-\alpha/4r}{1+\alpha/4r}\right)^2 c^2\,\mathrm{d}t^2 - \left(1+\frac{\alpha}{4r}\right)^4 \left(\mathrm{d}r^2 + r^2\,\mathrm{d}\theta^2 + r^2\sin^2\theta\,\mathrm{d}\phi^2\right). \tag{27}$$

To first order in α/r, the line element in Eq. (27) is

$$\mathrm{d}s^2 = \left(1-\frac{\alpha}{r}\right)c^2\,\mathrm{d}t^2 - \left(1+\frac{\alpha}{r}\right)\left(\mathrm{d}r^2 + r^2\,\mathrm{d}\theta^2 + r^2\sin^2\theta\,\mathrm{d}\phi^2\right). \tag{28}$$

To show that $\overset{(2)}{\Gamma}{}^{i}_{44}{}^{*} \neq \overset{(2)}{\Gamma}{}^{i}_{44}$ in isotropic coordinates, we need to show that the relevant $\overset{(0)}{g}\,\partial\overset{(2)}{g}$ terms do not vanish. This means that we need to show that

$$\overset{(0)}{g}{}^{ij}\,\partial_j\overset{(2)}{g}_{44} \neq 0. \tag{29}$$

Using this criterion, one immediately sees that the alternative interpretation of Einstein's approximation does work in Schwarzschild–Droste coordinates, even though these coordinates do not satisfy Einstein's coordinate condition $\sqrt{-g} = 1$. Since the 44 component of the Schwarzschild metric is simply $1 - \alpha/r$ (see Eq. (20)), there are no second- or higher-order terms in g_{44}, and the quantities on the left-hand side of Eq. (29) vanish. In isotropic coordinates, however, they do not. To second order in α/r, the 44 component of the metric in Eq. (27) is given by (cf. Misner et al. 1973, p. 1097, Eq. (40.1)):

$$g_{44} = 1 - \frac{\alpha}{r} + \frac{1}{2}\frac{\alpha^2}{r^2}. \tag{30}$$

Inserting the α^2/r^2 term of Eq. (30) into Eq. (29), and using that $\overset{(0)}{g}{}^{11} = -1$, we find that

$$\overset{(0)}{g}{}^{ij}\,\partial\overset{(2)}{g}_{44} = \overset{(0)}{g}{}^{11}\frac{\mathrm{d}}{\mathrm{d}r}\left(\frac{1}{2}\frac{\alpha^2}{r^2}\right) = \frac{\alpha^2}{r^3}. \tag{31}$$

This last expression clearly does not vanish, which shows that $\overset{(2)}{\Gamma}{}^{i}_{44}{}^{*} \neq \overset{(2)}{\Gamma}{}^{i}_{44}$ in isotropic coordinates.

6. Deriving the Equation of Motion

In Einstein's general theory the first step toward predicting the perihelion shift consists of deriving the equation of a timelike geodesic for the given line element, for the theory postulates that test particles freely falling in a gravitational field will trace out timelike geodesics of the metric $g_{\mu\nu}$ solving the field equations, and for purposes at hand we may treat the planets as test bodies moving in the gravitational field of the sun.

Almost half of section 2 of Einstein's perihelion paper is devoted to evaluating the leading terms in the geodesic equation. Using Eq. (6) to replace the Christoffel symbols in the geodesic equation with minus the components of the gravitational field, we arrive at equation E.7 (summation over repeated indices being understood):

$$\frac{\mathrm{d}^2x^\mu}{\mathrm{d}s^2} = \Gamma^\mu_{\rho\sigma}\frac{\mathrm{d}x^\rho}{\mathrm{d}s}\frac{\mathrm{d}x^\sigma}{\mathrm{d}s}. \tag{32}$$

We need to examine the order of magnitude of the various terms on the right-hand side of Eq. (32).

First, depending on whether $\mu = 4$ or $\mu = 1, 2, 3$, $\mathrm{d}x^\mu/\mathrm{d}s$ is of order 1 or of order $(\alpha/r)^{1/2}$, respectively. This can be seen as follows. Since $\overset{(0)}{g}_{\mu\nu} = \eta_{\mu\nu}$, $\mathrm{d}x^4/\mathrm{d}s \approx 1$ to the lowest order of approximation ($x^4 = ct$), and $\mathrm{d}x^i/\mathrm{d}s \approx v^i/c$, where v^i is the ith component of the ordinary three velocity. It follows from the virial theorem for gravitationally bound systems that in the weak-field, slow-motion approximation we are considering here, $(v^i/c)^2$ is of the same order of magnitude as α/r.[34]

We now turn to the components of the gravitational field. The leading terms in $\Gamma^\alpha_{\mu\nu}$ are of order α/r^2 (see Eq. (21)). Terms of this order of magnitude occur for components with either no or two indices equal to 4. Terms with one or three indices equal to 4 are zero to this first order of approximation. Only the Γ^i_{44} components will be needed to a second-order approximation with terms of order α^2/r^3 (see Eq. (23)).

Given these considerations, Eq. (32) to the lowest order of approximation becomes (cf. equation E.7a):

$$\frac{\mathrm{d}^2x^\mu}{\mathrm{d}s^2} = \overset{(1)}{\Gamma}{}^\mu_{44}. \tag{33}$$

For $\mu = 4$, the right-hand side vanishes, and it follows that to this order of approximation the arc length s can be set equal to $x^4 = ct$. For $\mu = i$, the right-hand side can be written as $(\alpha/2)\,\partial_i(1/r)$ (cf. Eq. (21)). With the help of these relations, the i components of Eq. (33) can be rewritten as:

$$\frac{\mathrm{d}^2x^i}{\mathrm{d}t^2} = -\frac{\partial\Phi_N}{\partial x^i}, \tag{34}$$

where $\Phi_N \equiv -\alpha c^2/2r$. Inserting $\alpha = 2GM/c^2$, one sees that Φ_N is the ordinary Newtonian potential $-GM/r$. Hence, Eq. (34) is just Newton's second law, and to this order of approximation there will be no perihelion motion.

Einstein moves on to the next order of approximation of Eq. (32):

$$\frac{\mathrm{d}^2x^\mu}{\mathrm{d}s^2} = \overset{(2)}{\Gamma}{}^\mu_{44}\frac{\mathrm{d}x^4}{\mathrm{d}s}\frac{\mathrm{d}x^4}{\mathrm{d}s} + 2\overset{(1)}{\Gamma}{}^\mu_{4i}\frac{\mathrm{d}x^4}{\mathrm{d}s}\frac{\mathrm{d}x^i}{\mathrm{d}s} + 2\overset{(1)}{\Gamma}{}^\mu_{ij}\frac{\mathrm{d}x^i}{\mathrm{d}s}\frac{\mathrm{d}x^j}{\mathrm{d}s}. \tag{35}$$

For $\mu = 4$, we have $\overset{(1)}{\Gamma}{}^4_{44} = \overset{(1)}{\Gamma}{}^4_{ij} = 0$. The second-order terms in Γ^4_{44} are of order α^2/r^3, a factor $\sqrt{\alpha/r}$ smaller than products of the form $\overset{(1)}{\Gamma}{}^\mu_{4i}\,\mathrm{d}x^i/\mathrm{d}s$.

So, for $\mu = 4$, only the second term on the right-hand side of Eq. (35) comes into play. Inserting $\overset{(1)}{\Gamma}{}^4_{4i} = -\frac{1}{2}\,\partial_i \overset{(1)}{g}_{44}$ and $\overset{(1)}{g}_{44} = -\alpha/r$ into this equation, one easily verifies that to this order of approximation (cf. equation E.9),

$$\frac{\mathrm{d}x^4}{\mathrm{d}s} = 1 + \frac{\alpha}{r}. \tag{36}$$

For $\mu = k$, we have $\overset{(1)}{\Gamma}{}^k_{4i} = 0$, so now only the first and the third term on the right-hand side of Eq. (35) come into play. Notice that the α/r corrections to $\overset{(1)}{\Gamma}{}^k_{44}$ in $\overset{(2)}{\Gamma}{}^k_{44}$ in the first term are of the same order of magnitude as the products $\overset{(1)}{\Gamma}{}^k_{ij}\,\mathrm{d}x^i/\mathrm{d}s\;\mathrm{d}x^j/\mathrm{d}s$ in the third term. With the help of Eqs. (21), (23), and (36), the k components of Eq. (35) can be written as

$$\frac{\mathrm{d}^2x^k}{\mathrm{d}s^2} = -\frac{\alpha}{2}\frac{x^k}{r^3}\left\{1 + \frac{\alpha}{r} + \left(2\delta_{ij} - 3\frac{x^i x^j}{r^2}\right)\frac{\mathrm{d}x^i}{\mathrm{d}s}\frac{\mathrm{d}x^j}{\mathrm{d}s}\right\}. \tag{37}$$

Using $\delta_{ij}\,\mathrm{d}x^i/\mathrm{d}s\;\mathrm{d}x^j/\mathrm{d}s \equiv u^2$ and $x^i\,\mathrm{d}x^i/\mathrm{d}s = r\,\mathrm{d}r/\mathrm{d}s$, we can rewrite Eq. (37) as (cf. equation E.7b)

$$\frac{\mathrm{d}^2x^k}{\mathrm{d}s^2} = -\frac{\alpha}{2}\frac{x^k}{r^3}\left(1 + \frac{\alpha}{r} + 2u^2 - 3\left(\frac{\mathrm{d}r}{\mathrm{d}s}\right)^2\right). \tag{38}$$

The first conclusion Einstein draws from Eq. (38) is that the area law holds to this order of approximation when time is measured in terms of proper time along the orbit. He does not pause to justify this claim, presumably because he thought the matter too obvious. The correctness of the claim simply follows from the linear dependence of the acceleration on x^i. In polar coordinates (r, ϕ), the area law can be written as

$$r^2\frac{\mathrm{d}\phi}{\mathrm{d}s} = B, \tag{39}$$

where B is some constant. Einstein now takes advantage of the fact that the factor in front of the expression in parentheses on the right-hand side of Eq. (38) is of order α/r^2. This means that the expression in parentheses itself only needs to be evaluated to the lowest order of approximation, in which, as we saw above, the familiar Newtonian results are recovered. In particular, Einstein can use Newtonian energy conservation

$$\tfrac{1}{2}u^2 + \Phi_N/c^2 = A, \tag{40}$$

where A is another constant. Since $\Phi_N/c^2 = -\alpha/2r$ (see Eq. (34)), Eq. (40) is equivalent to $\alpha/r = u^2 - 2A$. Substituting this expression for α/r into Eq. (38), along with $u^2 = (\mathrm{d}r^2 + r^2\,\mathrm{d}\phi^2)/\mathrm{d}s^2$ and $r\,\mathrm{d}\phi/\mathrm{d}s = B/r$, one obtains

$$\frac{\mathrm{d}^2 x^k}{\mathrm{d}s^2} = -\frac{\alpha}{2}\frac{x^k}{r^3}\left(1 - 2A + 3\frac{B^2}{r^2}\right). \tag{41}$$

This equation can be cast into the form of the Newtonian law Eq. (34). To this end, Einstein first divides both sides of Eq. (41) by the factor $1 - 2A$, rescales the proper time by the factor $\sqrt{1 - 2A}$ and absorbs the factor $1/\sqrt{1 - 2A}$ into the area law constant B.[35] Eq. (41) then turns into

$$\frac{\mathrm{d}^2 x^k}{\mathrm{d}s^2} = -\frac{\alpha}{2}\frac{x^k}{r^3}\left(1 + 3\frac{B^2}{r^2}\right). \tag{42}$$

This equation can be written as (cf. equation E.7c[36])

$$\frac{\mathrm{d}^2 x^k}{\mathrm{d}s^2} = -\frac{\partial \Phi_E}{\partial x^k}, \quad \Phi_E \equiv -\frac{\alpha}{2r}\left(1 + \frac{B^2}{r^2}\right). \tag{43}$$

Eq. (43) shows that Einstein's general theory revives, in a sense, Clairaut's 1745 hypothesis that the perihelion advance is due to the presence of an extra $1/r^4$ term in the gravitational force law. The sense is a rather attenuated one, however, since the area law constant B in Eq. (43) is not a universal constant but depends on the orbit.

With Eq. (43), finding the perihelion motion has become an exercise in Newtonian mechanics with a somewhat different force law. The perihelion motion is derived from energy conservation, $\frac{1}{2}u^2 + \Phi_E = A$, and from the area law Eq. (39).[37] When u^2 is expressed in polar coordinates, energy conservation can be rewritten as:

$$\frac{\mathrm{d}r^2 + r^2\,\mathrm{d}\phi^2}{\mathrm{d}s^2} = 2A - 2\Phi_E = 2A + \frac{\alpha}{r} + \frac{\alpha B^2}{r^3}. \tag{44}$$

Eliminating $\mathrm{d}s$ in favor of $\mathrm{d}\phi$ with the help of the area law and introducing $x \equiv 1/r$, one arrives at equation E.11:

$$\left(\frac{\mathrm{d}x}{\mathrm{d}\phi}\right)^2 = \frac{2A}{B^2} + \frac{\alpha}{B^2}x - x^2 + \alpha x^3. \tag{45}$$

As Einstein points out, the only difference with the Newtonian equation is the presence of an x^3 term on the right-hand side of Eq. (45).

In the next section, we will see how Eq. (45) can readily be integrated to find the perihelion motion. In the remainder of this section, we will discuss a simple alternative derivation of Eq. (45). This discussion serves two purposes. First, it will make Eq. (45) more perspicuous. Second, it may help us appreciate why Einstein derived Eq. (45) in what by comparison may seem a rather cumbersome fashion.

To first order in α/r, the Schwarzschild line element in the usual spherical coordinates (ct, r, θ, ϕ) is given in Eq. (19). As we saw in Section 5, this line element is equivalent to Einstein's first-order metric E.4b (cf. our Eq.(18)), the only difference being that Einstein used Cartesian instead of spherical coordinates. Starting from the line element in Eq. (19), the time-like geodesics representing planetary orbits follow from the variational principle

$$\delta \int \left\{ \left(1 - \frac{\alpha}{r}\right) c^2 \dot{t}^2 - \left(1 + \frac{\alpha}{r}\right) \dot{r}^2 - r^2 \left(\dot{\theta}^2 + \sin^2\theta \dot{\phi}^2\right) \right\} \mathrm{d}s = 0, \quad (46)$$

where the dots represent differentiation with respect to arc length s. The motion can be arranged to take place in a plane of fixed angle θ. With the choice $\theta = \pi/2$, one of the Euler–Lagrange equations is the area law

$$r^2 \frac{\mathrm{d}\phi}{\mathrm{d}s} = B, \quad (47)$$

where B is a constant. Another is

$$\left(1 - \frac{\alpha}{r}\right) c\dot{t} = C, \quad (48)$$

where C is another constant. Inserting these relations into the identity

$$1 = \left(1 - \frac{\alpha}{r}\right) c^2 \dot{t}^2 - \left(1 + \frac{\alpha}{r}\right) \dot{r}^2 - r^2 \dot{\phi}^2, \quad (49)$$

and introducing $x \equiv 1/r$, results in

$$\left(\frac{\mathrm{d}x}{\mathrm{d}\phi}\right)^2 (1 - \alpha^2 x^2) = \frac{C^2 - 1}{B^2} + \frac{\alpha}{B^2} x - x^2 + \alpha x^3. \quad (50)$$

If the $\alpha^2 x^2$ term on the left-hand side is neglected, Eq. (50) becomes the exact equation of motion for the Schwarzschild line element Eq. (20). Without the $\alpha^2 x^2$ term, Eq. (50) is also the same as equation E.11, our Eq. (45). At first sight, it looks as if the constant terms in the two equations are different. On closer examination, however, we see that they are both equal to zero. Consider the constant term in Eq. (50) first. From Eq. (48) it follows that to first order in α/r, $c\dot{t} = C(1 + \alpha/r)$. Comparing this relation with

$c\dot{t} = (1 + \alpha/r)$ (Eq. (36)), one sees that C has be to equal to 1. Hence, the constant term in Eq. (50) is zero. To see that $A = 0$ in Eq. (45), start from the identity in Eq. (49). Insert Eq. (36) and $\dot{r}^2 = 2A - 2\Phi_E - r^2\dot{\phi}^2$ (cf. Eq. (44)) into this identity. The constant term on the right-hand side of the identity will be $1 - 2A$; all other terms contain at least one factor α/r. Since the left-hand side is 1, A must be zero. So, apart from the $\alpha^2 x^2$ term, Eq. (45) and Eq. (50) are indeed the same. As an aside, we want to mention that this last calculation also demonstrates that Einstein's rescaling maneuver in Eqs. (41)–(42) was, in fact, redundant. To lowest order of approximation—and recall that the lowest order was all that was needed in the expressions in parentheses in Eqs. (41) and (42)—the constant A in Eq. (40), expressing Newtonian energy conservation, is equal to the constant A in Eq. (44), expressing energy conservation in the relativistic setting.

Returning to the main point, we see that Einstein's first-order metric directly leads to Eq. (45) or, equivalently, Eq. (50) without computing any Christoffel symbols. It is true that in the Cartesian coordinates that Einstein used the derivation does not run as smoothly as in the spherical coordinates used above. In particular, the area law does not drop out as one of the Euler–Lagrange equations. However, nothing prevents us from switching from Cartesian to spherical coordinates once the solution to the field equations has been found, even though spherical coordinates do not satisfy the coordinate condition imposed in finding the solution. Since Einstein, as we saw, switches from Cartesian to polar coordinates, the coordinate condition $\sqrt{-g} = 1$ would not have stopped him from deriving the equation for $(dx/d\phi)^2$ the way we did in Eqs. (46)–(50), had he thought of that possibility. What would have stopped him, though, is a tacit assumption that we made, namely, that for finding the perihelion motion it suffices to compute the metric to first order. As we argued in Section 5, Einstein felt he needed to take into account the effect of second-order terms in the metric, something he did by solving the second-order field equations Eq. (13) for—in our skeletal notation—$\overset{(2)}{\Gamma} = (\overset{(0)}{g} + \overset{(1)}{g})\,\partial \overset{(1)}{g} + \overset{(0)}{g}\,\partial \overset{(2)}{g}$. We also saw in Section 5 that it was a coordinate-dependent matter whether the components $\overset{(2)}{\Gamma}{}^i_{44}$, the only components playing a role in the equation for $(dx/d\phi)^2$ (see Eq. (37)), actually depend on second-order terms in the metric or not. In Cartesian and Schwarzschild–Droste coordinates they do not; in isotropic coordinates they do.

To emphasize that we have to be careful about neglecting second-order terms, we will go through the argument in Eqs. (46)–(50) in isotropic coordinates. The Schwarzschild line element in isotropic coordinates to first order in α/r is given in Eq. (28). In the variational problem for this line element the counterpart of Eq. (47) is $(1+\alpha/r)r^2\dot{\phi} = B$ and the counterpart of Eq. (48) is $(1-\alpha/r)c\dot{t} = D$, where D is a constant. Using these results in the counterpart of Eq. (49) gives

$$\left(\frac{\mathrm{d}x}{\mathrm{d}\phi}\right)^2 = \frac{D^2(1+\alpha x)}{B^2(1-\alpha x)} - \frac{(1+\alpha x)}{B^2} - x^2. \tag{51}$$

Expanding in αx and neglecting $\alpha^2 x^2$ terms reduces Eq. (51) to

$$\left(\frac{\mathrm{d}x}{\mathrm{d}\phi}\right)^2 = \frac{D^2-1}{B^2} + \frac{\alpha(2D^2-1)}{B^2}x - x^2. \tag{52}$$

Eq. (52) is of Newtonian form and, consequently, predicts a null perihelion shift.

Writing in the *Philosophical Magazine* in 1920, Prof. A. Anderson of University College, Galway, concluded from his analysis of motion in Einstein's general theory that "Mercury, unfortunately, is left with the advance of his perihelion unexplained" (Anderson 1920a, p. 628). Anderson's mistake lay in trying to draw conclusions from an unfortunate approximation similar to the one above.[38] In fact, the move from Eq. (51) to Eq. (52) is illegitimate. It follows from Eq. (51) that the first-order metric in isotropic coordinates gives a perihelion shift for Mercury of 4/3 the final GTR value, or some 57″ of arc per century (see Misner et al. 1973, pp. 1110–1116).[39]

The perihelion advance is, of course, an intrinsic effect, depending on the line element but not on the coordinate system in which the line element is expressed. But approximation procedures that use coordinate language are not intrinsic, and what is a good approximation in one coordinate system may be a disastrous approximation in another. In the case in point, the Schwarzschild–Droste coordinate system has the fortunate feature that the perihelion advance is a "first order" effect—neglecting α^2/r^2 terms in moving from the exact line element in Eq (20) to the approximate line element in Eq. (19) and neglecting α^2/r^2 terms in the equation of motion Eq. (50) that follows from Eq. (19) does not disturb the prediction of the advance. By contrast, in isotropic coordinates the perihelion shift is a "second order" effect—neglecting α^2/r^2 terms in the isotropic expression for the Schwarzschild line element in Eq. (27) to arrive at Eq. (28) leads to a perihelion advance that is too large by one third, and neglecting α^2/r^2 terms in the equation of motion Eq. (51) that follows from Eq. (28) leads to the prediction of no advance.

The upshot of these considerations is that the quick derivation of equation E.11, our Eq. (45), in Eqs. (46)–(50) is essentially just a lucky shot. It only gives the right answer in certain coordinate systems. Without having the exact solution in hand, there is no way to tell whether the terms that are neglected do or do not give a nonnegligible contribution to the perihelion advance. Einstein's own more labyrinthine derivation seems to be far more reliable.

7. The Perihelion Shift Formula

Two main methods for deriving the shift of the perihelion of a planet with an elliptical orbit are found in the textbooks. One starts from the second-order equation of motion. This equation is found by differentiating Eq. (45) with respect to ϕ:

$$2u''u' = \frac{\alpha}{B^2}u' - 2uu' + 3\alpha u^2 u', \tag{53}$$

where we introduced the notation $u \equiv x \equiv 1/r$ and $u' \equiv \mathrm{d}u/\mathrm{d}\phi$. If $u' = 0$, the motion is circular, so we can assume that $u' \neq 0$, with the result that

$$u'' + u = \frac{\alpha}{2B^2} + \frac{3\alpha}{2}u^2. \tag{54}$$

A perturbation approach is then used to solve Eq. (54) to first order and to show that the shift per revolution is

$$\frac{6\pi G^2 M^2}{B^2 c^2}. \tag{55}$$

Einstein's own method starts from the first-order equation Eq. (45). The idea is to find the perihelion shift by computing the deviation from π in the angle between aphelion and perihelion. To this end the equation for $\mathrm{d}x/\mathrm{d}\phi$, or rather its reciprocal $\mathrm{d}\phi/\mathrm{d}x$, is integrated between the values x takes on at aphelion and perihelion, respectively (see Einstein 1915c, p. 838):

$$\phi = \int_{\alpha_1}^{\alpha_2} \frac{\mathrm{d}x}{\sqrt{(2A/B^2) + (\alpha/B^2)x - x^2 + \alpha x^3}}, \tag{56}$$

where ϕ is the angle between the radius vectors from the sun to the planet at the times of aphelion and perihelion, and α_1 and α_2 are the roots of the cubic equation $0 = (2A/B^2) + (\alpha/B^2)x - x^2 + \alpha x^3$, which correspond to the roots of the classical quadratic equation $0 = (2A/B^2) + (\alpha/B^2)x - x^2$. Classically, the orbit is some fixed ellipse. Hence, the roots α_1 and α_2 must be the reciprocals of the distance of the planet to the sun at aphelion and perihelion, respectively. Without the rescaling maneuver in Eqs. (41)–(42), the coefficient of x in the cubic equation above would be $\alpha(1-2A)/B^2$ instead of α/B^2. At first sight, it looks as if the roots α_1 and α_2 of this cubic equation would be different from the roots of the classical quadratic equation. As we saw in Section 6, however, A can be set equal to zero, so we get the same results with or without rescaling proper time.

The method Einstein uses here in his perihelion paper is essentially the same as the method he and Besso used in 1913. In the Einstein–Besso manuscript, the law of energy conservation and the area law are derived and written in polar coordinates, ds is eliminated from these equations, and the resulting equation for $\mathrm{d}\phi/\mathrm{d}r$ is integrated between the values r takes on at perihelion and aphelion. There are some minor differences,[40] but the basic strategy is exactly the same. A modern discussion of the method can be found in Møller (1977, pp. 496–497). We have been unable to track down a source from which Einstein may have learned the method, but we strongly suspect it was fairly standard at the time.

The upshot in the perihelion paper is an advance per orbit of (see equation E.13)

$$\frac{3\pi\alpha}{a(1-e^2)}. \tag{57}$$

Inserting $\alpha = 2GM/c^2$ into Eq. (57) and the classical relation $B^2 = GMa(1-e^2)$ into Eq. (55), we see that these two equations are equivalent. Introducing the orbital period T through the Kepler relation allows the advance to be alternatively expressed as (see equation E.14):

$$\frac{24\pi^3a^2}{T^2c^2(1-e^2)} \tag{58}$$

Equation E.14, our Eq. (58), is the last to appear in Einstein's perihelion paper, and when he says that "The calculation yields . . . 43″ per century" (Einstein 1915c, p. 839), he is presumably referring to the result of inserting the planetary data into Eq. (58). But why do the calculation from Eq. (58) rather than Eq. (57), and, for that matter, why introduce Eq. (58) at all? Perhaps Freundlich, who supplied Einstein with the data, saw the orbital period T as having a more direct observational significance than GM, or perhaps he was influenced by Gerber's formula, which has the same form as Eq. (58).

To conclude our discussion of Einstein's perihelion paper, it seems appropriate to emphasize Einstein's impressive accomplishment in this paper. Within the space of a few days Einstein produced, by means of the most ingenious reasoning, the essentials of the solution to one of the great puzzles of astronomy. And whatever qualms one might have felt about

Einstein's derivation were shortly to be swept away when Schwarzschild (1916) produced an exact solution to the field equations and produced it in a way that made it clear that it was the most general solution satisfying the requirements Einstein laid down (see C.1–C.4, p. 140 above). A nonproblematic derivation of the perihelion shift was now possible, which Schwarzschild duly supplied.[41] Droste's identical solution appeared in print after Schwarzschild's (Droste 1917), but since Droste proceeded independently, the solution should rightly be called the Schwarzschild–Droste solution. Birkhoff (1923) showed that the assumptions C.1 of stationarity and C.3 of time orthogonality can be dropped without loss of generality: any spherically symmetric solution of the exterior field equations must be a piece of Schwarzschild space-time.

8. Reactions to Einstein's Perihelion Explanation

Some of the negative reactions to Einstein's explanation are amusing, others are simply silly, and still others have serious substance. Taken together, they are revealing of the attitudes toward and the levels of understanding of the general theory of relativity in the decade following its introduction. These reactions can be conveniently grouped into five categories.

Anti-relativity invective. Gehrcke, a sponsor of the anti-relativity campaign in Germany, attempted to revive Gerber's theory (Gehrcke 1916), which was then republished in the *Annalen der Physik* (Gerber 1917). Outside Germany there was no coordinated anti-relativity campaign, but Charles Lane Poor, a professor of astronomy at Columbia University and former head of the astronomy department at Johns Hopkins, waged his own campaign, charging that Einstein's derivation of the perihelion advance was incoherent and that Newtonian gravitation together with a form of Seeliger's hypothesis would suffice to provide an explanation (Poor 1921, 1922, 1925, 1930). In a communication to the *Astronomische Nachrichten*, Poor focused his criticism of Einstein's derivation on the rescaling maneuver discussed above in Sections 6 and 7. Poor argued that if the unit of time is changed, then there must be a corresponding change in the unit of mass, and that as a result of the latter change the constant α becomes, in "relativistic units," $\hat{\alpha} = \alpha\left(1 + 3C^2/r^2\right)$. So according to Poor's viewpoint, Einstein's equation of motion should be written as $\ddot{x}^i = -\hat{\alpha}x^i/2r^3$, whose "solution is identically the same as that of *Newton*: a fixed ellipse" (Poor 1930, p. 170). Poor concluded that "the so-called relativity rotation of planetary orbits is only a mathematical illusion.... The *Newtonian* law has not been abolished: there is no *Einsteinian* law of gravitation" (ibid.).

That the *Astronomische Nachrichten* should publish such an article is a sad commentary on the politics of German science in 1930.

Challenging the data. Ernst Grossmann argued (1921) that a reanalysis of the astronomical data used by Newcomb leaves a residual advance for Mercury's perihelion of between 29″ and 38″ per century. Wiechert (1920) set the anomalous advance at 34″ per century and claimed as a result that Einstein's theoretical value was too high. Von Gleich (1923) also used the discrepancy between Einstein's prediction and the figures of Grossmann and Wiechert to raise serious doubts about general relativity. (See also Lecornu 1922.)

Misunderstandings. As already mentioned, Anderson (1920a) claimed that no perihelion advance is predicted by Einstein's theory. His error was quickly discovered (see Pearson 1920), and Anderson himself published a correction (1920b). Le Roux (1921) complained that the perihelion advance "a bien été obtenu *à propos* de la théorie de la relativité, mais qu'il n'en est pas une conséquence et ne constitue même pas un argument en sa faveur" (Le Roux 1921, p. 1227). His complaint centered on the fact that the integration of Einstein's field equations for the static spherically symmetric case contains an undetermined constant of integration. Thus, the solution is not unique: "En réalité, il y a une infinité de solutions" (ibid., p. 1230). But an infinity of solutions is exactly what one wants, since the intended interpretation is the exterior field of a central body which may take on a continuum of mass values. This interpretation is confirmed by taking the weak-field, slow-motion approximation and verifying that the Newtonian equations of motion are obtained just in case the constant of integration equals twice the value of the central mass, something Einstein had already done in his perihelion paper.

A misunderstanding with more serious consequences was fostered by Allvar Gullstrand, professor of ophthalmology at Upsalla and the recipient of the 1911 Nobel Prize in physiology and medicine. Gullstrand claimed (1922, 1923) that the Schwarzschild metric was not the unique static solution of Einstein's field equations for a central mass and that the general solution contains, in addition to α, another parameter β that affects the value of the perihelion shift. As a result, Gullstrand thought that Einstein's perihelion explanation was merely an artifact of the coordinate system Einstein had employed. Under pressure from Kretschmann (1923a, 1923b), Gullstrand was forced to retreat, but the damage was already done by his 1921 report to the Nobel Committee for physics and by an updated version in 1922 which concluded that acceptance of Einstein's special and general theories of relativity was "a matter of faith."[42] This is undoubtedly part of the reason why Einstein's Nobel award does not mention his work on gravitation and was given "for his services to theoretical physics and especially his discovery of the law of the photoelectric effect" (quoted in Pais 1982, p. 510). In addition to Gullstrand's report, two other countervailing forces were at work. On the one hand, the Nobel Institute was influenced by philosophical objections to Einstein's special theory, most probably by Bergson's in particular.[43] On the other hand, there was an attempt, spearheaded by C.W. Oseen, to turn Swedish physics from a primarily experimental orientation toward a more theoretical outlook.[44] Trying to discern how these three factors interacted is a nice exercise that will not be attempted here.[45]

Alternative theories. This category can be subdivided, somewhat subjectively, into the uninteresting and the relatively more interesting. In the former belongs Wiechert's 1916 attempt to retain the ether and to explain the perihelion motion on the basis of an electromagnetic theory of gravitation. In the latter category belong two theories of Ludwik Silberstein. Although he found Gerber's theory "untenable," Silberstein (1917) could not resist labeling Eq. (58) "(G)" for Gerber. As for Einstein's general theory itself, Silberstein was skeptical:

> ...notwithstanding its broadness and mathematical elegance, it [Einstein's general theory] certainly offers many serious difficulties in its very foundations, while none of its predictions of new phenomena, as the deflection of a ray by the sun, have thus far been verified. And even the fact that Einstein's new theory gives Gerber's formula, and therefore the *full* excess of 43″ for Mercury, does not seem to be decisive in its favor. As far as I can understand from [Harold] Jeffreys' investigation [of Seeliger's hypothesis], it would rather alleviate the astronomer's difficulties if the Sun by itself gave only a *part* of these 43 seconds. (Silberstein 1917, p. 504)

Under these circumstances Silberstein thought it worthwhile to investigate how much of the anomalous advance could be accounted for on the basis of Einstein's "old" theory of relativity. Silberstein proceeded to rediscover a version of De Sitter's 1911 result; namely, that by introducing a factor of γ^n, $\gamma \equiv \sqrt{1 - v^2/c^2}$, into the special-relativistic force law, the perihelion shift formula becomes $4n\pi^3 a^2/T^2c^2(1 - e^2)$. If the entire excess of Mercury's perihelion is to be attributed to the sun, then $n = +6$. Silberstein confessed that he did not know why the value of n is just 6. But, he added, in self-defense, "as little do we know 'why' the exponent of r [in Newton's law] is 'just' or exceedingly nearly equal [to] -2," and, besides, "[s]uch a naturalistic method of improving Newton's law of gravitation seems a great deal safer than those based on fantastic constructions or rash generalizations" (1917, p. 509). In periods of scientific revolution, however, small improvements in old laws may in fact be less safe than fantastic constructions, at least if an Einstein is the constructor.

In 1918 Silberstein published his own theory of gravitation, which he called "general relativity without the equivalence hypothesis" (Silberstein 1918). Einstein's progress toward his general theory had been guided from the start by the principle of equivalence, but what Einstein took to be a foundation stone, Silberstein perceived as sand. The principle of equivalence was vulnerable, according to Silberstein, because "of its very special nature and the great number of assumptions which it tacitly implies" (ibid., p. 95). In addition, serious doubts arise about its acceptability because it leads directly to the gravitational red shift prediction which is contradicted by the "obstinately negative results quite recently obtained by St. John at the Mount Wilson Observatory" (ibid., p. 95). And the prediction of the bending of light, Silberstein noted, "still awaits its verification" (ibid., p. 96). Even the one "conspicuous and fascinating success" of Einstein's theory—the deduction of the 43″—is tainted by the fact that the secular motion of the perihelion "is most vitally conditioned by . . . g_{44}, which—to everybody's regret—has thus discredited itself at the Mount Wilson Observatory" (ibid., p. 96). Silberstein's own alternative theory, based on the postulate that space-time has a fixed constant curvature, regardless of the nature and distribution of the gravitational sources, gives a secular perihelion motion of $-4\pi^3a^2/T^2c^2(1-e^2)$, i.e., a retrograde motion of one-sixth Einstein's value. This feature was not regarded by Silberstein as a defect of his theory since he was able to refer to the forthcoming work by Jeffreys (1918), which attempted to show that a version of Seeliger's hypothesis would bring Silberstein's predictions into harmony with the observed secular motions of the inner planets. In the following year, however, Jeffreys abandoned his Seeligerizing.

Objection to the completeness of Einstein's explanation. The Italian paradoxer Burali-Forti (1922–1923) complained that Einstein's derivation did not really explain Mercury's perihelion motion because it did not contain an account of the perturbations of the other planets, showing that these would add up to the five hundred and seventy-some seconds of arc per century. One could respond that the fact that general relativity yields the Newtonian equations of motion in the weak-field, slow-motion approximation makes it plausible that the general-relativistic perturbations can be well approximated by the Newtonian ones. However, Einstein had shown this only for the Schwarzschild metric, whereas the relevant metric for computing, say, the perturbation on Mercury's orbit caused by Jupiter certainly will not be of Schwarzschild form. Without giving any reason, Burali-Forti claimed that as a result the Einstein value for Jupiter's perturbation will be different from the Newtonian value, leading to a discord between the GTR and the observed 570″ of arc per century.

If Burali-Forti had wanted to create a paradox, he could have argued as follows. In the case of the Schwarzschild metric, where there is a stationary, nonrotational timelike vector field, the notion of the perihelion shift can be given an invariant meaning. But in the N-body case, needed to give the full general-relativistic explanation of the observed 570″ of shift, stationarity and the other nice features will presumably be lost, and thus it becomes difficult to say precisely what the perihelion shift means.[46]

Positive reactions to Einstein's resolution of the perihelion anomaly were swift in coming, and they came from influential sources. De Sitter declared that "Seeliger's explanation of the anomalous motion of the perihelion of Mercury by the attraction of nebulous matter in the neighborhood of the Sun now becomes superfluous" (De Sitter 1916, p. 728). This declaration was contained in a lengthy three part review article detailing the principles and consequences of Einstein's general theory. Appearing in the *Monthly Notices of the Royal Astronomical Society*, the article served during the war years as the chief source of knowledge of Einstein's theory for scientists in England. It was read by Eddington, who was to become the most effective of the early champions of general relativity.

Harold Jeffreys did not give up on Seeliger's hypothesis until 1919, after Eddington had reported the verification of Einstein's light deflection prediction. Jeffreys (1919) argued that the numerical agreement between experimental values and Einstein's predictions for Mercury's perihelion and the deflection of light could not be counted as confirmation of the general theory until other potential causes of these effects had been eliminated.

> Suppose, for instance, that a true cause was known that would produce a motion of 10″ per century in the perihelion of Mercury; then Einstein's theory would predict the total excess motion to be 53″ per century, which differs from the observed 41″ by more than the permissible error of observation. Such a discovery would be fatal to a theory such as Einstein's, which contains no arbitrary constituent capable of adjustment to suit empirical facts. Now a sufficient amount of gaseous matter within the orbit of Mercury would be capable of producing the first two of Einstein's effects [perihelion advance and bending of light]—the first by its gravitation, and the second by its refraction; and such a matter is known to exist, causing the solar corona and the zodiacal light. It is therefore desirable to inquire whether its quantity is sufficient to invalidate the theory. (Jeffreys 1919, pp. 138–139)

Jeffreys proceeded to argue that an analysis of the observations of the solar corona and the luminosity of the zodiacal light shows that whatever intra-Mercurial matter exists will not appreciably affect either the secular motion of Mercury's perihelion or the displacement of star images taken during a solar eclipse. Thus ended the serious Seeligerizing in England.

The interaction among the classical tests of Einstein's general theory is a topic that deserves detailed study, but one facet of the interaction is already evident: the negative results of the early red shift measurements led some, such as Silberstein, to question Einstein's perihelion explanation, while the success reported by the English eclipse expeditions helped others, such as Jeffreys, to accept the explanation.

9. Conclusion

In the decades following 1915, the solar red shift measurements stubbornly refused to conform to Einstein's prediction, and the deflection of light measured by several eclipse expeditions exceeded the theoretical value. In this context, the resolution of the perihelion anomaly served as the main observational anchor of the general theory, but this anchor has always been a potential Achilles heel. As explained in Sections 4 and 5, it was doubly important in 1915 that Einstein obtain the 43″ per century or close to it, first to overcome worries about the validity of approximations and second to render superfluous Seeliger's hypothesis. The fact that the theory does give the 43″ leaves no room for maneuver if additional sources of perturbation of Mercury's perihelion are found, as Harold Jeffreys noted. What is so stunning about the explanation of the 43″ is that it was achieved without the leeway of any adjustable parameter, but it is exactly this feature of the theory that makes it vulnerable. In the 1960s Dicke tried to pierce this Achilles heel (see Dicke and Goldenberg 1967, 1974), claiming that optical measurements revealed a solar oblateness that would account for 3–5″ of the advance and would thus throw the general-relativistic prediction into doubt. The controversy that ensued had no clean resolution, but insofar as a consensus developed, it is in favor of orthodox relativity. At the same time, the two other "classical tests" have been perfected and stand solidly behind the theory, and new tests, such as the radar delay measurements, together with a deeper appreciation of the range of possible alternative theories of gravity that are eliminated by actual or feasible experiments have greatly strengthened the case that Einstein's theory of gravitation will prove to be as durable as Newton's.[47] It would thus be ironic indeed if the perihelion problem were to prove to be an Achilles heel. While claiming no powers of prognostication in this matter, we must state our opinion that such an irony would be most unseemly.

ACKNOWLEDGMENTS. We would like to thank Jean Eisenstaedt, Clark Glymour, David Hillman, A.J. Kox, John Norton, Jürgen Renn, and John Stachel for their helpful comments.

NOTES

[1] As the basis for this chapter, we used a paper written by one of us (JE) in collaboration with Clark Glymour, which was presented at the Second International Conference for the History of General Relativity held in Luminy, France, in September 1988. The so-called Einstein–Besso manuscript, discovered shortly afterward by the editors of *The Collected Papers of Albert Einstein*, made it clear that this 1988 paper contained some serious errors, especially in its analysis of Einstein's perihelion paper of November 1915. In this new version we have tried to correct those errors.

[2] Einstein found Seeliger's critique so caustic that he wrote to Sommerfeld: "Tell your colleague Seeliger that he has a horrible disposition. I had a taste of it recently in a reply to the astronomer Freundlich" (Hermann 1968, p. 37; translated by H. and R. Stuewer).

[3] For an account of the early tests of the gravitational red shift, see Earman and Glymour (1980a).

[4] An analysis of the early eclipse tests and their role in the reception of Einstein's general theory is given in Earman and Glymour (1980b).

[5] With a possible exception mentioned in Section 9 below.

[6] If an earth-based coordinate system is used, the observed is some 5500″ per century.

[7] The agreement, however, was not unanimous; see Section 8 below.

[8] A detailed study of the work of Le Verrier and Newcomb is to be found in Cohen (1971).

[9] The 42.95″ figure is due to a round-off error, while the more popular 43.03″ is due to the use of nonstandard values for the astronomical unit and the velocity of light; see Nobili and Will (1986). We are grateful to Professor Will for bringing this matter to our attention.

[10] They neglected the component of the sun's force exerted in a direction perpendicular to the radius vector from the earth to the moon; see Waff (1976).

[11] Roseveare (1982) also notes that Ritz's theory predicts a deflection of 1.31″ for starlight grazing the sun, a prediction contradicted by the majority of solar eclipse observations.

[12] This so-called Einstein–Besso manuscript will be published in Klein et al. (forthcoming).

[13] We are grateful to Jürgen Renn for drawing our attention to this calculation. The "scratch notebook" will be published in appendix A of Klein et al. (1993).

[14] See Norton 1984, 1987 and Earman and Norton 1987.

[15] In more detail, Einstein's three reasons for rejecting the Einstein–Grossmann theory are as follows:

> (1) I proved that the gravitational field for a uniformly rotating system does not satisfy the field equations.
> (2) The motion of the perihelion of Mercury yielded 18″ instead of 45″ per century.
> (3) The covariance requirement in my paper of last year did not yield the Hamiltonian function. It permits, if appropriately generalized, an arbitrary H. This led to the conclusion that the covariance with respect to "adapted" coordinate systems is a failure. (Hermann 1968, pp. 32–33; translation by H. and R. Stuewer)

[16] See Pais 1982, pp. 257–261.

[17] Footnote 2 in Einstein's perihelion paper reads: "E. Freundlich recently wrote a noteworthy article on the impossiblity of satisfactorily explaining the anomalies in the motion of mercury on the basis of Newtonian theory" (translation by Doyle, 1979a).

[18] See Norton 1984 for a detailed account of how Einstein reached his final field equations.

[19] The notation E.*n* will be used to refer to equation number *n* in Einstein's perihelion paper.

[20] Einstein's exposition is somewhat curious at this juncture. He writes down the field equations E.1 and then introduces the coordinate condition E.3; but, as we just saw and as is clear from the discussion in Einstein's previous paper (Einstein 1915b), equation E.1 actually presupposes equation E.3.

[21] In coordinate language, which Einstein was using, condition (C.1) means that $\partial g_{\mu\nu}/\partial x^4 = 0\,(\mu, \nu = 1, 2, 3, 4)$, and (C.3) means that $g_{i4} = g_{4i} = 0\,(i = 1, 2, 3)$. A metric is static when it satisfies both (C.1) and (C.3).

[22] See Fokker 1955. This reference is taken from Pais (1982), whose account of Einstein's explanation of the perihelion anomaly should also be consulted. Pais continues that what Einstein told de Haas "is even more profoundly significant: when he saw that his calculations agreed with the unexplained astronomical observations, he had the feeling that something actually snapped in him..." (Pais 1982, p. 253).

[23] Since we will discuss some fine points of this passage, the reader may want to look at the German:

> Wir setzen nun im folgenden voraus, dass sich die $g_{\mu\nu}$ von den in (4a) angegebenen Werten nur um Grössen unterscheiden, die klein sind gegenüber der Einheit. Diese Abweichungen behandeln wir als kleine Grössen "erster Ordnung," Funktionen *n*ten Grades dieser Abweichungen als "Grössen *n*ter Ordnung." Die Gleichungen (1) und (3) setzen uns in den Stand, von (4a) ausgehend, durch sukzessive Approximation das Gravitationsfeld bis auf Grössen *n*ter Ordnung genau zu berechnen. Wir sprechen in diesem Sinne von der "*n*ter Approximation"; die Gleichungen (4a) bilden die "nullte Approximation."

[24] For instance, Einstein writes: "Now that we have found $g_{\mu\nu}$ to first order, we can also compute the components $T^{\alpha}_{\mu\nu}$ of the gravitational field to first order" (Einstein 1915c, p. 834). "$T^{\alpha}_{\mu\nu}$" clearly is a typographical error and should be $\Gamma^{\alpha}_{\mu\nu}$.

[25] The opening of the sentence quoted in the preceding note—"Now that we have found $g_{\mu\nu}$ to first order..."—is one of those passages. We will have occasion to quote the other passage below.

[26] See the editorial note on the Einstein–Besso manuscript in Klein et al. (forthcoming).

[27] By $\overset{(1)}{\Gamma} = \overset{(0)}{g}\,\partial\overset{(1)}{g}$ we mean that to the lowest order of approximation $\Gamma^{\alpha}_{\mu\nu}$ is a sum of terms $\overset{(0)}{g}{}^{\alpha\beta}\,\partial_{\mu}\overset{(1)}{g}_{\beta\nu}$, etc. The terms $\overset{(1)}{g}\,\partial\overset{(1)}{g}$ and $\overset{(0)}{g}\,\partial\overset{(2)}{g}$ represent similar terms. These terms are smaller than the ones in $\overset{(1)}{\Gamma}$ by a factor of the order of $\overset{(1)}{g}$.

[28] Incidentally, Droste's paper also contains a clear exposition of the approximation procedure Einstein used both in 1913 and in 1915. Droste says he learned this method "from oral communications of Professor Lorentz" (Droste 1915, p. 999). Droste explicitly says that the solution for $g_{\mu\nu}$ "is obtained in the form of a power series" (ibid.).

[29] Actually, the coordinate system used in Eq. (20) is due to Droste (1917). Droste's radial coordinate r_D is related to Schwarzschild's radial coordinate r_S by $r_D = \left(r_S^3 + \alpha^3\right)^{1/3}$; see Eisenstaedt (1982) for details.

[30] This equation should have $1/r^3$ in the first term on the right-hand side instead of $1/r^2$.

[31] We are grateful to John Norton for drawing our attention to this passage.

[32] Cf. the second equation on p. 835 in Einstein 1915c. There is a minus sign missing on the right-hand side of Einstein's equtaion.

[33] Of course, the r in Eq. (27) is not the same as that in Eq. (20).

[34] Einstein clearly states this on p. 836 of his paper, where he writes: "...dass die Produkte $\frac{dx_\sigma}{ds}\,\frac{dx_r}{ds}$ mit Rücksicht auf (8) als Grössen erster Ordnung anzusehen sind." Equation 8 contains the Newtonian law for energy conservation. The subscript "r" is a misprint and should be "τ."

[35] We will show below that A can be set equal to zero, which renders this rescaling maneuver, for which Einstein would be criticized by Charles Lane Poor (see Section 8), superfluous.

[36] A factor $1/r$ is missing in the expression for Φ.

[37] Eq. (39) still holds even though the meaning of B and s has changed meanwhile. Because of the redefinition of the area law constant in Eq. (42), B in Eq. (39) should be replaced by $B\sqrt{1-2A}$. The square root is absorbed into ds on the left-hand side of Eq. (39) to rescale the proper time as in Eq. (42).

[38] It should be mentioned that in an otherwise useless paper, Anderson gives what is perhaps the first prediction of black hole formation in general relativistic space-times:

> ...if, in accordance with the suggestion of Helmholtz, the body of the sun should go on contracting, there will come a time when it will be shrouded by darkness, not because it has no light to emit, but because its gravitational field will be impermeable to light. (Anderson 1920a, p. 627)

For an analysis of Anderson's prediction, see Eisenstaedt (1982).

[39] We are grateful to Professor H. Goenner for bringing this point to our attention and for correcting an error in an earlier version of our paper.

[40] The derivation of energy conservation is different, there is no switch from r to $1/r$, and the roots are expressed in terms of the coefficients of the relevant (fourth-order) equation rather than simply set equal to the distance of the planet from the sun at perihelion and aphelion. Further details will be provided in the editorial apparatus for the Einstein–Besso manuscript in Klein et al. (forthcoming).

[41] Schwarzschild wrote:

> It is always pleasant to have at one's disposal a rigorous solution of simple form; it is more important that the calculation produce, at the same time, the unequivocal determination of the solution. Einstein's treatment still leaves some doubt, and, as is shown below, this uniqueness could be proved only with difficulty by his method. This solution therefore manages to let Einstein's result shine through in increased purity. (Translation from Doyle 1979b)

[42] As quoted by Friedman (1981, p. 3) from the *Protokoll, Nobelkommitteen for fysik*, September 6, 1922. See also Eisenstaedt 1982.

[43] As remarked by A.I. Miller, Luminy Conference. Cf. Pais 1982, p. 510.

[44] As remarked by S. Sigurdsson, Luminy Conference.

[45] See Friedman 1981 for an overview of the issues.

[46] We thank Professor J. Winicour for bringing this point to our attention.

[47] For a review of the recent experimental work, see Will 1981, 1984; a popular exposition is given in Will 1986.

References

Abraham, Max (1912a). "Zur Theorie der Gravitation." *Physikalische Zeitschrift* 13: 1–4.

—— (1912b). "Das Elementargesetz der Gravitation." *Physikalische Zeitschrift* 13: 4–5.

—— (1912c). "Das Gravitationsfeld." *Physikalische Zeitschrift* 13: 793–797.

Adler, Ronald, Bazin, Maurice, and Schiffer, Menahem (1975). *Introduction to General Relativity*, 2d ed. New York: McGraw-Hill.

Anderson, A. (1920a). "On the Advance of the Perihelion of a Planet and the Path of a Light Ray in the Gravitational Field of the Sun." *Philosophical Magazine* 39: 626–628.

—— (1920b). "Advance of the Perihelion of a Planet." *Philosophical Magazine* 40: 670.

Birkhoff, Garrett (1923). *Relativity and Modern Physics*. Cambridge, MA: Harvard University Press.

Burali-Forti, Cesare (1922–1923). "Flessione dei raggi luminosi stellari e spostamento secolare del perielio di Mercurio." *Accademia delle Scienze (Turin). Classe di Scienze Fisische, Mathematiche e Naturali* 58: 149–151.

Clairaut, Alexis-Claude (1745). "Du système du monde dans les principes de la gravitation universelle." *Mémoires de l'Académie Royale des Sciences (Paris)* 58: 329–364.

Clemence, Gerold M. (1947). "The Relativity Effect in Planetary Motions." *Reviews of Modern Physics* 19: 361–364.

Cohen, Paul I. (1971). *Relativity and the Excess Advances of Perihelia in Planetary Orbits*. MS Thesis, University of Pennsylvania.

De Sitter, Willem (1911). "On the Bearing of the Principle of Relativity on Gravitational Astronomy." *Monthly Notices of the Royal Astronomical Society* 71: 388–415.

—— (1913). "The Secular Variations of the Elements of the Four Inner Planets." *Observatory* 36: 296–303.

—— (1916–1917). "On Einstein's Theory of Gravitation, and its Astronomical Consequences." *Monthly Notices of the Royal Astronomical Society* I: 76: 699–728; 77: 481 (Errata); II: 77: 155–184, 481 (Errata); III: 78: 3–28.

Dicke, Robert H. and Goldenberg, H. Mark (1967). "Solar Oblateness and General Relativity." *Physical Review Letters* 18: 313–316.

—— (1974). "The Oblateness of the Sun." *Astrophysical Journal Supplement Series* 27: 131–182.

Doyle, Brian (1979a). English translation of Einstein 1915c, in Lang and Gingerich 1979.

—— (1979b). English translation of Schwarzschild 1916, in Lang and Gingerich 1979.

Droste, Johannes (1915). "On the Field of a Single Center in Einstein's Theory of Gravitation." *Koninklijke Akademie van Wetenschappen te Amsterdam. Proceedings of the Section of Sciences* 17: 998–1011.

—— (1917). "The Field of a Single Center in Einstein's Theory of Gravitation and the Motion of a Particle in that Field." *Koninklijke Akademie van Wetenschappen te Amsterdam. Proceedings of the Section of Sciences* 19: 197–215.

Earman, John and Glymour, Clark (1980a). "The Gravitational Redshift as a Test of Einstein's General Theory: History and Analysis." *Studies in the History and Philosophy of Science* 11: 175–214.

—— (1980b). "Relativity and the Eclipses: The British Eclipse Expeditions of 1919 and Their Predecessors," *Historical Studies in the Physical Sciences* 11: 49–85.

Earman, John and Norton, John (1987). "What Price Space-Time Substantivalism? The Hole Story." *British Journal for the Philosophy of Science* 38: 515–525.

Einstein, Albert (1912a). "Lichtgeschwindigkeit und Statik des Gravitationsfeldes." *Annalen der Physik* 38: 355–369.

—— (1912b). "Zur Theorie des Statischen Gravitationsfeldes." *Annalen der Physik* 38: 443–458.

—— (1915a). "Zur allgemeinen Relativitätstheorie." *Königlich Preussische Akademie der Wissenschaften (Berlin). Sitzungsberichte*: 778–786.

—— (1915b). "Zur allgemeinen Relativitätstheorie (Nachtrag)." *Königlich Preussische Akademie der Wissenschaften (Berlin). Sitzungsberichte*: 799–801.

—— (1915c). "Erklärung der Perihelbewegung des Merkur aus der allgemeinen Relativitätstheorie." *Königlich Preussische Akademie der Wissenschaften (Berlin). Sitzungsberichte*: 831–839.

—— (1915d). "Die Feldgleichungen der Gravitation." *Königlich Preussische Akademie der Wissenschaften (Berlin). Sitzungsberichte*: 844–847.

Einstein, Albert and Grossmann, Marcel (1913). *Entwurf einer verallgemeinerten Relativitätstheorie und einer Theorie der Gravitation.* Leipzig: Teubner. Reprinted with added "Bemerkungen," *Zeitschrift für Mathematik und Physik* 62(1914): 225–261.

Eisenstaedt, Jean (1982). "Histoire et singularités de la solution de Schwarzschild (1915–1923)" *Archive for the History of the Exact Sciences* 27 (No. 2): 157–198.

Fokker, Adriaan D. (1955). "Albert Einstein, 14 Maart 1878–18 April 1955." *Nederlands Tijdschrift voor Natuurkunde* 21: 125–129.

Friedman, Robert Marc (1981). "Nobel Physics Prize in Perspective." *Nature* 292: 793–798.

Freundlich, Erwin (1915). "Über die Erklärung der Anomalien im Planeten-System durch die Gravitationswirkung interplanetarer Massen." *Astronomische Nachrichten* 201: 49–56.

Gehrcke, Ernst (1916). "Zur Kritik und Geschichte der neueren Gravitationstheorien." *Annalen der Physik* 51: 119–124.

Gerber, P. (1898). "Die raumlich und zeitliche Ausbreitung der Gravitation." *Zeitschrift fur Mathematik und Physik* 43: 93–104.

—— (1902). "Die Fortpflanzungsgeschwindigkeit der Gravitation." *Programmabhandlung des städtischen Realgymnasiums zu Stargard in Pommerania.*

—— (1917). "Die Fortpflanzungsgeschwindigkeit der Gravitation." *Annalen der Physik* 52: 415–441.

Gleich, Gerold von (1923). "Die allgemeine Relativitätstheorie und das Merkurperihel." *Annalen der Physik* 72: 221–235.

Grossmann, Ernst (1921). "Die Bewegung des Merkurperihels." *Zeitschrift für Physik* 5: 280–284.

Gullstrand, Allvar (1922). "Allgemeine Lösung des statischen Einkörperproblems in der Einsteinschen Gravitationstheorie." *Arkiv för Matematik, Astronomi O. Fysik* 16 (No. 8): 1–15.

—— (1923). "Das statische Einkörperproblem in der Einstein'schen Theorie." *Arkiv för Matematik, Astronomi O. Fysik* 17 (No. 3): 1–5.

Hall, Asaph (1894). "A Suggestion in the Theory of Mercury." *Astrophysical Journal* 14: 49–51.

Hermann, Armin, ed. (1968). *Albert Einstein/Arnold Sommerfeld Briefwechsel.* Basel: Schwabe & Co.

Jeffreys, Harold (1916). "The Secular Perturbations of the Four Inner Planets." *Monthly Notices of the Royal Astronomical Society* 77: 112–118.

——— (1918). "The Secular Perturbations of the Inner Planets." *Philosophical Magazine* 36: 203–205.

——— (1919). "On the Crucial Test of Einstein's Theory of Gravitation." *Monthly Notices of the Royal Astronomical Society* 80: 138–154.

Klein, Martin, Kox, A.J., Renn, Jürgen, and Schulmann, Robert, eds. (1993). *The Collected Papers of Albert Einstein*. Vol. 3, *The Swiss Years: Writings, 1909–1911*. Princeton: Princeton University Press.

——— (forthcoming). *The Collected Papers of Albert Einstein*. Vol. 4, *The Swiss Years: Writings, 1912–1914*. Princeton: Princeton University Press.

Kretschmann, Erich (1923a). "Eine Bermerkung zu Hrn. A. Gullstrands Abhandlung: 'Allgemeine Lösung des statischen Einkörperproblems in der Einsteinschen Gravitationstheorie'." *Arkiv för Matematik, Astronomi, O. Fysik* 17 (No. 2): 1–4.

——— (1923b). "Das statische Einkörperproblem in der Einstein'schen Theorie." *Arkiv för Matematik, Astronomi O. Fysik* 17 (No. 25): 1–4.

Lang, Kenneth R. and Gingerich, Owen, eds. (1979). *Source Book in Astronomy and Astrophysics*. Cambridge, MA: Harvard University Press.

Lecornu, Léon (1922). "Quelques remarques sur la relativité." *Comptes Rendus* 174: 337–342.

Le Roux, J. (1921). "Sur la théorie de la relativité et la mouvement séculaire du périhélie de Mercure." *Comptes Rendus* 172: 1227–1230.

Le Verrier, Urbin J.J. (1859). "Théorie du mouvement de Mercure." *Annales de l'Observatorie impériel de Paris* 5: 1–196.

Mie, Gustav (1913) "Grundlagen einer Theorie der Materie." *Annalen der Physik* 40: 25–63.

Minkowski, Hermann (1908). "Die Grundgleichungen für die elektromagnetischen Vorgänge in bewegten Körpern." *Nachrichten von der Königlichen Gesellschaft der Wissenschaften zu Göttingen*: 53–111.

Misner, Charles W., Thorne, Kip S., and Wheeler, John Archibald (1973). *Gravitation*. San Francisco: W.H. Freeman.

Møller, Christian (1972). *The Theory of Relativity*, 2d ed. Oxford: Oxford University Press.

Newcomb, Simon (1882). "Discussion and Results of Observations on Transits of Mercury from 1677 to 1881." *Astronomical Papers Prepared for the Use of the American Ephemeris and Nautical Almanac* 1: 367–487. Washington, D.C.: U.S. Govt. Printing Office.

——— (1895). *The Elements of the Four Inner Planets and the Fundamental Constants of Astronomy*. Washington, D.C.: U.S. Govt. Printing Office.

——— (1912). "Researches in the Motion of the Moon." *Astronomical Papers Prepared for the Use of the American Ephemeris and Nautical Almanac* 9: 226–227. Washington, D.C.: U.S. Govt. Printing Office.

Nobili, Anna M. and Will, Clifford M. (1986). "The Real Value of Mercury's Perihelion Advance." *Nature* 320: 39–41.

Nordström, Gunnar (1912) "Relativitätsprinzip und Gravitation." *Physikalische Zeitschrift* 13: 1126–1129.

——— (1913). "Zur Theorie der Gravitation vom Standpunkt des Relativitätsprinzips." *Annalen der Physik* 42: 533–554.

——— (1914). "Die Fallgesetze und Planetenbewegungen in der Relativitätstheorie." *Annalen der Physik* 43: 1101–1110.

Norton, John (1984). "How Einstein Found His Field Equations." *Historical Studies in the Physical Sciences* 14: 253–316.

——— (1987). "Einstein, the Hole Argument, and the Objectivity of Space." In *Measurement, Realism, and Objectivity*, J. Forge, ed. Dordrecht: D. Reidel.

Pais, Abraham (1982). *Subtle Is the Lord.* Oxford: Oxford University Press.

Pavanini, G. (1912). "Prime conseguenze d'una recente teoria della gravitazione." *Reale Accademia dei Lincei (Rome). Rendiconti* 21: 648–655.

——— (1913). "Prime conseguenze d'una recente teoria della gravitazione: le disuguanlianze secolari." *Reale Academia dei Lincei (Rome). Rendiconti* 22: 369–376.

Pearson, E.S. (1920). "Advance of the Perihelion of a Planet." *Philosophical Magazine* 40: 342–344.

Poincaré, Henri (1906). "Sur la dynamique de l'électron." *Rendiconti del Circulo Mathematiche di Palermo* 21: 494–550.

——— (1953). "Les limites de la loi de Newton." *Bulletin astronomique (Paris)* 17: 121–269.

Poor, Charles Lane (1921). "The Motions of the Planets and the Relativity Theory." *Science* 54: 30–34.

——— (1922). *Relativity versus Gravitation.* New York: Putnam.

——— (1925). "Relativity and the Motion of Mercury." *Annals of the New York Academy of Sciences* 29: 285–319.

——— (1930). "Relativity and the Law of Gravitation." *Astronomische Nachrichten* 238: 165–170.

Ritz, Walter (1909). "Die Gravitation." *Scientia* 5: 241–255.

Roseveare, N.T. (1982). *Mercury's Perihelion from Le Verrier to Einstein.* Oxford: Clarendon Press.

Schwarzschild, Karl (1916). "Über das Gravitationsfeld eines Massenpunktes nach der Einsteinschen Theorie." *Königlich Preussische Akademie der Wissenschaften (Berlin). Sitzungsberichte*: 189–196.

Seelig, Carl (1956). *Albert Einstein, A Documentary Biography.* London: Staples Press.

Seeliger, Hugo von (1906). "Das Zodiakallicht und die empirischen Glieder in Bewegung der innern Planeten." *Königlich Bayerische Academie der Wissenschaften (Munich). Sitzungsberichte* 36: 595–622.

——— (1916). "Über die Gravitationswirkung auf die Spektrallinien." *Astronomische Nachrichten* 202: 83–86.

Shapiro, Irwin I., Counselman, C.C., and King, R.W. (1976). "Verification of the Principle of Equivalence for Massive Bodies." *Physical Review Letters* 36: 555–558.

Silberstein, Ludwik (1917). "The Motion of the Perihelion of Mercury." *Monthly Notices of the Royal Astronomical Society* 77: 503–510.

—— (1918). "General Relativity without the Equivalence Hypothesis." *Philosophical Magazine* 36: 94–128.

Speziali, Pierre, ed. (1972). *Albert Einstein/Michele Besso Correspondance*. Paris: Hermann.

Waff, Craig B. (1976). *Universal Gravitation and the Moon's Apogee: The Establishment and Reception of Newton's Inverse Square Law, 1687–1749*. Ph.D. Dissertation, Johns Hopkins University.

Weinberg, Steven (1972). *Gravitation and Cosmology*. New York: Wiley.

Whitrow, Gerald J. and Morduch, G.E. (1960). "General Relativity and Lorentz-Invariant Theories of Gravitation." *Nature* 188: 790–794.

—— (1965). "Relativistic Theories of Gravitation." *Vistas in Astronomy* 6: 1–67.

Wiechert, Emil (1916). "Perihelbewegung des Merkur und die allgemeine Mechanik." *Physikalische Zeitschrift* 17: 442–448.

—— (1920). "Bemerkungen zu einer elektrodynamischen Theorie der Gravitation." *Astronomische Nachrichten* 211: 275–284.

Will, Clifford M. (1981). *Theory and Experiment in Gravitational Physics*. Cambridge: Cambridge University Press.

—— (1984). "The Confrontation between General Relativity and Experiment: An Update." *Physics Reports* 113: 345–422.

—— (1986). *Was Einstein Right?* New York: Basic Books.

Pieter Zeeman's Experiments on the Equality of Inertial and Gravitational Mass

A.J. Kox

1. Introduction

In the years 1917–1918 the Dutch experimental physicist Pieter Zeeman (1865–1943) performed a series of experiments to test the equality of inertial and gravitational mass for radioactive and anisotropic bodies. Zeeman, the discoverer of the Zeeman effect and one of the first recipients of the Nobel Prize for physics, had a reputation for designing and carrying out precision experiments, and his work on the equality of inertial and gravitational mass is typical of his choice of problems and his style of work. In this chapter I will discuss these experiments; in addition I will comment on the reasons that led Zeeman to perform them and on their reception by the physics community.[1]

2. Zeeman and Relativity

Zeeman is, of course, best known for his work on the Zeeman effect, which was discovered by him in the fall of 1896. Less known is that he was also interested in what one might call experimental relativity. His interest was at first focused on special relativity, but later included general relativity as well.

In the field of special relativity, Zeeman performed a series of experiments during the years 1914–1923 on the propagation of light in moving transparent media. He started by repeating an experiment first performed by the Frenchman Hippolyte Fizeau in the middle of the nineteenth century, namely the determination of the speed of light in running water. Zeeman was motivated by an apparent discrepancy between the theoretical prediction and the best available experimental data.

Thirty years after Fizeau's work, Albert Michelson and Edward Morley had repeated the experiment with much greater accuracy (see Michelson and Morley 1886). Like Fizeau, they had found excellent agreement with the existing theoretical prediction, due to Augustin Fresnel. His expression for the speed of light in moving media was derived in the framework of an ether theory and included his well-known "dragging coefficient." In 1895, however, Hendrik Antoon Lorentz used his electron theory to derive a new expression in which a dispersion term occurred, in addition to the dragging coefficient (see Lorentz 1895, section 71). For the speed of light in running water, c', he found:

$$c' = \frac{c}{\mu} \pm w\left(1 - \frac{1}{\mu^2}\right) \mp \frac{w}{\mu} T \frac{\mathrm{d}\mu}{\mathrm{d}T}, \tag{1}$$

where w is the speed at which the water is running, T is the period of vibration, and μ is the index of refraction; the upper sign corresponds to light and water moving in the same direction. The second term is Fresnel's dragging coefficient, the third term the dispersion term. Several years after 1895, the same result was found in the framework of the special theory of relativity (see, e.g., von Laue 1907). When Lorentz's new and supposedly more accurate expression was compared with Michelson and Morley's data, however, worse agreement was found than with the original formula. Zeeman's aim was to provide new and better data, so that it could be decided whether Lorentz's formula was correct or not. To that end he first repeated Fizeau's experiment and subsequently measured the speed of light in moving glass and quartz. In spite of many experimental difficulties, he succeeded with a high degree of accuracy and found his data in agreement with Lorentz's theoretical prediction.[2]

Although it would be going too far to say that Zeeman's work provided an experimental confirmation of special relativity, if only because it did not discriminate between electron theory and relativity, his results were received with enthusiasm. Einstein wrote that Zeeman's experiment filled "a still-existing unpleasant gap."[3]

Apparently inspired by the success of the Fizeau experiments, Zeeman decided to apply his experimental skills to another precision experiment in the field of relativity, namely the testing of the equality of gravitational and inertial mass, with the explicit goal of providing a more solid foundation for general relativity. He set himself two tasks: first to reach a higher degree of accuracy than in earlier experiments, and second to extend the measurements to crystals (to determine a possible anisotropy in the gravitational interaction) and, more importantly, to radioactive substances.

3. Inertial and Gravitational Mass

The history of experimental tests of the equality—or rather the universal proportionality—of gravitational and inertial mass goes back to Newton, who performed pendulum experiments to investigate a possible dependence of the period of a pendulum on the composition of the bob. It is easy to show that, for a bob with gravitational mass m_g and inertial mass m_i, the period T of a pendulum of length l is given by

$$T = 2\pi\sqrt{l/g}\,\sqrt{m_i/m_g}. \tag{2}$$

By comparing the periods of pendulums of equal lengths but different compositions, differences in the ratio of inertial and gravitational mass for different substances can be determined. As Newton reports in his *Principia*, he found the same ratio for all substances he investigated. His accuracy was approximately $1 : 10^3$. A more accurate repetition of Newton's experiment was carried out by Bessel around 1830 (see Bessel 1832). He reached an accuracy of $1 : 5 \times 10^4$.

The next major step forward was taken in the 1880s by the Hungarian physicist Lorand (or Roland) von Eötvös (Eötvös 1890). He used a different and much more accurate method, based on the measurement of the torque exerted on a beam at each end of which a mass was hung. The beam was suspended on a wire attached to its center. If the gravitational accelerations of the two bodies were different, the horizontal component of the centripetal acceleration of the earth would exert a slight torque on the beam, which would reverse in sign if the apparatus were turned through 180°. The effect was greatest when the beam was oriented in the east–west direction.[4] The great advantage of this kind of experiment was that it was a null experiment, allowing for much greater precision than in the case of pendulum experiments. Eötvös succeeded in achieving an accuracy of $1 : 2 \times 10^7$.

The outcome of the Eötvös experiment not only had connections with general relativity[5] but also with special relativity, in particular through the mass–energy equivalence. This was pointed out, in particular, by Einstein in 1912 (Einstein 1912, p. 1062). He gave the following argument (which he ascribed to Langevin): if the loss of energy and thus of inertial mass suffered by decaying radioactive substances would not be accompanied by a proportional decrease in gravitational mass, the acceleration of a body in a gravitational field would depend on its composition.

Interestingly, not very long after the appearance of Einstein 1912, in a letter to Wilhelm Wien of July 10, 1912, Einstein raised the question of whether it would be possible to test the universal proportionality of gravitational and inertial mass for radioactive bodies with the help of a pendulum experiment of sufficient accuracy. From a calculation, he had concluded that the relative difference in period between a uranium pendulum and a lead pendulum would be approximately 2×10^{-4}, so that an accuracy of $1 : 10^5$ would be sufficient. In a postscript to the letter and as an afterthought, he very surprisingly proposed the same experiment that Eötvös had performed more than two decades earlier. Apparently, Einstein did not know of the experiment at that time.[6] Someone, perhaps Wien, must have subsequently drawn Einstein's attention to Eötvös' work: when he discussed the equivalence principle the next year in the introductory section of Einstein and Grossmann 1913, he cited Eötvös' work as experimental evidence for this fundamental principle.

It is thus not surprising that the idea came up to test the proportionality of gravitational mass and inertial mass for radioactive substances, all the more so because a test that would confirm the universal proportionality would also support the mass–energy equivalence in an indirect way. The first person to do so was Leonard Southerns in 1910 (Southerns 1910). Southerns worked in J.J. Thomson's laboratory and followed up on earlier experiments by Thomson himself.[7] He used specially constructed pendulums in a Newton-like experiment and came to the conclusion that the ratio of the two masses was equal for lead and uranium up to $1 : 2 \times 10^5$.

4. Zeeman's Experiment

Zeeman's experiment is essentially analogous to the one by Eötvös. One difference is that his torsion balance was much smaller than the one used by Eötvös, and the weights were smaller as well.[8] As Zeeman writes in the paper he published on his experiments in the fall of 1917 (Zeeman 1917), the physics laboratory in Amsterdam was unsuitable for performing the experiment with the accuracy he had in mind. For instance, the regularly passing streetcars produced vibrations that interfered with the experiment. At first Zeeman tried to construct a special housing for the apparatus, consisting of a container of thick oil in which a second container was floating, but to no avail. It was impossible to suppress the disturbing influence of vibrations with a period of 300 or 400 seconds that turned out to be present day and night. Zeeman wrote: "It was therefore hopeless to work with the torsion balance in Amsterdam, and I resolved to continue my experiments in the cellar of a country house near Huis ter Heide" (Zeeman 1918, p. 547). The house was in fact Zeeman's own country house in the province of Utrecht in which the Zeeman family usually spent its summers. The house had been built with part of the Nobel money won in 1902 and stood on very solid sandy soil. Tests showed that there were no disturbing vibrations, and, as Zeeman relates, even stomping on the floor had no effect at all.

In the summer of 1917 Zeeman started his measurements, first in the cellar, but later also in the hallway of the house. The latter series was kept short, however:

> Several excellent series of observations were obtained. As they extended, however, over the whole day and the principal entrance of the house was then put out of use, I restricted these observations to a rather limited number of days. (Zeeman 1918, p. 547)

The outcome of the experiments was that for anisotropic bodies the influence of the orientation on the ratio of inertial and gravitational mass was less than $1 : 3 \times 10^7$. For uranyl nitrate a difference of less than $1 : 5 \times 10^6$ was found. Results for uranium oxide that seemed to indicate a difference in inertial and gravitational mass were rejected on the grounds that the samples were probably contaminated with iron, so that magnetic effects came into play. The samples were not tested for the presence of iron, however.

The above figures show that only for radioactive substances did Zeeman reach a much higher degree of accuracy than had been the case in previous experiments; in particular, the accuracy of Eötvös' results from 1891 was not significantly exceeded.

Unfortunately, Zeeman was ignorant of other, much more accurate work by Eötvös and his group. In the first decade of the century, Eötvös and his collaborators Pekár and Fekete had performed a series of new experiments that greatly improved on Eötvös' earlier work. The results of the investigation had been submitted as a prize essay to the the Göttingen Philosophical Faculty. It concerned the Benecke Prize for 1906, which was to be awarded for a detailed test of the proportionality of inertial and gravitational mass. In its report, the jury summarized the experimental results[9] and pointed out that the essay—the only one submitted—fully deserved being awarded the prize, in spite of certain shortcomings. Still, the work by Eötvös and his collaborators remained unknown to most people, including Zeeman, until a full publication appeared in 1922, three years after Eötvös' death (Eötvös et al. 1922). Publication had been postponed because even more accurate experiments were planned. In the end nothing came of these, and the original prize essay was published in a somewhat abbreviated form. In these experiments the proportionality of inertial and gravitational mass was established up to $1 : 2 \times 10^8$, markedly more accurate than Zeeman's data. For a radioactive substance, the difference is $1 : 4 \times 10^6$, the same order of magnitude as Zeeman's result.

The situation is intriguing, because in the published paper from 1922 no mention is made either of Southerns' work or of Zeeman's experiments. It seems that Zeeman's work had not made a great impact. This impression is reinforced by the fact that most textbooks, both from that time and more modern ones, mention Eötvös' pioneering work, and sometimes also the 1922 publication, but do not cite Zeeman. Wolfgang Pauli, for instance, only mentions Eötvös (including the 1909 jury report) and Southerns in his review paper (Pauli 1921). Von Laue, on the other hand, mentions Eötvös and Zeeman together in his textbook (von Laue 1921), but it should be kept in mind that its first edition appeared before the publication of Eötvös' later experiments. In modern textbooks Zeeman is often not cited either: Misner et al. 1973 and Weinberg 1972 only list Eötvös. And when in the sixties Roll, Krotkov, and Dicke performed an improved Eötvös experiment with an accuracy of $1 : 10^{11}$, in their discussion of previous experiments Zeeman's work was not mentioned at all (see Roll et al. 1964)). Given the accuracy of Zeeman's results this lack of attention is not very surprising insofar as nonradioactive substances are concerned. For his results for radioactive substances it can only be concluded that the work was apparently not considered sufficiently significant. An additional factor for the lack of attention may have been that Zeeman's work was published during World War I, when international scientific contacts were severely disrupted.

There is perhaps another reason why Zeeman's work on the equality of gravitational and inertial mass did not create the same enthusiasm as his experiments on the dragging coeffient. The experiments did not really test general relativity, as the three "classic" tests did, by testing a prediction made by the theory; they simply tested its foundations by reestablishing an equality that had already been established with great precision and in which most physicists tended to believe anyway. Not long after Zeeman's work, the results of the 1919 eclipse measurements gave such strong support to general relativity that the issue of the universal proportionality of gravitational and inertial mass as a foundation of general relativity became of minor importance.[10]

In this respect, Einstein's reaction to Zeeman's results is illustrative. In January 1918 he wrote Zeeman a letter, in which he first thanked Zeeman for sending him some reprints, including some further work on the Fizeau experiment, which Einstein characterized as "your beautiful papers on the dragging coefficient."[11] He then continued:

> Of your new investigations, I am mainly interested in the ones on the inertial and gravitational mass of uranyl nitrate. For the investigations on weight and inertia of crystallic substances I miss the theoretical viewpoints that have led to the formulation of the problem.[12]

Einstein is right in questioning Zeeman's theoretical viewpoints, as is borne out by the text of the opening paragraph of Zeeman 1918:

> Our ideas concerning gravitation have been so radically changed by Einstein's theory of gravitation that questions of the utmost interest in older theories are now simply discarded or at least appear in a changed perspective. We cannot try anymore to form an image of the mechanism of the gravitational action between two bodies, and we must return to the older theories in order to justify the suspicion, that the structure of substances might influence their mutual attraction. In most crystalline substances the velocity of propagation of light, the conduction for heat and the dielectric constant are different in different directions, and we might then suspect that the lines of gravitative force spread out from a crystal unequally in different directions. (Zeeman 1918, p. 542)

This quotation can hardly be called a convincing motivation for the investigation of anisotropic substances and betrays, moreover, a lack of understanding of the foundations of general relativity. But that was of course not unusual in those days, especially among experimenters.

From the further history of Zeeman's work on inertial and gravitational mass one might conclude that the experiments fell short of Zeeman's own expectations as well. At the end of his paper, Zeeman announced a planned improvement of his apparatus that would increase its accuracy at least tenfold. From correspondence with one of the laboratory technicians it becomes clear that in the summer of 1918 a new instrument was constructed and that it was tested with satisfactory results, but it apparently did not lead to a new series of experiments. Perhaps the desired accuracy was never reached; another possibility is that Zeeman was discouraged by the lack of response to his earlier work. In any case, the work was ended, and no further publication resulted.

ACKNOWLEDGMENTS. I am grateful to the Hebrew University of Jerusalem for permission to quote from Einstein's unpublished letters.

NOTES

[1] This chapter is partly based on material present in the Zeeman Archive, which was discovered in the fall of 1989. It is now located in the Rijksarchief Noord-Holland at Haarlem, The Netherlands. See Kox 1992 for a recent biographical sketch of Zeeman, based on the new material. Several publications on Zeeman's work, including one on the discovery of the Zeeman effect, are in preparation.

[2] See Zeeman 1927 for a detailed review of the experiments.

[3] "Eine bisher unangenehm fühlbare Lücke," Albert Einstein wrote to Zeeman on August 15, 1915. Zeeman Archive, Haarlem.

[4] See, for instance, Weinberg 1972, pp. 11–13, for an elementary discussion, and von Laue 1921, section 1, for a more sophisticated calculation.

[5] The statement that all bodies fall with the same acceleration is sometimes called the "weak equivalence principle" and is an essential part of the more general equivalence principle. See, e.g., Will 1981, section 2.4 for a discussion of Eötvös-like experiments as tests of the weak equivalence principle.

[6] See Illy 1989 for a detailed discussion of the letter and its historical context.

[7] It should be noted that these experiments were motivated by purely classical ether-theoretical considerations concerning the relation between energy and mass, in particular by the argument that the existence of electromagnetic mass suggests a general relation between potential energy and mass. The relativistic relation between mass and energy is not mentioned at all.

[8] Eötvös' balance was 25–30 cm long, and he used weights of 30 g; Zeeman's weights were about 1.5 g and were about 10 cm apart.

[9] The jury cited an accuracy of $1 : 2 \times 10^7$ for the single radioactive substance that was tested and $1 : 2 \times 10^8$ for other materials. See *Nachrichten von der Königlichen Gesellschaft der Wissenschaften zu Göttingen. Geschäftliche Mitteilungen* 1909, pp. 37–41.

[10] Ironically, the actual accuracy of the eclipse results never warranted the crucial importance attached to them; see, e.g., Will 1981, section 7.1, for a discussion of light-bending experiments and their accuracy.

[11] "Ihre wundervolle Abhandlungen über den Mitführungs-Koeffizienten," wrote Albert Einstein to Zeeman, January 18, 1918. Zeeman Archive, Haarlem.

[12] "Von Ihren neuen Untersuchungen interessieren mich hauptsächlich diejenigen über die träge und schwere Masse des Uranylnitrats. Bei den Untersuchungen über die Schwere und Trägheit kristallinischer Substanzen fehlen mir die theoretischen Gesichtspunkte, welche die Fragestellung veranlasst haben" (ibid.).

References

Bessel, Friedrich W. (1832). "Versuche über die Kraft, mit welcher die Erde Körper von verschiedener Beschaffenheit anzieht." *Annalen der Physik und Chemie* 25: 401–417.

Einstein, Albert (1912). "Relativität und Gravitation. Erwiderung auf eine Bemerkung von M. Abraham." *Annalen der Physik* 38: 1059–1064.

Einstein, Albert and Grossmann, Marcel (1913). *Entwurf einer verallgemeinerten Relativitätstheorie und einer Theorie der Gravitation.* Leipzig: Teubner. Also *Zeitschrift für Mathematik und Physik* 62 (1914): 225–259.

Eötvös, Roland von (1890). "Über die Anziehung der Erde auf verschiedene Substanzen." *Mathematische und naturwissenschaftliche Berichte aus Ungarn* 8: 65–68.

Eötvös, Roland von, Pekár, Desiderius, and Fekete, Eugen (1922). "Beiträge zum Gesetz der Proportionalität von Trägheit und Gravitation." *Annalen der Physik* 68: 11–66.

Illy, József (1989). "Einstein und der Eötvös-Versuch: Ein Brief Albert Einsteins an Willy Wien." *Annals of Science* 46: 417–422.

Kox, A.J. (1992). "Pieter Zeeman (1865–1943): Meester van het experiment." In *Een brandpunt van geleerdheid in de hoofdstad: De Universiteit van Amsterdam rond 1900 in vijftien portretten.* J.C.H. Blom et al., eds. Hilversum: Verloren; Amsterdam: Amsterdam University Press, pp. 213–228.

Lorentz, Hendrik A. (1895). *Versuch einer Theorie der electrischen und optischen Erscheinungen in bewegten Körpern.* Leiden: Brill.

Michelson, Albert A. and Morley, Edward W. (1886). "Influence of the Motion of the Medium on the Velocity of Light." *American Journal of Science* (3)31: 377–386.

Misner, Charles W., Thorne, Kip S., and Wheeler, John A. (1973). *Gravitation.* New York: Freeman.

Pauli, Wolfgang (1921). "Relativitätstheorie." In *Encyklopädie der mathematischen Wissenschaften, mit Einschluss ihrer Anwendungen.* Vol. 5, *Physik*, part 2, pp. 539–775. Arnold Sommerfeld, ed. Leipzig: Teubner, 1904–1922. Issued November 15, 1921.

Roll, P.G., Krotkov, R., and Dicke, R.H. (1964). "The Equivalence of Inertial and Passive Gravitational Mass." *Annals of Physics* 26: 442–517.

Southerns, Leonard (1910). "A Determination of the Ratio of Mass to Weight for a Radioactive Substance." *Royal Society of London. Proceedings A* 84: 325–344.

von Laue, Max (1907). "Die Mitführung des Lichtes durch bewegte Körper nach dem Relativitätsprinzip." *Annalen der Physik* 23: 989–990.

—— (1921). *Die Relativitätstheorie.* Vol. 2, *Die allgemeine Relativitätstheorie und Einsteins Lehre von der Schwerkraft.* Braunschweig: Vieweg.

Weinberg, Steven (1972). *Gravitation and Cosmology: Principles and Applications of the General Theory of Relativity.* New York: Wiley.

Will, Clifford M. (1981). *Theory and Experiment in Gravitational Physics.* Cambridge: Cambridge University Press.

Zeeman, Pieter (1917). "Enkele proeven over de zwaartekracht. De trage en zware massa van kristallen en radioactieve stoffen." *Koninklijke Akademie van Wetenschappen te Amsterdam. Wis- en Natuurkundige Afdeeling. Verslagen van de Gewone Vergaderingen* 26 (1916–1917): 451–462. (Meeting of September 29, 1917.)

—— (1918). "Some Experiments on Gravitation. The Ratio of Mass to Weight for Crystals and Radioactive Substances." *Koninklijke Akademie van Wetenschappen te Amsterdam. Proceedings of the Section of Sciences* 20 (1917–1918): 542–553. (Translation of Zeeman 1917.)

—— (1927). "Expériences sur la propagation de la lumière dans des milieux liquides ou solides en mouvement." *Archives Néerlandaises des Sciences Exactes et Naturelles* (IIIA) 10: 131–220.

Part III

Variational Principles in General Relativity

Variational Derivations of Einstein's Equations

S. Kichenassamy

1. Introduction

Variational derivations of the relativistic gravitational field equations proceed from the adaptation of the Hamilton principle of classical mechanics to field theories, as is illustrated in the case of the electromagnetic field (Larmor 1900, p. 167; Schwarzschild 1903). Since its use by Einstein and Grossmann (1914) to determine the restricted covariance group of the *Entwurf* equations (Einstein and Grossmann 1913), now abandoned, and by Hilbert (1915) to obtain the correct field equations in the presence of an electromagnetic field, the variational principle remains an "unusually impressive mathematical" tool (Sommerfeld 1952, p. 208) in the search for field equations in the presence of sources, as well as in the source-free case.

> The derivation of the field equations from a variational principle has the advantage that the compatibility of the resulting equations is assured [when an extremum can exist] and that the identities connected with general covariance, the "Bianchi identities" as well as the conservation laws result in a systematic manner. (Einstein 1955, p. 154)

Field equations were derived, taking as gravitational variables (Kichenassamy 1986)

(1) the metric g_{ij} and its derivatives (Hilbert- or g-variation);

(2) the metric g_{ij} and the connection $\Gamma^{i}{}_{jk}$ (and its first derivatives), considered as independent (g–Γ variation) or constrained by the Ricci identity $\nabla_k g_{ij} = 0$ ($\mathcal{C}$-variation);

(3) the connection alone, with or without a soldering metric (affine variation), as in gauge theories or purely affine theories.

In this chapter, we comment on some of the earlier derivations, hoping that this will throw some light on the emergence of the present formulation of general relativity. Indeed, most of the papers discussed here originated in the difficulties encountered by Einstein and others to understand the details of Hilbert's 1915 paper. Tricky points include the meaning of Hamilton's principle in field theories and the nature of the variation involved. This chapter is therefore organized as follows:

- Section 2 briefly recalls the specific features of Hamilton's principle, which distinguish it from other integral variational principles of mechanics.
- Section 3 describes the three types of variation considered in the various papers (the functional, label, and Lie variations) and discuss their implications.
- Section 4 includes comments on the papers of Einstein and Grossmann (1914), Einstein (1914), Lorentz (1915), Hilbert (1915), Einstein (1916a), Lorentz (1916), Einstein (1916b), and Palatini (1919). Some modern trends are also indicated.
- Section 5 gives a brief conclusion.

2. Integral Variational Principles of Mechanics

Integral variational principles are used to derive the equations of motion of a system by requiring the action integral to be stationary. The action integral has the form

$$I = \int_{t_1}^{t_2} L\big(t, q^a(t), \dot{q}^a(t)\big)\, \mathrm{d}t, \tag{1}$$

where $q^a(t)$ are the generalized position coordinates of the system, depending on the parameter t, and $\dot{q}^a(t) := \mathrm{d}q^a(t)/\mathrm{d}t$. One can distinguish two principles of this kind: Hamilton's principle and the principle of least action. Historically, the latter came first (see Sommerfeld 1952, p. 181).

2.1 Hamilton's Principle

Hamilton's principle says that we obtain the equations of motion by requiring that $\delta I = 0$ for any variations δq^a satisfying the following two conditions:

(1) they are taken at constant time

$$\delta t = 0; \tag{2}$$

(2) they vanish at the endpoints:

$$(\delta q^a = 0, \quad \delta \dot{q}^a = 0)_{t=t_1,t_2}. \tag{3}$$

Condition (2) ensures that q^a on the original path and $q^a + \delta q^a$ on the varied path belong to the same time t, so that

$$\delta \dot{q}^a = \frac{\mathrm{d}}{\mathrm{d}t}\delta q^a. \tag{4}$$

When these conditions are satisfied, the requirement that $\delta I = 0$ leads to the Euler–Lagrange equations

$$E_a = \frac{\partial L}{\partial q^a} - \frac{\mathrm{d}}{\mathrm{d}t}\frac{\partial L}{\partial \dot{q}^a} = 0.$$

The term $(\frac{\partial L}{\partial \dot{q}^a}\delta q^a)_{t_1}^{t_2}$ vanishes on account of (3).

When $\partial L/\partial t = 0$, L is *invariant* under the translation $t \to t + u$, and we have conservation of energy ($H =$ Energy):

$$\frac{\mathrm{d}H}{\mathrm{d}t} = 0, \quad H := \frac{\partial L}{\partial \dot{q}^a}\dot{q}^a - L.$$

Even when L explicitly depends on t, the function H can be defined by the same relation and is called the Hamiltonian associated to L. L is sometimes called Hamilton's function, not to be confused with the Hamiltonian.

When constraints reduce the number of degrees of freedom of the system, one may keep all the variables q^a independent through the use of Lagrange multipliers; these may be interpreted as the reaction to the constraints, as in hydrodynamics (the pressure versus the incompressibility).

2.2 Principle of Least Action of Maupertuis

The principle of least action can be derived from $\delta I = 0$ for *conservative* systems with $L = T - V$ (kinetic energy minus potential energy) when we keep condition (3) and replace (2) by

$$\delta E := \delta(T + V) = 0. \tag{5}$$

It must be noted that q^a and $q^a + \delta q^a$ no longer correspond to the same time; we have

$$\delta \dot{q}^a = \frac{\mathrm{d}}{\mathrm{d}t}\delta q^a - \dot{q}^a \frac{\mathrm{d}}{\mathrm{d}t}\delta t,$$

and $\delta I = 0$ reduces to the *Maupertuis principle*:

$$\delta \int_{t_1}^{t_2} 2T\, dt = 0.$$

2.3 Remark

As the evolution of a dynamical system is obtained not from its present state, but from both its past and future states, these integral variational principles were for a long time considered to be "teleological" rather than "causal," i.e., "shaped by a purpose" or even "best expressing the Wisdom of the Creator" (see Sommerfeld 1952, section 37; for a more elaborate philosophical discussion, see Gueroult 1934 and Dugas 1950 [part III, chapter 5]). From the beginning of this century onward, however, they have simply been looked upon as an important mathematical tool, refined more than once by Weierstrass (1927), Hilbert (1906), Tonelli (1921–1923) and others (see any standard treatise on the calculus of variations, e.g., Gelfand and Fomin 1963). Note that Hilbert (1899) gave the first existence proof for an extremum of a definite integral (see Bolza 1904, pp. 245–263), and that the calculus of variations is the object of three of his famous problems: 5th (role of Lie groups without differentiability assumptions), 19th (analyticity of solutions of "regular" problems), and 23rd (developments of its methods).

3. Hamilton's Principle in General-Relativistic Field Theories

Hamilton's principle has been extended to field theories on a space-time manifold M with metric g_{ij} by substituting the coordinates x^i (Hilbert's world parameters) ($i = 0, 1, 2, 3$) for t, and the field variables $\Phi_A(x)$ (A standing for a collection of indices) for the position coordinates $q^a(t)$. The action I is replaced by the functional $I_U(\Phi)$ over the domain $U \subset M$:

$$I_U(\Phi) = \int_\Omega \hat{L}\big(x, \Phi(x)\big)\, \mathrm{d}^4x, \tag{6}$$

where $\Omega \subset \mathbf{R}^4$ is the image of U by the homeomorphism ψ assigning coordinates x^k to the points of U; where $\Phi(x) := \{\Phi_A(x)$ and its derivatives up to a certain order}; where $d^4x := dx^0 \wedge dx^1 \wedge dx^2 \wedge dx^3$; and where $\hat{L}$ is used rather than L to indicate that we are dealing with a density.

Now, what is meant by the variations of $\Phi_A(x)$? We may consider at least three different kinds: functional variations, label variations, and Lie variations.

3.1 The Functional Variation δ

The functional variation δ is the one used in the calculus of variations. It is obtained by considering a smooth one-parameter family of field configurations $\Phi_A(x, \lambda)$ such that $\Phi_A(x, 0) = \Phi_A(x)$ and

$$\Phi_A(x, \lambda) = \Phi_A(x) + \lambda \phi_A(x) + \cdots,$$

where $\phi_A(x)$ are arbitrary. The functional variation $\delta\Phi_A(x)$ is now defined as

$$\delta\Phi_A(x) = \frac{d}{d\lambda}\Phi_A(x, \lambda)_{|\lambda=0}, \tag{7}$$

where x^i is held fixed (just like t in Hamilton's principle). From (7), it is clear that

$$\delta\Phi_{A,i}(x) = \big(\delta\Phi_A(x)\big)_{,i}, \tag{8}$$

where $_{,i} := \partial/\partial x^i$.

The variation of $I_U(\Phi)$, assumed to depend on derivatives $\Phi_{A,i}$ and $\Phi_{A,ij}$, is given by

$$\delta I_U = \int_\Omega (\hat{L}^A \delta\Phi_A + \partial_i \delta\hat{Q}^i)\, d^4x, \tag{9}$$

with

$$\delta\hat{Q}^i = \Big[\partial\hat{L}/\partial\Phi_{A,i} - 2(\partial\hat{L}/\partial\Phi_{A,ij})_{,j}\Big]\delta\Phi_A + \Big(\frac{\partial\hat{L}}{\partial\Phi_{A,ij}}\delta\Phi_A\Big)_{,j}.$$

Under the conditions

$$\big(\delta\Phi_A(x),\ \delta\Phi_{A,i}(x)\big)_{|\partial\Omega} = 0, \tag{10}$$

the requirement $\delta I_U = 0$ leads to the field equations

$$\hat{L}^A = \partial\hat{L}/\partial\Phi_A - (\partial\hat{L}/\partial\Phi_{A,i})_{,i} + (\partial\hat{L}/\partial\Phi_{A,ij})_{,ij} = 0.$$

3.2 The Label-Variation Δ

The label-variation Δ is induced by a local coordinate transformation or a *relabeling* of points of U:

$$x^i(P) \;\to\; x^{i'}(P) = x^i(P) + \rho v^i(x) + \cdots,$$

i.e., a *transformation of* $\mathbf{R}^4$ inducing an isomorphism of the tangent spaces T_{x^i} and $T_{x^{i'}}$. This variation Δ depends on the nature of the geometric object. For a scalar L, defined as a one-component quantity independent of the orientation of the local frame,

$$\Delta L = 0, \tag{11}$$

from which we may derive, by standard methods, the label-variation for other quantities. Thus, for a covector A_i,

$$\Delta A_i = -A_k v^k{}_{,i} \tag{12}$$

and we have

$$\Delta A_{i,j} = (\Delta A_i)_{,j} - A_{i,k} v^k{}_{,j}. \tag{13}$$

As $\Delta \Phi_A$ is merely due to the relabeling of the points of the underlying topological space $\mathcal{M}$, all physical descriptions should be independent of it. This label-invariance is the basis for the coordinate covariance principle, considered to be trivial by Kretschmann (1917) and admitted to be so by Einstein (1918).

Consequently, one should require that I_U be a number and therefore that $\hat{L}$ be a scalar density:

$$\hat{L}(x')J = \hat{L}(x),$$

where $J := \det|J^{i'}_k|$ with $J^{i'}_k := \partial x^{i'}/\partial x^k$. In other words, $L = \hat{L}/\sqrt{-g}$ is a scalar:

$$L\big(x', \Phi'(x')\big) = L\big(x, \Phi(x)\big) \;\text{ or }\; \Delta L = 0.$$

This has many consequences, such as the coordinate covariance of the field equations. It also provides some insight into the structure of the Lagrangian L.

Coordinate covariance. As I_U is label-invariant, $\int_\Omega \hat{L}^A \delta\Phi_A \, \mathrm{d}^4x$ should be label-invariant, and it follows that if $\delta\Phi_A$ is a tensor of type (k, l), $\hat{L}^A/\sqrt{-g}$ should be of type (l, k). Hence, we have coordinate covariance of the field equations.

Structure of L. Under a local coordinate transformation

$$x^{i'} = x^i + \rho v^i + \cdots,$$

a scalar L should satisfy

$$\partial L/\partial v^i{}_{,j} = 0, \quad \partial L/\partial v^i{}_{,jk} = 0, \ldots. \tag{16}$$

This has important consequences for the structure of L. We give two examples below:

(1) Hilbert–Mie type electromagnetic Lagrangian L_{em}.

This Lagrangian is of the form $L_{\text{em}} = L(g^{ij}, A_i, A_{i,j})$, where g^{ij} and A_i are the gravitational and electromagnetic potentials. Using the variations (12), (13), and

$$\Delta g^{ij} = g^{ih} v^j{}_{,h} + g^{hj} v^i{}_{,h},$$

we have:

$$\frac{\partial L}{\partial v^i{}_{,jk}} = \left(\frac{\partial L}{\partial A_{j,k}} + \frac{\partial L}{\partial A_{k,j}}\right) A_i = 0, \tag{17}$$

i.e., the components $A_{i,j}$ enter L only through $M_{ij} = A_{j,i} - A_{i,j}$, and from

$$\partial L/\partial v^i{}_{,k} = 0$$

we find

$$2g^{jk}\partial L/\partial g^{ij} - A_i \partial L/\partial A_k - M_{ij}\partial L/\partial M_{kj} = 0$$

or

$$\hat{T}_i{}^k = -\frac{\partial \hat{L}}{\partial g^{ij}} g^{jk} = \tfrac{1}{2}\left(\hat{L}\delta_i{}^k - \frac{\partial \hat{L}}{\partial A_k} A_i - \frac{\partial \hat{L}}{\partial M_{kj}} M_{ij}\right). \tag{18}$$

This relation between $\partial \hat{L}/\partial g^{ij}$ and Mie's electromagnetic energy tensor emerging from L_{em} (which is a scalar and which is only a function of g^{ij} and not of its derivatives), led Hilbert to describe it as "a circumstance which first brought my attention to the very close and unavoidable relation between Einstein's theory of general relativity and Mie's electrodynamics, and which gave me the conviction that the theory here developed was indeed correct" (Hilbert 1915, p. 404).

(2) Einstein–Grossmann type Lagrangian L_{EG}.

This Lagrangian is of the form

$$\hat{L}_{\text{EG}} = \sqrt{-g}\, L(g^{ij}, g^{ij}{}_{,k}),$$

where L is only a linear scalar. Using the variations (16) and

$$\Delta g^{ij}{}_{,k} = (\Delta g^{ij})_{,k} - g^{ij}{}_{,h} v^h{}_{,k}, \tag{19}$$

we get

$$\tfrac{1}{2}\Delta L = s_i{}^k v^i{}_{,k} + L^j{}_h{}^k v^h{}_{,jk}, \tag{20}$$

with

$$s_i{}^k := \frac{\partial L}{\partial g^{hi}} g^{hk} + L_{ji}{}^h g^{jk}{}_{,h} - \tfrac{1}{2} L_{jh}{}^k g^{jh}{}_{,i},$$
$$L^j{}_i{}^k := g^{hj} L_{hi}{}^k, \quad L_{ij}{}^k := \frac{\partial L}{\partial g^{ij}{}_{,k}}.$$

Under linear transformations ($v^i{}_{,jk} = 0$), $\Delta L = 0$ leads to

$$s_i{}^k = 0. \tag{21}$$

Can L be a scalar under some more general class of transformations ($v^i{}_{,jk} \neq 0$ for a class of v^i)? In that case, we should have, since $\partial L/\partial v^i{}_{,jk} = 0$ for such v^i s, that

$$L^{jik} + L^{kij} = 0,$$

where $L^{jik} := g^{ih} L^j{}_h{}^k$. Hence, L satisfies

$$L^{ijk} = L^{jik} \text{ and } L^{jik} = -L^{kij}, \tag{22}$$

the first equality following from g^{ij} being a symmetric tensor. Such an object must be zero by the "braid lemma." Indeed,

$$L^{jik} = -L^{kij} = -L^{ikj} = L^{jki} = L^{kji} = -L^{ijk} = -L^{jik} = 0. \tag{23}$$

Therefore, L can be at most a linear scalar.

3.3 The Lie-Variation $\mathcal{L}$

The Lie-variation $\mathcal{L}$ is induced by a one-parameter group of diffeomorphisms of (M, g_{ij}) generated by a vector field v^i on M. Let h_ε be such a diffeomorphism of M:

$$h_\varepsilon : \; x^i \mapsto h_\varepsilon(x^i) = x^i(\varepsilon) = x^i + \varepsilon v^i + \cdots.$$

The Lie-variation of Φ_A is

$$\mathcal{L}(v)\Phi_A := \frac{\mathrm{d}}{\mathrm{d}\varepsilon}(h^*_{-\varepsilon}\Phi_A)_{\varepsilon=0}, \tag{24}$$

where $h^*_{-\varepsilon}$ is the natural pull-back associated to the diffeomorphism. We thus have, for a scalar Φ,

$$\mathcal{L}(v)\Phi = v^k\Phi_{,k}; \tag{25}$$

for a covector A_i,

$$\mathcal{L}(v)A_i = v^k A_{i,k} + A_k v^k{}_{,i} \tag{26}$$

and

$$\mathcal{L}(v)A_{i,j} = \left(\mathcal{L}(v)A_i\right)_{,j}. \tag{27}$$

As the Lie-variation defines an equivalence of manifold structures $(M, \Phi_A) \to (M, h\Phi_A)$, one may require the *general invariance* of physical descriptions under the diffeomorphism pseudogroup. In that case, points of the underlying topological space lose their individuality and the geometry of the space-time manifold reduces to the geometry of the metric field structure (M, g_{ij}), in the spirit of the generalized *Erlangen* program (cf. Veblen and Whitehead 1932). This was, in fact, the endpoint of Einstein's search for general covariance (in late 1915 and early 1916), as is argued in some very important papers by Stachel (1979, 1989) and Norton (1984) on the "hole argument." The general invariance may be considered as generalizing the Lorentz invariance, and thus corresponds to Einstein's relativity principle (1918). However, it must be stressed that space-time can no longer be thought of as an arena for physics, as in special relativity.

On the other hand, it is well known (see, e.g., Hilbert 1915, Noether 1918) that the general invariance of I_U leads to the covariance of the field equations under diffeomorphisms, which can be seen in the same way as in the case of Δ-variation, by showing that $\partial_i(\mathcal{L}Q^i)$ vanishes whenever v^i and its derivatives on $\partial\Omega$ vanish; to the generalized "Bianchi" identities, which follow from $\hat{L}^A\mathcal{L}\Phi_A = 0$; and to the conservation laws related to assumed symmetries of the fields. In short:

(1) The functional variation δ leads, through the requirement that I_U be stationary, to the field equations.

(2) The label-variation Δ helps define the coordinate covariance, and leads to conditions on the structure of the Lagrangian.

(3) The Lie-variation $\mathcal{L}$ determines the general invariance of I_U, and leads to conservation identities and conservation laws.

With the above considerations on the three kinds of variation and their different roles, we are now in a position to comment on some of the early variational derivations of the field equations of general relativity.

4. Comments on Some Variational Derivations of Einstein's Field Equations

4.1 Covariance Group of the *Entwurf* Equations

The route followed by Einstein toward general relativity is now fairly well known, thanks to many historical investigations by Mehra (1973), Earman and Glymour (1978a, 1978b), Stachel (1979, 1989), Pais (1982), and Norton (1984). By 1913, Einstein felt compelled to abandon the postulate of general covariance, and proposed with Marcel Grossmann (Einstein and Grossmann 1913) the *Entwurf* equations. In Einstein and Grossmann 1914, they derived these field equations from the Lagrangian

$$\hat{L}_{\mathrm{EG}} = \tfrac{1}{4}\sqrt{-g}\, g^{ij} g_{hk,i}\, g^{hk}{}_{,j}$$

with the help of Paul Bernays (Einstein and Grossmann 1914, p. 119, footnote). Paul Bernays was a mathematician of the Göttingen school. The year before, he had written a paper in which he gave a competent exposition of the special theory of relativity, even though he rejected it on philosophical grounds (Bernays 1913). Much later, he would coauthor *Grundlagen der Mathematik* with Hilbert (Bernays and Hilbert, 1934/1939).

The idea behind Einstein and Grossmann's 1914 paper and a subsequent paper by Einstein alone (Einstein 1914), which has a more general Lagrangian $L = L(g^{ij}, g^{ij}{}_{,k})$, was to restrict the "covariance" group of the *Entwurf* field equations to a "justified" group, more general than the linear group. The coordinate transformations in this "justified" group were to relate "adapted" coordinate systems determined by the condition

$$B_i := \left(\frac{\partial \hat{L}}{\partial g^{ki}{}_{,h}} g^{kj}\right)_{,jh} = 0. \tag{29}$$

This condition is obtained from $\frac{1}{2}\Delta\hat{L} = \hat{L}^j{}_i{}^h v^i{}_{,jh}$ (modulo $S_i{}^k$, which is set equal to zero) through integration by parts, the divergence terms cancelling on the assumption that $(v^i, v^i{}_{,k})_{|\partial\Omega} = 0$. We can make two remarks:

(1) Condition (29) may be satisfied by $L^j{}_i{}^h + L^h{}_i{}^j = 0$, in which case, as we showed in Section 3.2, L is independent of $g^{ij}{}_{,k}$.

(2) The field equations $\delta\hat{L}/\delta g^{ij} = 0$ are indeed covariant under the linear group, for $\Delta_{\text{lin}}\delta I_U = 0$. But with respect to the larger "justified group," when it exists, $\Delta_{\text{just}}\delta I_U \neq 0$, since L is not a scalar under this larger group of transformations; and there is no apparent reason why $\Delta_{\text{just}}\delta\hat{L}/\delta g^{ij}$ should be zero, i.e., why the field equations should be covariant.

Cattani and De Maria (1989) have given an analysis of part of the 1915 epistolary controversy between Einstein and Levi-Civita on the restricted covariance of the *Entwurf* field equations. It appears essentially that the main observation of Levi-Civita concerned the nontensor character of those field equations. This correspondence, his visit to Göttingen in late June and early July 1915, and a number of points of dissatisfaction with the introduction of "adapted" coordinates, led Einstein to believe that a "more far-reaching covariance, where possible *general* covariance, must be demanded" (Einstein to Lorentz, January 1, 1916; as quoted in Norton 1984, p. 299).

We should note that Lorentz brought "simplicity and clearness" to some parts of Einstein's variational paper (Einstein and Grossmann 1914), basing it on a principle similar to that of Hamilton, so much so, in fact, that "Hamilton's name may properly be connected with it" (Lorentz 1915, p. 229; see also de Donder 1921, pp. 21–31)). Two essential features of Lorentz's paper were:

(1) the variations of the field variables are quite arbitrary and not related to any other procedure, such as a coordinate transformation;
(2) the Lagrangian of a system depending on several field variables is the sum of the Lagrangians corresponding to each of them.

On the other hand, when Lorentz related the covariance of the field equations to the "invariant" character of $\hat{L}\,\mathrm{d}^4x$ under any change of coordinates, he simply remarked, without further comment, that the covariance of the gravitational equations "is a consequence of the invariancy [*sic*] of [$\delta\int\hat{L}\,\mathrm{d}^4x$], which Einstein has proved by an ingenious mode of reasoning" (Lorentz 1915, p. 245).

4.2 Hilbert's "Die Grundlagen der Physik"

Hilbert's derivation of the field equations is based on two axioms:

Axiom 1. The Lagrangian is the world-function (i.e., a scalar) $L = L(g_{kl}, g_{kl,i}, g_{kl,ij}, A_i, A_{i,j})$, where g_{ij} and A_i are the gravitational and electromagnetic potentials, which gives the field equations $\delta\hat{L}/\delta g_{ij} = 0$ and $\delta\hat{L}/\delta A_i = 0$;

Axiom 2. (Axiom of general invariance.) L is invariant with respect to an arbitrary transformation of coordinates;

while the Einstein Lagrangians are "in no sense invariants, nor do they contain electric potentials." Although not explicitly stated, it is clear from the text that Hilbert used the functional variation δ to get field equations, the label-variation Δ to ascertain the scalar character of L, and the Lie-variation $\mathcal{L}$ to derive conservation identities. Hilbert considered the introduction of the electromagnetic energy-momentum tensor as the derivative of L_{em} with respect to the gravitational potentials g^{ij} (see Section 3.2) as the triumph of axiomatics. It seems that this result, following simply from L being a scalar, has never been well understood. Not only did Einstein and Lorentz make no use of it, there was also some debate in the 1960s as to why the Eulerian $(\partial\hat{L}/\partial g^{ij})$ and the canonical energy-momentum tensors in field theories are equivalent, an equivalence here noted by Hilbert.

With the help of a third axiom, $L = R + L_{\text{em}}$, where R is the curvature scalar, and assuming that L_{em} contains no derivatives of g^{ij}, Hilbert obtains, using now standard procedures, the field equations

$$R_{ij} - \tfrac{1}{2}Rg_{ij} = (T_{ij})_{\text{em}},$$
$$\frac{\partial\hat{L}_{\text{em}}}{\partial A_i} - \left(\frac{\partial\hat{L}}{\partial A_{i,j}}\right)_{,j} = 0,$$

and shows that "four of them are always a consequence of the remaining $n-4$, in the sense that between the n differential equations and their total derivatives, four combinations, independent from each other, are always identically satisfied" (Hilbert 1915, p. 397). The four relations Hilbert is talking about here are, of course, just those following from Noether's theorem on pseudogroups (Noether 1918).

4.3 Einstein's "The Foundations of the General Relativity Theory"

In his 1916a, Einstein ignored Hilbert's paper and gave the "Hamiltonian function" for the gravitational field as

$$\begin{cases} L = g^{ij}\Gamma^k{}_{il}\Gamma^l{}_{jk}, \\ \sqrt{-g} = 1. \end{cases} \tag{30}$$

He allowed the condition $\sqrt{-g} = 1$ as "a hypothesis as to the physical nature of the continuum under consideration, and at the same time a convention as to the choice of coordinates" (Einstein 1916a, p. 130). The field equations, in the presence of "matter," become:

$$-(\Gamma^k{}_{ij})_{,k} + \Gamma^k{}_{ih}\Gamma^h{}_{jk} = -k(T_{ij} - \tfrac{1}{2}Tg_{ij}). \tag{31}$$

It is easy to verify that the left-hand side is the Ricci tensor for $\sqrt{-g} = 1$, so that these equations are almost generally covariant, i.e., they are covariant under unimodular transformations, transformations with $J = 1$.

Why did Einstein not consider the Hilbert Lagrangian in this 1916 paper? Was he technically handicapped because of the second-order derivatives in the Lagrangian, as one might suspect from Hilbert's remark: "Every boy in the streets of Göttingen understands more about four-dimensional geometry than Einstein. Yet, in spite of that, Einstein did the work and not the mathematicians" (Reid 1986, p. 142)? Why did he insist on the condition $\sqrt{-g} = 1$? Was it because it implied, as Einstein had shown in his November 11, 1915 paper (Einstein 1915), that $T^k{}_k = 0$ and because it would have been satisfactory if all matter were electromagnetic in nature (see Norton 1984, p. 308)? Was he hoping to find a new physical argument to support his assumption against the mathematician's view of nature? Or was it simply because the details of Hilbert's paper were still obscure to him, as it appears from his letters to Hilbert in May–June 1916 (see, especially, Einstein to Hilbert, May 30, 1916, EA 13-102; see Norton 1984, p. 315)?

I will now show that the condition $\sqrt{-g} = 1$ is not that innocent. Consider the constrained Lagrangian

$$\hat{L} = R\sqrt{-g} + \lambda^k(\sqrt{-g})_{,k}, \tag{32}$$

where λ^k is a Lagrange multiplier, and the second term is not a scalar density. The gravitational field equations become

$$\begin{cases} (\sqrt{-g})_{,k} = 0 \\ R_{ij} - \frac{1}{2}Rg_{ij} + \frac{1}{2}(\nabla_k\lambda^k)g_{ij} = 0, \end{cases} \tag{33}$$

the second equation being manifestly covariant. We now have two options: to recover Eqs. (31) by requiring $\nabla_k\lambda^k = 0$, i.e., by requiring that M admits a one-parameter volume-preserving group generated by λ^k, or to interpret the extra term as a reaction to the constraint $\sqrt{-g} = \text{const}$.

4.4 Lorentz's "On Einstein's Theory of Gravitation"

In a series of four papers, H.A. Lorentz considers the variational derivations of Einstein and Hilbert and gives his own version of the gravitational equations in the presence of incoherent matter and electromagnetic field. I will comment only on the gravitational part (see also Kox 1988 and Janssen 1992).

In the first paper, Lorentz gave the geometric interpretation of the curvature scalar as the sum of mean curvatures in the directions $\lambda_a{}^i$, where the λ_a form an orthogonal basis (see, e.g., Ricci 1904). (Recall that the mean curvature $m_{(a)}$ in the direction $\lambda_a{}^i$ is defined by

$$m_{(a)} = \frac{R_{ij}\lambda_a{}^i\lambda_a{}^j}{g_{ij}\lambda_a{}^i\lambda_a{}^j}$$

and $R = \sum_a m_{(a)}$.) In the third paper, he adopted $L_{\text{grav}} = R$, and he showed that the term

$$M_{ij} = \frac{\delta\hat{L}}{\delta\hat{g}^{ij}}$$

in the field equations (with $\hat{g}^{ij} = g^{ij}\sqrt{-g}$) is, in fact, equal to the Ricci tensor R_{ij} in a particular coordinate system in which $g_{ij} = \eta_{ij}$ and $g_{ij,k} = 0$, and equal to

$$\frac{\delta\hat{L}}{\delta g^{ij}} = \sqrt{-g}\,(R_{ij} - \tfrac{1}{2}Rg_{ij})$$

in an arbitrary system of coordinates. In the last part of his third paper, Lorentz considered an additional electromagnetic field, but unlike Hilbert he did not obtain the corresponding energy-momentum tensor as the derivative with respect to g_{ij}. Lie variation is used to get conservation identities. At the end of the third paper and in the fourth paper, different candidates for the gravitational energy-momentum pseudotensor are proposed and compared to Einstein's.

4.5 Einstein's "Hamilton's Principle and the General Theory of Relativity"

At the very beginning of his paper (Einstein 1916b), Einstein admitted that "the general theory of relativity has recently been given in a particularly clear form by H.A. Lorentz and D. Hilbert, who have deduced its equations from one single principle of variation" (Einstein 1916b, p. 167). His aim is now to "do the same thing," making "as few specializing assumptions as possible, in marked contrast to Hilbert's treatment of the subject,"and with "complete liberty in the choice of the system of coordinates" (ibid.).

However, using the relation

$$\hat{R} = \hat{E} + \hat{F}, \tag{34}$$

with

$$\hat{E} = -\hat{g}^{ij}(\Gamma^k{}_{ih}\Gamma^h{}_{jk} - \Gamma^k{}_{ij}\Gamma^h{}_{kh}),$$
$$\hat{F} = (\hat{g}^{ij}\Gamma^k{}_{ij} - \hat{g}^{ik}\Gamma^j{}_{ij})_{,k} = \partial_i \hat{Q}^i,$$

he adopted the first-order gravitational Lagrangian $\hat{E}(g^{ij}, g^{ij}{}_{,k})$, equivalent to $\hat{R}$, so that the total Lagrangian is

$$\hat{L} = \hat{E} + \hat{M} = \hat{R} + \hat{M} - \hat{F},$$

the matter part $\hat{M} = \hat{M}(g^{ij}, q_i, q_{i,j})$ depending only on g^{ij} as in Hilbert's paper. The field equations are

$$-\hat{E}_{ij} = \left(\frac{\partial \hat{E}}{\partial g^{ij}{}_{,k}}\right)_{,k} - \frac{\partial \hat{E}}{\partial g^{ij}} = \frac{\partial \hat{M}}{\partial g^{ij}}, \tag{35}$$

$$\left(\frac{\partial \hat{M}}{\partial q_{i,k}}\right)_{,k} - \frac{\partial \hat{M}}{\partial q_i} = 0. \tag{36}$$

Since

$$\delta \int_\Omega \hat{L}\, \mathrm{d}^4x = \delta \int_\Omega (\hat{E} + \hat{M})\, \mathrm{d}^4x = \delta \int_\Omega (\hat{R} + \hat{M})\, \mathrm{d}^4x$$

under the conditions $\delta g^{ij}|_{\partial\Omega} = \delta g^{ij}{}_{,k}|_{\partial\Omega} = 0$, and since R and M are scalars, we can infer that

$$\int_\Omega (\hat{E}_{ij} + \hat{M}_{ij})\delta g^{ij}\, \mathrm{d}^4x = \int_\Omega [(\hat{R}_{ij} - \tfrac{1}{2}R\hat{g}_{ij}) + \hat{M}_{ij}]\delta g^{ij}\, \mathrm{d}^4x,$$

where $\hat{M}_{ij} = \partial \hat{M}/\partial g^{ij}$, so that we get

$$\hat{E}_{ij} = \hat{R}_{ij} - \tfrac{1}{2}R\hat{g}_{ij} = -\hat{M}_{ij}, \tag{37}$$

which is manifestly covariant.

By performing a local infinitesimal coordinate transformation such that "Δx^i differ from zero only in the interior of a given domain [Ω], but in infinitesimal proximity to the boundary [$\partial\Omega$] they vanish," Einstein then convinced himself that "the value of the boundary integral [$\int_\Omega \hat{F}\, \mathrm{d}^4x$] does not change" (Einstein 1916b, p. 171), and he concluded that

$$\Delta \int \hat{R}\, \mathrm{d}^4x = \Delta \int \hat{E}\, \mathrm{d}^4x = 0. \tag{38}$$

Since $\hat{E}$ and $\hat{F} = \partial_i \hat{Q}^i$ are not scalar densities, however, one should be careful in applying Stokes' theorem. Substituting $x^i \to x^{i'} = x^i + \Delta x^i$, we have, by virtue of (14),

$$J(\hat{E}' + \partial_{i'} \hat{Q}^{i'}) = \hat{E} + \partial_k \hat{Q}^k + \Delta \hat{E} - \left(\hat{Q}^l \Delta x^k{}_{,l} + \hat{g}^{ij} \Delta x^k{}_{,ij} - \hat{g}^{ik} \Delta x^h{}_{,ih}\right)_{,k}. \quad (39)$$

Since, on the other hand, $J \hat{R}(x') = \hat{R}(x) = \hat{E} + \partial_k \hat{Q}^k$, we have in Ω:

$$\Delta \hat{E} - \left(\hat{Q}^l \Delta x^k{}_{,l} + \hat{g}^{ij} \Delta x^k{}_{,ij} - \hat{g}^{ik} \Delta x^h{}_{,ih}\right)_{,k} = 0. \quad (40)$$

This means that

(1) $\Delta \hat{E}$ is not vanishing, but is compensated by terms originating in $\hat{F} = \partial_i \hat{Q}^i$, since $\hat{Q}^i$ is not a vector density;

(2) under the conditions imposed on Δx^k and its derivatives on $\partial\Omega$,

$$\Delta \int_\Omega \hat{E}\, \mathrm{d}^4 x = 0,$$

as asserted in (38), since, by virtue of (40), $\Delta \hat{E}$ is equal to a divergence. The first-order Lagrangian of general relativity, $\hat{E}$, therefore has the following peculiar property: *although it is not a scalar density, $\hat{E}$ yields an equivalent Lagrangian under a local coordinate transformation.*

One can then use the second part of (38), as Einstein did, to derive "the conservation of momentum and energy" (Einstein 1916b, p. 172), expressed by

$$(\hat{M}_i{}^h + \hat{t}_i{}^h)_{,h} = 0, \quad (41)$$

where

$$\hat{M}_i{}^h = \hat{M}_{ik} g^{hk}, \quad (42)$$

$$\hat{t}_i{}^h = -\left(\frac{\partial \hat{E}}{\partial g^{ij}{}_{,k}} g^{jh}{}_{,k} + \frac{\partial \hat{E}}{\partial g^{ij}} g^{jh}\right) = \tfrac{1}{2}\left(\hat{E}\delta_i{}^h - \frac{\partial \hat{E}}{\partial g^{jk}{}_{,h}} g^{jk}{}_{,i}\right). \quad (43)$$

The second equality in (43) comes from the vanishing of

$$\hat{s}_i{}^h = \frac{\partial \hat{E}}{\partial g^{ij}} g^{jh} + \frac{\partial \hat{E}}{\partial g^{ij}{}_{,k}} g^{jh}{}_{,k} + \tfrac{1}{2}\left(\hat{E}\delta_i{}^h - \frac{\partial \hat{E}}{\partial g^{jk}{}_{,h}} g^{jk}{}_{,i}\right), \quad (44)$$

which follows from the fact that $\hat{E}$ is a linear scalar density (cf. Equation (21)). Equation (40), combined with the field equations (35), leads to the conservation of the energy-momentum tensor:

$$\nabla_k \hat{M}_i{}^k = 0.$$

Einstein then concluded:

> It is to be emphasized that the (generally covariant) laws of conservation (21) and (22) [here (40) and (45)] are deduced from the field equations (7) of gravitation [here (36)], in combination with the postulate of general covariance (relativity) *alone*, without using the field equations for material phenomena. (Einstein 1916b, p. 173)

It should be noted that the conservation laws (40) and (45) are a consequence of the label-variation Δ of the Lagrangian and of the gravitational field equations (35), while the Lie-variation $\mathcal{L}$ leads to four identities between those field equations, as was first shown by Hilbert. Did Einstein really distinguish between these two kinds of variation in the years 1914–1918? When did he *fully* understand the implications of the failure of the "hole argument," i.e., the necessity of the general invariance of the space-time manifold structure? Obviously, these considerations are not at variance with Stachel and Norton's historical analyses of Einstein's struggle with the "hole argument."

4.6 Palatini's "Invariant Derivation of Gravitational Equations from Hamilton's Principle"

After noting that Hilbert (1915) and Weyl (1917) used "non-invariant formulae" (Palatini 1919, p. 203) in their derivations, Palatini showed that

$$\delta\hat{R} = R_{ij}\delta\hat{g}^{ij},$$

$\hat{g}^{ij}\delta R_{ij}$ contributing only a divergence $\nabla_k\hat{t}^k$, by virtue of $\nabla_k g_{ij} = 0$. The Palatini variation is in fact a device to get the field equations by varying only g^{ij} as in Hilbert's g-variation (see Ferraris et al. 1982; Goenner 1979; Stephenson 1958).

4.7 General Remarks on Other Variational Derivations

Constrained variation. The Palatini device was historically followed by P-variation, i.e., the independent variation of g^{ij} and $\Gamma^i{}_{jk}$. Although g- and P-variation yield equivalent results for the Lagrangian of general relativity, they generally give different field equations for more general Lagrangians (see, e.g., Kichenassamy 1986). The use of Lagrange multipliers to cope with the Ricci constraint $\nabla_k g_{ij} = 0$ removes this discrepancy and brings a wealth of useful information (Kichenassamy 1986). The general gravitational Lagrangian is

$$\hat{L} = \hat{A}(g, \Gamma, \partial\Gamma) + \hat{B}(g, \Phi_A, \nabla\Phi_A) + \hat{C}(g, \partial g, \Gamma). \qquad (46)$$

The equation $\delta\hat{C} = 0$ restores the constraint. In a Lorentzian manifold, in which case $A = R$ and $\partial\hat{B}/\partial\Gamma^i{}_{jk} = 0$, we get the Einstein equations (see, e.g., Ray 1975). The cases in which $\partial\hat{B}/\partial\Gamma^i{}_{jk} \neq 0$ (minimum gravitational coupling) include more elaborate "matter" configurations.

The constraints may also restrict the independence of "matter variables." In this context a very important literature is now developing on perfect and dissipative fluids.

From Lagrangian to Hamiltonian. With a view toward quantizing general relativity, much work has been done to derive the Hamiltonian from the gravitational Lagrangian. An important problem arises from the fact that the boundary integral cannot be made to vanish by assuming that δg^{ij}, but not $\delta g^{ij}{}_{,k}$, vanish on the boundary $\partial\Omega$. As a consequence, the Lagrangian has to be modified before the transition to Hamiltonian formalism can be made.

Extensions of general relativity. Variational methods are used to obtain generalizations of Einstein's theory: unified field theories, Einstein–Cartan theory, and gravitational gauge theories (see Kichenassamy 1986).

5. Conclusion

From the various contributions to the variational formulation of general relativity that I discussed in this chapter, the following picture emerges. On the one hand, we see how Einstein, through a hard struggle, with misconceptions and misleading techniques, was slowly led to a fundamental model of physical description, with the help of Grossmann, Lorentz, Ehrenfest, Levi-Civita, Sommerfeld, and others. On the other hand, we have Hilbert, well acquainted with Einstein's work, who could anticipate the exact mathematical structure of this model. Unfortunately, Hilbert was preoccupied more with the foundations of physics than with physics itself, whereas in Einstein's case one can seriously doubt whether he fully understood the implications of his theory in the period 1915–1918 that I considered here.

Anyhow, I hope to have shown in this chapter that the variational approach leads to some valuable insights into the foundations of general relativity, and that it raises some interesting historical questions about how Einstein coped with his theory.

ACKNOWLEDGMENTS. It is a pleasure to thank Jean Eisenstaedt who generously provided me with a number of helpful papers and documents.

REFERENCES

Bernays, Paul (1913). *Über die Bedenklichkeit der neueren Relativitätstheorie.* Göttingen: Vandenhoeck and Ruprecht.

Bernays, Paul and Hilbert, David (1934/1939). *Grundlagen der Mathematik.* Vols. 1 and 2. Berlin: Springer.

Bolza, Oskar (1904). *Lectures on the Calculus of Variations.* New York: G.E. Streichert. Reprinted in 1940.

Cattani, Carlo and De Maria, Michelangelo (1989). "The 1915 Epistolary Controversy between Einstein and Tullio Levi-Civita." In *Einstein and the History of General Relativity.* Don Howard and John Stachel, eds. Boston: Birkhäuser, pp. 175–200.

De Donder, Theophile (1921). *La Gravifique Einsteinienne.* Paris: Gauthier-Villars.

Dugas, René (1950). *Histoire de la Mécanique,* Préface de Louis de Broglie. Neuchatel: Editions du Griffon.

Earman, John and Glymour, Clark (1978a). "Lost in the Tensors: Einstein's Struggles with Covariance Principles 1912–1916." *Studies in History and Philosophy of Science* 9: 251–278.

—— (1978b). "Einstein and Hilbert: Two Months in the History of General Relativity." *Archive for History of Exact Sciences* 19: 291–308.

Einstein, Albert (1914). "Die formale Grundlage der allgemeinen Relativitätstheorie." *Königlich Preussische Akademie der Wissenschaften* (Berlin). *Sitzungsberichte*: 1030–1085.

—— (1915). "Zur allgemeinen Relativitätstheorie (Nachtrag)." *Königlich Preussische Akademie der Wissenschaften (Berlin). Sitzungsberichte*: 799–801.

—— (1916a). "Die Grundlage der allgemeinen Relativitätstheorie." *Annalen der Physik* 49: 769–822. Quotations are from the English translation in Lorentz et al. 1923, pp. 111–164. Page numbers refer to translation.

—— (1916b). "Hamiltonsches Prinzip und allgemeine Relativitätstheorie." *Königlich Preussische Akademie der Wissenschaften* (Berlin). *Sitzungsberichte*: 1111–1116. Quotations are from the English translation in Lorentz et al. 1923, pp. 167–173. Page numbers refer to translation.

—— (1918). "Prinzipielles zur allgemeinen Relativitätstheorie." *Annalen der Physik* 55: 241–244.

—— (1955). *The Meaning of Relativity.* 5th ed. Princeton: Princeton University Press.

Einstein, Albert and Grossmann, Marcel (1913). *Enwurf einer verallgemeinerten Relativitätstheorie und einer Theorie der Gravitation.* Leipzig: Teubner. Reprinted with added "*Bemerkungen*" in *Zeitschrift für Mathematik und Physik* 62: 225–261.

—— (1914). "Kovarianzeigenschaften der Feldgleichungen der auf die verallgemeinerte Relativitätstheorie gegründeten Gravitationstheorie." *Zeitschrift für Mathematik und Physik* 63: 215–225.

Ferraris, Marco, Francaviglia, Mauro, and Reina, Cesare (1982). "Variational Formulation of General Relativity from 1915 to 1925; 'Palatini's Method' Discovered by Einstein in 1925." *General Relativity and Gravitation.* Vol. 14, no. 3: 243–254.

Gelfand, Izrail' M. and Fomin, Sergefi V. (1963). *Calculus of Variations.* Richard A. Silvermann, trans. Englewood Cliffs, New Jersey: Prentice-Hall.

Goenner, Hubert F. (1979). "On the Equivalence of the Palatini and Hilbert Methods of Variation." *Tensor* New Series 33: 307–312.

Gueroult, M. (1934). *Dynamique et métaphysique leibniziennes, suivi d'une "Note sur le principe de la moindre action chez Maupertuis."* Paris: Les Belles-Lettres.

Hilbert, David (1899). "Über das Dirichlet'sche Princip." *Jahresberichte* 8: 184–188.

——— (1906). "Zur Variationsrechnung." *Mathematische Annalen* 62: 351–370.

——— (1915). "Die Grundlagen der Physik." *Königlichen Gesellschaft der Wissenschaften zu Göttingen, Nachrichten:* (I) (1915), 395–407; (II) (1916), 53–76.

Janssen, Michel (1992). "H.A. Lorentz's Attempt to Give a Coordinate-Free Formulation of the General Theory of Relativity." In: *Studies in the History of General Relativity.* Jean Eisenstaedt, A.J. Kox, eds. Boston: Birkhäuser, pp. 344–363.

Kichenassamy, S. (1986). "Lagrange Multipliers and Theories of Gravitation." *Annals of Physics N.Y.* 168: 404–424.

Kox, A.J. (1988). "Hendrik Antoon Lorentz, the Ether and the General Theory of Relativity." *Archive for History of Exact Sciences* 38: 66–78. Also in *Einstein and the History of General Relativity.* Don Howard and John Stachel, eds. Boston: Birkhäuser, 1989, pp. 201–212.

Kretschmann, Erich (1917). "Über den physikalischen Sinn der Relativitätspostulate, A. Einsteins neue und seine ursprüngliche Relativitätstheorie." *Annalen der Physik* 53: 575–614.

Larmor, Joseph J. (1900). *Aether and Matter.* Cambridge: Cambridge University Press.

Lorentz, Hendrik A. (1915). "Het beginsel van Hamilton in Einstein's theorie der Zwaartekracht." *Koninklijke Akademie van Wetenschappen te Amsterdam. Verslagen van de Gewone Vergaderingen der Wis- en Natuurkundige Afdeeling* 23: 1073-1089; "On Hamilton's Principle in Einstein's Theory of gravitation." *Koninklijke Akademie van Wetenschappen te Amsterdam. Proceedings of the Section of Sciences* 19: 751-767. Quotations are from the English version.

——— (1916). "Over Einstein's theorie der zwaartekracht." *Koninklijke Akademie van Wetenschappen te Amsterdam. Verslagen van de Gewone Vergaderingen der Wis- en Natuurkundige Afdeeling* (I) 24, (1916): 1389–1402; (II) 24, (1916): 1759–1774; (III) 25, (1916): 468–486; (IV) 25, (1917): 1380–1396; "On Einstein's Theory of gravitation." *Koninklijke Akademie van Wetenschappen te Amsterdam. Proceedings of the Section of Sciences* (I) 19, (1916): 1341–1354; (II)19, (1916): 1354–1369; (III) 20, (1916): 2–19; (IV) 20, (1917): 20–34. Quotations are from the English version.

Lorentz, Hendrik A., Einstein, A., Minkowski, H., and Weyl, H., eds. (1923). *The Principle of Relativity: A Collection of Original Memoirs on the Special and General Theory of Relativity*, with Notes by A. Sommerfeld. W. Perrett and G.B. Jeffery, trans. London: Methuen; reprint New York: Dover, 1952.

Mehra, Jagdish (1973). "Einstein, Hilbert, and the Theory of Gravitation." In *The Physicist's Conception of Nature*. Jagdish Mehra, ed. Boston and Dordrecht: D. Reidel, pp. 92–178.

Noether, Emmy (1918). "Invariante Variationsprobleme." *Königliche Gesellschaft der Wissenschaften zu Göttingen. Mathematisch-Physikalische Klasse. Nachrichten.*: 235–257.

Norton, John (1984). "How Einstein Found His Field Equations, 1912–1915" *Historical Studies in the Physical Sciences* 14: 253–316. Also in *Einstein and the History of General Relativity*. Don Howard and John Stachel, eds. Boston: Birkhäuser, 1989, pp. 101–159.

Pais, Abraham (1982). *"Subtle Is the Lord..." The Science and Life of Albert Einstein*. Oxford: Clarendon Press; New York: Oxford University Press.

Palatini, Attilio (1919). "Deduzione invariantiva delle equazioni gravitazionali dal principio di Hamilton." *Rendiconti del Circolo Matematico di Palermo* 43: 203–212.

Ray, John R. (1975). "Palatini Variational Principle." *Nuovo Cimento* 25B: 706–710.

Reid, Constance (1986). *Hilbert–Courant*. New York: Spinger-Verlag.

Ricci, Gregorio (1904). "Formole Fundamentali nella Teoria Generali di Varieta et della lors curvatura." *Rendiconti Accademia dei Lincei* ser. 5, 11: 355–362.

Schwarzschild, Karl (1903). "Zur Elektrodynamik." *Königliche Gesellschaft der Wissenschaften zu Göttingen. Mathematisch-Physikalische Klasse. Nachrichten.* I, 126–131; II, 132–141; III, 245–278.

Sommerfeld, Arnold (1952). *Lectures on Theoretical Physics. Vol. 1. Mechanics*. New York: Academic Press.

Stachel, John (1979). "The Genesis of General Relativity." In *Einstein Symposion Berlin*. H. Nelkowski, A. Hermann, H. Poser, R. Schrader, R. Seiler, eds. (Lecture Notes in Physics, Vol. 100.) Berlin, Heidelberg, New York: Springer-Verlag, pp. 428–442.

—— (1989). "Einstein's Search for General Covariance, 1912–1915." In *Einstein and the History of General Relativity*. Don Howard and John Stachel, eds. Boston: Birkhäuser, pp. 63–100. Based on a paper delivered to the Ninth International Conference on General Relativity and Gravitation, Jena, 1980.

Stephenson, G. (1958). "Quadratic Lagrangians and General Relativity" *Nuovo Cimento* 9: 263–269.

Tonelli, Leonida (1921–1923). *Fondamenti di Calcolo delle Variazioni*, Vol. 1 and 2. Bologna: Zanichelli.

Veblen, Oswald and Whitehead, John H.C. (1932). *The Foundations of Differential Geometry*. Cambridge: Cambridge University Press.

Weierstrass, Karl (1927). *Vorlesungen über Variationsrechnung*. Rudolf Rothe, ed. Leipzig: Akademische Verlagsgesellschaft.

Weyl, Hermann (1917). "Zur Gravitationstheorie." *Annalen der Physik* 54: 117–145.

Levi-Civita's Influence on Palatini's Contribution to General Relativity

Carlo Cattani

1. Introduction

In the years 1913–1918, Albert Einstein made every effort to give a rigorous variational formulation of his gravitational field equations and to find solutions of these equations in some special cases. In this way, he hoped to clarify the physical significance of his new gravitational theory. During the years 1915–1925, the Italian mathematician Tullio Levi-Civita and his follower Attilio Palatini made a number of fundamental contributions to the new theory that were very important for such clarification. The most important of these were:

(1) Levi-Civita's proof, in 1915, of the mathematical inconsistency of Einstein's early gravitational theory (see Cattani and De Maria 1989b),

(2) the correct definition of the conservation of the gravitational stress-energy tensor through the contracted Bianchi identities (Levi-Civita 1917a),

(3) the correct variational formulation of the final version of the gravitational field equations (Palatini 1919),

(4) a profound discussion of gravitational energy (see Cattani and De Maria 1989a, and my chapter with De Maria in this volume),

(5) a series of investigations into problems of static gravitation (Levi-Civita 1917b, 1917/19; Palatini 1918, 1919/20, 1921a, 1921b, 1923a, 1923b).

In each of these cases, Levi-Civita and Palatini challenged such authorities as Einstein, David Hilbert, Hermann Weyl, and Hendrik A. Lorentz, and proposed original points of view that came to be internationally recognized.

Palatini is best remembered for his variational method (Palatini 1919), but the main focus of his research was, in fact, on finding solutions of the gravitational field equations. This part of his work is quite unknown. Palatini was introduced to general relativity by Levi-Civita, his former teacher, who, for at least ten years (from 1913 until 1923), had a strong influence on his scientific work. Their correspondence[1] clearly shows how Levi-Civita stimulated Palatini's interest in general relativity, making him aware of the progress in the field, how he suggested new problems to work on, and how he acted as an adviser for Palatini's work. In 1923, after a bright initial career, Palatini's star slowly started to fade. No longer in direct contact with Levi-Civita, he was unable to produce any interesting new results in general relativity. In this chapter, we will trace how Palatini entered the European debate on variational formulations of general relativity and how he subsequently contributed, with Levi-Civita, to the study of static solutions of the gravitational field equations.

2. Early European Debates on Variational Principles in General Relativity, 1913–1916

Before discussing Palatini's 1919 variational method, I briefly want to discuss the earlier uses (and misuses) of variational principles in general relativity (see also my chapter with De Maria in this volume).

In Einstein and Grossmann 1914 and Einstein 1914, Einstein had already used a variational principle to derive the field equations of the so-called *Entwurf* theory (Einstein and Grossmann 1913). This theory was marred both by the restricted covariance of its gravitational field equations and by some mathematical mistakes in one of its crucial proofs.[2] In their variational derivation of the field equations, Einstein and Grossmann made use of a noninvariant Lagrangian density H, thus obtaining the equations of limited covariance of the *Entwurf* theory. In his October 1914 article, Einstein tried to make the covariance properties of the *Entwurf* equations more explicit, using a generalized version of his previous variational approach. Although he believed he had found a more satisfactory derivation of the gravitational field equations in this way, namely, a derivation "in a purely covariant theoretical form" (Einstein 1914, p. 1030), he was forced, one year later, to abandon the *Entwurf* theory. It has been shown that Levi-Civita played an important role in convincing Einstein of the mathematical defectiveness of his *Entwurf* theory (Cattani and De Maria 1989b), leading him to redirect his steps toward general covariance.

Levi-Civita showed that Einstein, at a crucial point in his 1914 reasoning, confused invariance and covariance. Looking at the transformation properties of both the action integral and the Lagrangian function, Einstein found that the invariance of these quantities and the *covariance* of the gravitational field equations were limited to the so-called "justified" transformations, i.e., transformations satisfying the condition that a certain vector $\mathcal{B}_\mu$, defined as

$$\mathcal{B}_\mu = \frac{\partial^2}{\partial x^\sigma x^\alpha}\left(g^{\nu\alpha}\frac{\partial H\sqrt{-g}}{\partial g_\sigma^{\mu\nu}}\right),$$

vanishes (Einstein 1914, p. 1070; Einstein and Grossmann 1914, p. 218). Contrary to what Einstein thought, however, it does not follow that the field equations are *invariant* under these justified transformations. With four conditions $\mathcal{B}_\mu = 0$ on the coordinates, one cannot have both *independence* and *covariance* of the field equations. In the course of his correspondence with Einstein between March and May of 1915,[3] Levi-Civita discovered this gap in the argument. Einstein was eventually forced to accept the validity of Levi-Civita's objections, and, in particular, to admit that his proof was incomplete.[4]

In the same year, Lorentz published a paper criticizing Einstein's variational formulation of the field equations, and proposing "a more correct approach" (Lorentz 1915, p. 1073). He derived the gravitational field equations without specifying the form of the Lagrangian H, simply assuming H to be a function only of the metric tensor and its first order derivatives. Hence, H is still not invariant under general transformations. It should be stressed that Lorentz was perfectly aware of the limited covariance of the field equations he obtained in this way (which are just the *Entwurf* equations). He accepted this result without question, however, and considered Einstein's proof of it (which Levi-Civita had shown to be wanting) as an example of his "ingenious mode of reasoning" (Lorentz 1915, p. 1089).

Only in November 1915 did Einstein write a sequence of papers (Einstein 1915a, 1915b, 1915c) in which he finally succeeded in obtaining the correct generally covariant field equations. He did not use a variational method, however, and, in order to obtain conservation of energy, he was forced to adopt the hypothesis that all matter is electromagnetic and the coordinate condition $\sqrt{-g} = 1$.

In the meantime, on November 20, 1915, Hilbert had presented the first of a two-part paper (Hilbert 1915), in which he tried to unify gravitational theory with the electromagnetic theory of matter advanced by Gustav Mie (Mie 1912). In this paper, Hilbert showed that Einstein's gravitational field equations can easily be obtained from a variational principle, at least in the presence of an electromagnetic field (Hilbert 1915). Hilbert assumed that the quantities characterizing the field are the ten gravitational potentials $g_{\mu\nu}$ and the four electrodynamic potentials q_{μ}, which are the fundamental variables of a generally invariant "world" function H. In his opinion,

> Einstein gave the fundamental original idea of general invariance a simple expression. However, for Einstein, the Hamilton principle (only) plays a subordinate role, and his function H is not at all generally invariant. Moreover, the electrical potentials are not included [in the theory]. (second footnote to Hilbert 1915, (I), p. 396)

Splitting the Lagrangian into two parts—the curvature scalar R for the gravitational field and a Lagrangian M for the electrodynamic field—and evaluating the corresponding functional derivatives, Hilbert obtained Einstein's gravitational field equations as well as the equations of motion of electromagnetic matter in a curved space-time.

In 1916, Lorentz published a long paper in four parts (Lorentz 1916), in which he used a variational principle to obtain the gravitational field equations, the "stress-energy complex," and the conservation laws for the gravitational field. Contrary to his previous article, Lorentz now identified the Lagrangian density as the curvature scalar R. By coupling R with several kinds of matter variables, Lorentz obtained the correct gravitational field equations. He made some mathematically unwarranted assumptions, however, such as the assumption that the infinitesimal variations of the metric tensor have tensor character. Moreover, the generality of his proof was limited by a special choice of coordinates.

In November 1916, Einstein once again used a variational method in general relativity. He claimed that his approach was more general than and "especially in contrast" with Hilbert's, since he rejected Mie's hypothesis about the fully electromagnetic nature of matter (Einstein 1916b, p. 1111). He conceded, however, that Lorentz and Hilbert had given general relativity an especially transparent form by deriving its equations from a single variational principle. Contrary to his earlier claims in the context of the *Entwurf* theory, Einstein now admitted that, in order to satisfy the principle of general relativity, the Lagrangian for the gravitational field "must be the linear invariant of the Riemann curvature tensor since there is no other invariant that has the required properties" (Einstein 1916b, p. 1113). In line with his previous variational approach (Einstein and Grossmann 1914; Einstein 1914), Einstein showed (Einstein 1916b) that the correct gravitational field equations could be obtained from a noninvariant Lagrangian,[5] and that by using a suitable infinitesimal coordinate transformation, the condition $\mathcal{B}_\mu = 0$ could still be satisfied.

Einstein's persisting uncertainties about coordinate conditions and general covariance led him, in 1917, to start a new polemic with Levi-Civita. Einstein's misconceptions (Einstein 1916b) about the covariance of the stress-energy tensor for the gravitational field along with some incorrect conclusions about gravitational waves (Einstein 1916a, 1918),[6] motivated Levi-Civita to study the correct analytical form of the gravitational stress-energy tensor (Levi-Civita 1917a). In Levi-Civita's opinion,

> the idea of a gravitational [stress-energy] tensor belongs to the majestic construction of Einstein. However, the definition proposed by the author is unsatisfactory. First of all, from the mathematical point of view, it lacks the invariant character it should have in the spirit of general relativity. (Levi-Civita 1917a, p. 381)

In his paper, Levi-Civita showed, among other things, that some differential relations holding for the gravitational field equations plus the conservation law for the stress-energy tensor for matter were identically fulfilled by the metric tensor (the so-called contracted Bianchi identities, still ignored by Einstein at the time). Since these identities are generally covariant, Levi-Civita pointed out, Einstein was mistaken in thinking that energy-momentum conservation calls for a restriction on the choice of coordinates (Levi-Civita 1917a).

3. Palatini's 1919 Variational Principle

In 1913, Attilio Palatini received his Bachelor's degree in mathematics from the University of Padova. His thesis, written under supervision of Levi-Civita, dealt with fluid mechanics. In the following years, under the scientific influence of Levi-Civita, Palatini wrote his more important articles on general relativity. His first article on relativity, in 1917, dealt with the perihelion motion of Mercury and the bending of light beams in gravitational fields (Palatini 1917). One year later, he investigated dynamical paths for a stationary metric (Palatini 1918). In 1919, encouraged by Levi-Civita, Palatini proposed a new approach to the variational formulation of gravitational field equations (Palatini 1919). In this brilliant article, he rebutted the early variational approaches of Einstein and Grossmann (Einstein and Grossmann 1914; Einstein 1914) and improved on the variational methods developed since (Einstein 1916b; Hilbert 1915; Lorentz 1916). In Palatini's words:

> Since Einstein's discovery of the gravitational equations, many efforts were made to derive them from a variational principle just as one derives the equations of Lagrange from Hamilton's principle in classical mechanics. This goal was accomplished by Einstein himself establishing, a new version of Hamilton's principle, though subsequently, more precise adjustments were made by Hilbert and Weyl. However, these authors do not conform to the spirit of the absolute differential calculus, because they obtained invariant equations (with respect to changes of the variables) via some formulæ lacking in such invariance. (Palatini 1919, pp. 203–204)

Like Hilbert, Palatini began with a first axiom:

> Any physical law is dependent on a sole universal function H. Such a function is invariant with respect to any coordinate transformation; it depends on the gravitational potentials $g^{\mu\nu}$, on the corresponding Christoffel symbols, and on those parameters $[q_\mu]$ characterizing any physical event as well. (Palatini 1919, p. 204)

Following Hilbert again, Palatini split H into two parts: the curvature scalar R and a function M representing matter, i.e., everything not intrinsically included in the space-time structure. Palatini felt that,

> at least from a speculative point of view, it seems desirable to attribute to all these manifestations (directly or not) the same electromagnetic origin. The explicit form of the Lagrangian cannot be given as a function of the parameters $[q_\mu]$, due to the complicated form of this dependence, therefore, some restrictions on the representation of the Lagrangian [for matter] M are needed. (Palatini 1919, p. 204)

Rather than making specific assumptions about the nature of matter, Palatini just assumed that matter can always be represented by an energy-momentum tensor $T_{\mu\nu}$, which macroscopically describes the stress-energy components for matter. Once the components $T_{\mu\nu}$ are given, M can be expressed as

$$M = \chi T_{\mu\nu} g^{\mu\nu},$$

where $T_{\mu\nu}$ are independent of the variations $\delta g^{\mu\nu}$ of the metric tensor,[7] i.e.,

$$\delta(\sqrt{-g}\, T_{\mu\nu}) = 0.$$

The next step was to evaluate the infinitesimal variation of the connection, as well as of the curvature tensor, with respect to arbitrary variations $\delta g^{\mu\nu}$ of the metric tensor that vanish on the boundary of the space-time integration domain.[8] He considered both the metric tensor and the affine (symmetric) connection as independent variables, and rederived Einstein's field equations together with a set of conditions on the affine connections that reduce them to the Christoffel symbols.

The main results of Palatini's variational method were the following (see Ferraris et al. 1982; see also Wald 1984, pp. 450–459). First, he preserved the tensor character of all equations at each step of his derivation. Second, he showed, for the first time, that the variations of the Christoffel symbols are the components of a tensor. Moreover, his method of varying the Riemann curvature tensor was independent of any particular choice of a symmetric affine connection. Finally, he was the first to give an example of a metric/affine variational principle though he was working in the framework of a theory traditionally phrased in terms of a metric variational principle. This important result, based on his choice of the connection as an independent variable, opened up a new and important field of research that would be fully exploited later in Einstein's unified field theories (Einstein 1925; see Ferraris et al. 1982).

4. Levi-Civita and Palatini's Contribution to Einstein's Statics, 1917–1923

The inspiration that Palatini drew from Levi-Civita in his work on variational principles is just an example of the fundamental role Levi-Civita played, during the years 1917–1919, in organizing an enthusiastic and ambitious research program in general relativity in Italy. This program was aimed at (a) strengthening the physical basis of general relativity and raising the standards of mathematical rigor in order to make the theory acceptable to a hostile scientific environment and (b) minimizing the gap between classical and relativistic mechanics. In particular, Levi-Civita wanted to show that relativity formed a natural progression from classical mechanics.

During this period, Levi-Civita therefore worked hard on rederiving classical results from a first order static approximation to Einstein's equations. He started in 1917 (Levi-Civita 1917b), studying Einstein's gravitational field equations for static[9] phenomena and the motion of a particle in a static field, with a view to linking Einstein's theory to Newton's. Starting from a given static distribution of matter[10] and neglecting the influence of a test particle on the gravitational field, Levi-Civita obtained the relative gravitational field equations and the relative motion of a test particle in a weak static field.[11] In this way, he was able to show how, in a first approximation, relativistic particle motion coincides with particle motion in ordinary mechanics (Levi-Civita 1917b).

In a long article, in nine parts, presented to the Accademia dei Lincei between December 1917 and January 1919 (Levi-Civita 1917/19), Levi-Civita continued his study of first order solutions to Einstein's field equations. He proposed an intrinsic approach in terms of independent congruences,[12] using a conformal transformation of the static metric, and also considered some conformal metrics as well as the so-called Weyl metric (Weyl 1917, 1918).[13] Levi-Civita was able to evaluate the geometrical structure of the space-time for the case of a cylindrical matter distribution. In particular, he showed that the longitudinal solutions[14] include the Einstein–Schwarzschild solutions (which are special cases of the Weyl metric). These solutions prompted Levi-Civita to study a particular field generated by a symmetric source, that is, a homogeneous cylinder. He also elaborated on the result of his paper (Levi-Civita 1917b, p. 464) that, in an empty static space-time, Einstein's equations reduce to seven equations which allow for considerable further simplification (Levi-Civita 1926, p. 381; Palatini 1921a, p. 464).[15] The resulting equations, as Levi-Civita put it, "reduce the Einsteinian statics to the three dimensions of the associated space. Their form is invariant with respect to the metric of this space" (Levi-Civita 1926, p. 381). Moreover, Levi-Civita showed that the square root of the time–time component of the metric is a harmonic function. The next step was to integrate a similar system of differential equations for a region of space in which the components of the stress-energy tensor do not all vanish. In doing so, one must assume that the metric differs by "very little" from the Euclidean type. Levi-Civita evaluated both the scalar potential (i.e., the time–time component of the metric) and the spatial metric to first order (Levi-Civita 1917b, 1926, pp. 383–392). Palatini went up to second order (Palatini 1921a). Levi-Civita and Palatini showed that the solutions depend on the arbitrary choice of a harmonic function and, in second order, on ordinary potentials that are a function of the density.[16]

Meanwhile, Palatini was slowly losing contact with Levi-Civita. Levi-Civita had moved to Rome in 1918, leaving Palatini behind in Padova. In 1920, on Levi-Civita's recommendation, Palatini was offered the chair of mechanics in Messina. Palatini at first was pleased with the offer, as can be gathered from his reply to the letter in which Levi-Civita informed him of the "unexpectedly"[17] good news. Palatini's letter, dated April 17, 1920, also bears testimony of the great esteem in which he held the addressee:

> My very highly honored Teacher.... Let me express my respectful affection[18] and remind you of my gratitude for your constant and encouraging help, as well as for your manifold and continuous advice while I was studying. Your trust has given me much support, especially recently, over the last few years, when I had to get over many crises. I

> will forever be very grateful [to you]. (Palatini to Levi-Civita, April 17, 1920, p. 1)

In 1921, Palatini actually moved to Messina. It did not take long before he started to feel very isolated. As he wrote to Levi-Civita, "I cannot move from here, because the journey is too uncomfortable.... I am very sorry because I could have had the pleasure of some talks with you" (Palatini to Levi-Civita, March 28, 1921, p. 4).

Despite this intellectual isolation, Palatini continued to study the solutions of Einstein's equations, without concealing his difficulties to his "teacher" Levi-Civita:

> Two years have elapsed since you suggested to me to solve the problem of cylindrical potentials within the Einstein theory. Again and again, I have strained myself to come up with the solution. As you know, I have studied this argument in second order approximation. Taking into account these results, I have recently reached some conclusions that I would like to submit to you for your judgement. But I must confess that I look upon my conclusions with distrust, because they are extremely simple. Although they might be easily criticized for their extreme simplicity, I cannot find anything wrong with them myself. If, however, it turns out that my considerations are right, their simplicity is not only a consequence of the fact that what I was seeking was concealed for a long time, but it also explains why all my previous attempts stubbornly led me to the ds^2 of Weyl. (Palatini to Levi-Civita, March 28, 1921, pp. 4–5)

Just one month before this letter, Palatini had delivered a paper (Palatini 1921b) dealing with the second order solutions for the cylindrical potentials he and Levi-Civita had been looking at. Palatini showed that in a static weak field with the harmonic function independent of one of the spatial coordinates, the spatial potentials depend only upon the remaining spatial coordinates. Furthermore, to second order, the approximated axial symmetric potentials derived by Palatini coincide with the exact solutions for the static spherical and symmetric potentials previously obtained by Levi-Civita (Levi-Civita 1917/19, VIII). These results encouraged Palatini to try to work toward the exact solution of the cylindrical potentials. In fact, the correspondence up to second order between symmetric potentials[19] and cylindrical potentials should, in Palatini's opinion, be valid in general, and the required metric[20] should depend only on a pair of harmonic functions of two spatial coordinates. In the letter to Levi-Civita quoted above, Palatini suggested a simple form for these two functions (Palatini to Levi-Civita, March 28, 1921, p. 5). He later realized that they could not have this simple form (see Palatini 1923a, p. 266).

In a letter to Levi-Civita of January 22, 1922, and in subsequent papers (Palatini 1923a, 1923b), Palatini returned to the problem. To Levi-Civita, he wrote:

> I went back to the gravitational equations again: I may be wrong to be so one-sided, but I am driven first by the desire to reap some benefits of my long study of Einstein's equations, and second by the fact that I do not find the necessary feedback here [in Messina]. In particular, as I said to you, I don't have access to [scientific] publications, so that I never know what is being done and said in the mathematical field and, in particular, in mechanics. (Palatini to Levi-Civita, January 22, 1922, p. 1)

Once again, we see Palatini's dissatisfaction with the scientific and cultural life in Messina cropping up in this passage. In the same letter, Palatini writes:

> I live with the wish to leave Messina, for the reasons you know; besides, I am disgusted by the low and petty university environment.... I heard a rumor about new transfers: hopefully, there is something for me in it as well. (Palatini to Levi-Civita, January 22, 1922, p. 1)

I do not want to draw general conclusions about the consequences of the environment on Palatini's scientific work, but it is clear that he missed the strong direct influence of Levi-Civita.

Still, this same letter contains Palatini's evaluation of a symmetric potential belonging to the class of Schwarzschild solutions. Without any substantial changes, these considerations appeared in the article published in 1923 (Palatini 1923a), in which Palatini addressed the problem of longitudinal solutions within the Weyl solutions. As he reported to Levi-Civita, Palatini wanted to show that "in first approximation, from a relativistic point of view, matter distributed over a round ellipsoid behaves like matter distributed inside a sphere of some suitably chosen radius" (Palatini to Levi-Civita, January 22, 1922, p. 5), and, more generally, that solutions for a symmetric distribution of matter over a round ellipsoid are Weyl solutions. Hence, the longitudinal solution (of Schwarzschild) is a particular case of the Weyl solutions. Moreover, since the Weyl solutions correspond to symmetric solutions, Palatini was able to give an explicit expression for the potential[21] in the symmetric solutions of the Schwarzschild type. Further attempts by Palatini to demonstrate the correspondence between longitudinal and symmetric solutions in general can be found in follow-up letters to Levi-Civita,[22] who apparently was not satisfied by them. In the process, Palatini obtained the nonphysical case of matter distributed at infinity and sought Levi-Civita's advice. In this letter, Palatini is lamenting over his situation in Messina again: "I cannot continue to go deeply into this question. I implore you to give me some suggestions. Forgive me if I continue to disturb you. You can surely imagine how painful this loneliness is for me" (Palatini to Levi-Civita, March 28, 1922, p. 6).

In a letter from April 1922, Palatini asked Levi-Civita to choose between two titles he had come up with for his papers on the subject (Palatini 1923a, 1923b).[23] This is the last reference to the problem in the existing correspondence between the two men. Palatini's desire to move away from Messina is mentioned a few more times in the course of 1922. In April, Palatini wrote to Levi-Civita, "I have always kept you informed of my steps to go back to . . . the civilized world" (Palatini to Levi-Civita, April 28, 1922, p. 3), and, in August, he wrote, "I trust in your help and I still believe in a favorable solution, although, due to the circumstances, my hopes have been greatly reduced" (Palatini to Levi-Civita, August 2, 1922, p. 3).

Palatini would soon leave Messina to go to Parma, where he lived from 1922 to 1924, and then to Pavia, where he stayed the rest of his life (he died in 1949). By the time he left Messina, Palatini had exhausted most of his relativistic inspiration, certainly due in large measure to the lack of direct influence from Levi-Civita. With the exception of one paper in 1929 (Palatini 1929) and a review article in 1947 (Palatini 1947), Palatini did no further work in the field of relativity. After 1923, six years elapsed before Palatini published another article on general relativity. Despite this long public silence on the subject, Palatini in 1929 still boasted the formal rigor learned at "Levi-Civita's school," as the following comment on a paper by Einstein shows:

> With all due respect, Einstein is a muddler. Mixing up covariant components with invariant components has given rise to formal complications, which prevent both Einstein and Weitzenböch from reaching definitive results. (Palatini to Levi-Civita, March 23, 1929, p. 1)

Then Palatini became so involved in other problems (mainly of a private nature), that he was no longer able to meet the challenge of relativistic questions. As a result of the growing interest in the theory, these questions multiplied and required quick answers, while Palatini by his own admission "could not work any faster" (Palatini to Levi-Civita, May 6, 1927, p. 2). His 1929 article on Einstein's unified field theories would remain his last contribution to the field.

5. Conclusion

In this chapter, a very short segment of the Italian history of general relativity is reconstructed, singling out the role played by Levi-Civita in Palatini's contributions during the years 1919–1923. Levi-Civita's two main concerns were with variational formulations of the gravitational field equations and with approximate solutions of these equations in some special cases. On the latter problem, Levi-Civita and Palatini worked intensively for a period of nearly five years. The fruits of their labor were discussed in Section 4. The former problem led Levi-Civita into a short but sharp polemic with Einstein in 1915, which helped to convince Einstein that his *Entwurf* theory was untenable (see Cattani and De Maria 1989b); it led Palatini to a new fully invariant variational method. Before Palatini, the variational method for gravitational field equations was studied by Lorentz, whose 1915 paper on the subject was marred by its unspecified Lagrangian; by Einstein, facing problems with coordinate conditions and a noncovariant Lagrangian (Einstein 1916b); and by Hilbert, adopting rather restrictive hypotheses on the nature of matter (Hilbert 1915). Nonetheless, by 1916 their efforts had resulted in a self-consistent, covariant, and by and large satisfactory formulation of a variational principle for the gravitational field equations (Lorentz 1916, Einstein 1916b). Their approaches, however, were not considered satisfactory by Levi-Civita and Palatini, so it happened that Palatini, a relatively unknown Italian mathematician, was the first to show, in 1919, that the variations of the Christoffel symbols were covariant, thereby securing, at last, formal invariance at each and every step of the variational method. He also showed that a more general variational principle could be formulated, adopting as fundamental variables not just the components of the metric tensor but the components of the affine connection as well. This opened up an entirely new field of research. Palatini's contribution was immediately acknowledged by Weyl[24] and Pauli (Pauli 1921, p. 621, footnote 5). Einstein also gave credit to Palatini by giving the new variational principle his name (Einstein 1950, appendix II, p. 141).

ACKNOWLEDGMENTS. I am very grateful to M. De Maria for his critical reading of the manuscript, and to M. Janssen for many helpful suggestions.

NOTES

[1] A. Palatini to T. Levi-Civita, Levi-Civita Papers, Accademia dei Lincei, Rome.

[2] In their 1913 article, Einstein and Grossmann tried to argue that their gravitational theory cannot be generally covariant. For a critical discussion of these arguments, see, e.g., Cattani and De Maria 1989b and Norton 1984.

[3] A. Einstein to T. Levi-Civita Correspondence, Levi-Civita Papers, Family collection, Rome.

[4] Einstein to Levi-Civita, May 5, 1915. For a more detailed analysis, see Cattani and De Maria 1989b.

[5] The form of the noninvariant Lagrangian, H^*, obtained after boundary is, in terms of the Christoffel symbols Γ,

$$H^* = \sqrt{-g}\, g^{\mu\nu}(\Gamma_{\mu\alpha}{}^{\beta}\Gamma_{\nu\beta}{}^{\alpha} - \Gamma_{\mu\nu}{}^{\alpha}\Gamma_{\alpha\beta}{}^{\beta}).$$

[6] Cattani and De Maria 1989a. See also my chapter with De Maria in this volume.

[7] In the following years, Palatini tried to eliminate this restriction on matter, but in spite of many efforts he eventually had to recognize his failure in a letter to the Italian mathematician Roberto Marcolongo: "Once again I have tried to get rid of this condition even by using some of my drafts of the time I studied the derivation of the gravitational equations. I did not succeed and I do not pursue my attempt any longer, because I am convinced that is impossible to avoid it." (A. Palatini to R. Marcolongo, May 10, 1922, Marcolongo Papers, Department of Mathematics, University "La Sapienza," Rome.)

[8] By applying the variational principle to the integral

$$\int_\Sigma H\sqrt{-g}\,\mathrm{d}\tau = \int_\Sigma (R + \chi T_{\mu\nu}g^{\mu\nu})\sqrt{-g}\,\mathrm{d}\tau,$$

it follows that

$$\int_\Sigma \delta R\sqrt{-g} + R\delta\sqrt{-g} + \chi\delta\big(T_{\mu\nu}\sqrt{-g}\big)g^{\mu\nu} + \chi\big(T_{\mu\nu}\sqrt{-g}\big)\delta g^{\mu\nu}\,\mathrm{d}\tau = 0,$$

so that he obtains

$$\delta\int_\Sigma H\sqrt{-g}\,\mathrm{d}\tau = \int_\Sigma \big(R_{\mu\nu} - \tfrac{1}{2}Rg_{\mu\nu} + \chi T_{\mu\nu}\big)\delta g^{\mu\nu}\,\mathrm{d}\tau + \int_{\partial\Sigma}\sqrt{-g}\,f^\sigma\,\mathrm{d}\omega_\sigma = 0.$$

Since the last integral vanishes on the boundary and the $\delta g^{\mu\nu}$ are arbitrarily chosen, the gravitational field equations easily follow.

[9] A metric field that is time-independent (i.e., $\partial_0 g_{ik} = 0$, $(i, k = 0, \ldots, 3)$) is called *stationary*. A stationary metric field with vanishing time-space components (i.e., $g_{0k} = 0$) is called *static*.

[10] So that the four-dimensional line element reduces to

$$\mathrm{d}s^2 = c^2\,\mathrm{d}t^2 - \mathrm{d}l^2, \qquad \mathrm{d}l^2 = \gamma_{ik}\,\mathrm{d}x^i\,\mathrm{d}x^k,$$

where the coeffient c^2 and the coefficients of the spatial line element γ_{ik} are functions of the space coordinates only.

[11] Because of the time-independence of the three-dimensional metric, Levi-Civita was able to split the four-dimensional space-time into a three-plus-one structure: the relative three-dimensional space plus time. The ten gravitational equations reduce to only seven equations. In fact, the equations corresponding to mixed terms 01, 02, 03 vanish because in the static case there are no fluxes of energy.

[12] i.e., referring to four independent vector fields.

[13]

$$\mathrm{d}s^2 = c^2\,\mathrm{d}t^2 - e^{-2\nu}[e^{2\lambda}(\mathrm{d}r^2 + \mathrm{d}z^2) + r^2\,\mathrm{d}\mathring{x}_3^2],$$

where $\nu = \nu(r, z)$ is a symmetric potential, and λ is known when ν is given.

[14] where the metric tensor depends only on two coordinates, and the spatial metric is orthogonal with respect the third coordinate.

15

$$\widetilde{R_{ik}} + \frac{\nabla_i \nabla_k U}{U} = 0, \quad {}^{(3)}R = 0, \quad \sim \Delta_2 U = 0 \quad (i, k = 1, 2, 3).$$

16

$$U^2 = c^2(1 + 2\Gamma + 2\Gamma^2), \quad \gamma_{ik} = \delta_{ik}(1 - 2\Gamma + 2\Gamma^2 - \eta^*) + \eta^*_{ik},$$

where $\eta^* \stackrel{\text{def}}{=} \gamma^{ik}\eta^*_{ik}$ and the arbitrary harmonic function Γ with the coefficients η^*_{hk} fulfill a set of conditions, as shown by Palatini in 1921a, p. 476.

17 Palatini to Levi-Civita, April 17, 1920, p. 2.

18 Palatini's sincere gratitude and respect for Levi-Civita are expressed more or less explicitly in all subsequent letters. Palatini was always very thankful for Levi-Civita's help; he once started a letter with: "Many thanks for your suggestions which I will use for further study of the question I am working on" (Palatini to Levi-Civita, May 6, 1927, p. 1). In a subsequent letter to Levi-Civita, he wrote "I have written two papers that I am submitting to you, I want you to decide if they should be published or not in Lincei. . . . As your respectful pupil I am always ready to accept your verdict" (Palatini to Levi-Civita, May 28, 1928, pp. 1–2).

19 of the same kind as Weyl 1917.

20 That Palatini wrote in the form (Palatini to Levi-Civita, March 28, 1921, p. 5)

$$d\sigma^2 = e^{-2\nu}\left[e^{-2\lambda}(d\mu^2 + d\nu^2) + A^2\, dx^{3^2}\right].$$

21 Palatini to Levi-Civita, January 22, 1922, p. 2, Palatini 1923a, p. 265:

$$\nu = \tfrac{1}{2}\log\frac{(C^2+\tau)^{1/2} - \alpha/2}{(C^2+\tau)^{1/2} - \alpha/2}$$

C being a constant and τ the parameter of a family of round ellipsoids.

22 As he wrote to Levi-Civita, "I intend to investigate whether the longitudinal solutions are included in the binary Weyl solutions" (Palatini to Levi-Civita, March 28, 1922, p.2).

23 "Which title do you think more suitable? *Sopra i potenziali simmetrici che conducono alle soluzioni longitudinali delle equazioni gravitazionali* [or] *Legame tra le soluzioni longitudinali delle equazioni gravitazionali e le soluzioni binarie di Weyl-Levi civita?*" (Palatini to Levi-Civita, April 28, 1922, p. 9).

24 See Weyl 1921, p. 216, with the corresponding note 6 on p. 292. (See also Weyl 1952, p. 238, and the corresponding note 6 on p. 322.)

References

Cattani, Carlo, and De Maria, Michelangelo (1989a). "Gravitational Waves and Conservation Laws in General Relativity: A. Einstein and T. Levi-Civita, 1917 Correspondence." In *Proceedings of the Fifth M. Grossmann Meeting on General Relativity*, D.G. Blair and M.J. Buckingham, eds. Singapore: World Scientific, pp. 1335–1342.

——— (1989b). "The 1915 Epistolary Controversy between A. Einstein and T. Levi-Civita." In *Einstein and the History of General Relativity*, D. Howard and J. Stachel, eds. Boston: Birkhäuser, pp. 175–200.

Einstein, Albert (1914). "Die formale Grundlage der allgemeinen Relativitätstheorie." *Königlich Preussischen Akademie der Wissenschaften (Berlin). Sitzungsberichte*: 1030–1085.

—— (1915a). "Zur allgemeinen Relativitätstheorie." *Königlich Preussischen Akademie der Wissenschaften* (Berlin). *Sitzungsberichte*: (I) November 4, 778–786; (II) November 11, 799–801.

—— (1915b). "Erklärung der Perihelbewegung des Merkur aus der allgemeinen Relativitätstheorie." *Königlich Preussischen Akademie der Wissenschaften* (Berlin). *Sitzungsberichte*: November 18, 831–839.

—— (1915c). "Feldgleichungen der Gravitation." *Königlich Preussischen Akademie der Wissenschaften* (Berlin). *Sitzungsberichte*: November 25, 844–847.

—— (1916a). "Näherungsweise Integration der Feldgleichungen der Gravitation." *Königlich Preussischen Akademie der Wissenschaften* (Berlin). *Sitzungsberichte*: 688–696.

—— (1916b). "Hamiltonsches Prinzip und allgemeine Relativitästheorie." *Königlich Preussischen Akademie der Wissenschaften* (Berlin). *Sitzungsberichte*: 1111–1116.

—— (1918). "Über Gravitationswellen." *Königlich Preussischen Akademie der Wissenschaften* (Berlin). *Sitzungsberichte*: 154–167.

—— (1925). "Einheitliche Feldtheorie von Gravitation und Elektrizität." *Königlich Preussischen Akademie der Wissenschaften* (Berlin). *Sitzungsberichte*: 414–421.

—— (1950). *The Meaning of Relativity.* 3rd ed. Princeton: Princeton University Press.

Einstein, Albert and Grossmann, Marcel (1913). *Entwurf einer verallgemeinerten Relativitätstheorie und einer Theorie der Gravitation. I. Physikalischer Teil von Albert Einstein. II. Mathematischer Teil von Marcel Grossmann.* Leipzig and Berlin: B.G. Teubner. Reprinted with added "Bemerkungen," in *Zeitschrift für Mathematik und Physik* 62 (1914): 225–261.

—— (1914). "Kovarianzeigenschaften der Feldgleichungen der auf die verallgemeinerte Relativitätstheorie gegründeten Gravitationstheorie." *Zeitschrift für Mathematik und Physik* 63: 215–225.

Ferraris, Marco, Francaviglia, Mauro, and Reina, Cesare (1982). "Variational Formulation of General Relativity from 1915 to 1925 'Palatini's Method' Discovered by Einstein in 1925." *General Relativity and Gravitation*, Vol. 14, No. 3: 243–254.

Hilbert, David (1915). "Die Grundlagen der Physik." *Königlichen Gesellschaft der Wissenschaften zu Göttingen, Nachrichten:* (I) (1915), 395–407; (II) (1916), 53–76.

Levi-Civita, Tullio (1917a). "Sulla espressione analitica spettante al tensore gravitazionale nella teoria di Einstein." *Rendiconti Accademia dei Lincei* Ser. 5, Vol. XXVI: 381–391.

—— (1917b). "Statica einsteiniana." *Rendiconti Accademia dei Lincei* Ser. 5, Vol. XXVI: 458–470.

——— (1917/19). "ds^2 einsteiniani in campi newtoniani." *Rendiconti Accademia dei Lincei* $[\bullet]_1$ Nota I, vol. XXVI, (1917): 307–317; $[\bullet]_2$ Nota II, vol. XXVII, (1918): 3–12; $[\bullet]_3$ Nota III, vol. XXVII, (1918): 183–191; $[\bullet]_4$ Nota IV, vol. XXVII, (1918): 220–229; $[\bullet]_5$ Nota V, vol. XXVII, (1918): 240–248; $[\bullet]_6$ Nota VI, vol. XXVII, (1918): 283–292; $[\bullet]_7$ Nota VII, vol. XXVII, (1918): 343–351; $[\bullet]_8$ Nota VIII, vol. XXVIII, (1919): 3–13; $[\bullet]_2$ Nota IX, vol. XXVIII, (1919): 101–109.

——— (1926). *The Absolute Differential Calculus*. London and Glasgow: Blackie & Son Limited. Republished in 1977, New York: Dover.

Lorentz, Hendrik Antoon (1915). "Het beginsel van Hamilton in Einstein's theorie der Zwaartekracht." *Koninklijke Akademie van Wetenschappen te Amsterdam. Verslagen van de Gewone Vergaderingen der Wis- en Natuurkundige Afdeeling* 23: 1073–1089. English translation: "On Hamilton's Principle in Einstein's Theory of Gravitation." *Koninklijke Akademie van Wetenschappen te Amsterdam. Proceedings of the Section of Sciences* 19: 751–767.

——— (1916). "Over Einstein's theorie der zwaartekracht." *Koninklijke Akademie van Wetenschappen te Amsterdam. Verslagen van de Gewone Vergaderingen der Wis- en Natuurkundige Afdeeling* (I) 24, (1916): 1389–1402; (II) 24, (1916): 1759–1774; (III) 25, (1916): 468–486; (IV) 25, (1917): 1380–1396. English translation: "On Einstein's Theory of Gravitation." *Koninklijke Akademie van Wetenschappen te Amsterdam. Proceedings of the Section of Sciences* (I) 19, (1916): 1341–1354; (II)19, (1916): 1354–1369; (III) 20, (1916): 2–19; (IV) 20, (1917): 20–34.

Mehra, Jagdish (1974). *Einstein, Hilbert and the Theory of Gravitation*, Dordrecht: D. Reidel.

Mie, Gustav (1912). "Grundlagen einer Theorie der Materie." *Annalen der Physik* (I) 37, (1912): 511–534; (II) 39, (1912): 1–40; (III) 40, (1913): 1–66.

Norton, John (1984). "How Einstein Found His Field Equations: 1912–15." *Historical Studies in the Physical Sciences*, 14: 253–316; see also in *Einstein and the History of General Relativity*, D. Howard and J. Stachel, eds. Boston: Birkhäuser (1989), pp. 101–160.

Palatini, Attilio (1917). "Lo spostamento del perielio di Mercurio e la deviazione dei raggi luminosi secondo la teoria di Einstein." *Nuovo Cimento* (6) 14: 12–54.

——— (1918). "Moti einsteiniani stazionari." *Atti del Reale Istituto Veneto* 78: 589–606.

——— (1919). "Deduzione invariantiva delle equazioni gravitazionali dal principio di Hamilton." *Rendiconti del Circolo Matematico di Palermo* 43: 203–212; reprinted in an English translation by Roberto Hojman and Chandrasekher Mukku in *Cosmology and Gravitation*, Peter G. Bergmann and Venzo De Sabbata, eds. New York: Plenum Press, 1980, pp. 477–488. Quotations from this paper are given in Hojman and Mukku's translation.

——— (1919/20). "ds^2 einsteiniani in rappresentazione conforme con lo spazio euclideo." *Atti del Reale Istituto Veneto* 79: 321–325.

—— (1921a). "Sulle equazioni della statica einsteiniana in seconda approssimazione." *Rendiconti dell'Istituto Lombardo* (2) 54: 463–476.

—— (1921b). "L' analogo einsteiniano dei potenziali cilindrici in seconda approssimazione." *Rendiconti dell'Istituto Lombardo* (2) 54: 570–576.

—— (1923a). "Sopra i potenziali simmetrici che conducono alle soluzioni longitudinali delle equazioni gravitazionali di Einstein." *Rendiconti Accademia dei Lincei* (5) 32: 263–267.

—— (1923b). "Sopra i potenziali simmetrici che conducono alle soluzioni longitudinali delle equazioni gravitazionali di Einstein." *Nuovo Cimento* (7) 26.

—— (1929). "Intorno alla nuova teoria di Einstein." *Rendiconti Accademia dei Lincei* (6) 9: 633–639

—— (1947). "Teoria della relativitá." In *Enciclopedia delle Matematiche Elementari*, Vol. III, P. I, Hoepli, Milano: 779–819.

Pauli, Wolfgang (1921). "Relativitätstheorie." In *Enzyklopädie der Mathematischen Wissenschaften mit Einschluss ihrer Anwendungen, Vol. 5, Physik*, Part 2. Arnold Sommerfeld, ed. Leipzig: Teubner, 1904–1922, pp. 539–775.

Wald, Robert M. (1984). *General Relativity*. Chicago: The University of Chicago Press.

Weyl, Hermann (1917). "Zur Gravitationstheorie." *Annalen der Physik* 54: 117–145.

—— (1918). *Raum-Zeit-Materie*. Berlin: Julius Springer.

—— (1921). *Raum-Zeit-Materie*. 4th ed. Berlin: Julius Springer.

—— (1952). *Space, Time, Matter*. New York: Dover. Translation of Weyl 1921.

Part IV

The Reception and Development of General Relativity

The American Contribution to the Theory of Differential Invariants, 1900–1916

Karin Reich

Summary. The father of the theory of differential invariants was Gauss, who developed the first examples. The development of a calculus was due to Christoffel and, especially, Gregorio Ricci, who regarded it as a formal calculus comparable to the calculus of forms in algebraic invariant theory. Then came many contributions to the new theory of differential invariants. The first textbook was written by Joseph Wright. While European mathematicians favored a generalization of of the theory of differential invariants, the Americans also tied new links: they combined it with vector calculus and regarded vector and tensor calculus as the foundation of differential geometry. Unfortunately, Wilson and Moore's publication on this subject came too early to include general relativity and too late to influence Einstein.

1. Preliminaries

Tensor calculus in the form of absolute differential calculus is due to Gregorio Ricci (1853–1925); thus it was also called Ricci calculus. In 1901 Ricci published a comprehensive paper of nearly 80 pages together with his former student Tullio Levi-Civita (1873–1941). Their paper "Méthodes du calcul différentiel absolu et leurs applications" had appeared in a broadly recognized journal, the *Mathematische Annalen.* This paper played a crucial role for Einstein; by means of it he became familiar with tensor calculus, which is regarded as the fundamental mathematical tool in his general relativity. Along with relativity theory, Ricci, who had been a professor in

Padua since 1880, also became famous and was finally honored by gaining membership to several academies.

But what about Ricci's recognition before Einstein? Several historians think that Ricci was not a well-known scientist; this also meant that his tensor calculus was not widely known and respected. The following statements illustrate this: "Their [Ricci's and Levi-Civita's] seventy-six page memoir attracted little attention from mathematicians outside of Italy, while Einstein, in 1912, made it the starting point for his own work in general relativity" (Goodstein 1982/3, p. 247). J.E. Wright's textbook, *Invariants of Quadratic Differential Forms* (1908), was "perhaps the first book dealing with the calculus of Ricci and Levi-Civita" (Guth 1970, p. 203). This was also a common opinion at the time of Einstein's relativity; Wilson and Moore, for example, stated that "the few authors who cite Ricci do so in a manner which suggests strongly that his method was practically unknown" (Wilson and Moore 1916a, p. 274).

This impression was supported by the fact that Ricci had twice tried to win the royal prize of the Accademia dei Lincei, but in vain. He was rejected. Beltrami defeated his application in 1887, Luigi Bianchi in 1901. Beltrami and Bianchi were the leading Italian mathematicians of their time; both worked in the field of differential geometry. Bianchi's argument was that tensor calculus "could not conceivably be of any use, even for a differential geometer" (Roth 1942, p. 266). With their denials it was revealed that tensor calculus was not thought to be of essential importance in differential geometry.

Ricci himself had not believed that his calculus was a part of differential geometry. According to him, tensor calculus was a formal calculus that could be adopted by several fields, in differential geometry as well as in other mathematical or physical disciplines such as, for example, in elasticity theory or heat propagation. Ricci's starting point was invariant theory; he wanted to transfer the methods of algebraic invariant theory to differential invariants. His aim was a new theory of differential invariants, i.e., a calculus of differential forms and not a contribution to differential geometry.

Invariant theory was one of the most important mathematical disciplines of the 19th century; Sylvester had defined the terms *invariant, covariant*, and *contravariant* (Sylvester 1852) and had introduced the calculus of forms (Sylvester 1853). Invariant theory belonged to algebra; the idea of covariance was well established in algebra before Ricci started with his investigations. The history of tensor calculus has to be regarded as the history of differential invariants, cultivated in Italy and other European countries. The theory of invariants was also propagated in the U.S. The Americans contributed not only the largest number of papers but also the most eminent ones.

2. History of Differential Invariants

Gauss is thought to be the father of differential geometry (Wrede 1972, p. 197). This evaluation is based on his article "General Investigations of Curved Surfaces" (Gauss 1828). With this paper Gauss also became the father of differential invariants. Assuming the linear element of a surface to be $\mathrm{d}s^2 = g_{ik}\,\mathrm{d}x^i\,\mathrm{d}x^k$, $i, k = 1, 2$, he defined the curvature of his surface as

$$\frac{R_{1212}}{g} = \frac{\partial \Gamma_{221}}{\partial x^1} - \frac{\partial \Gamma_{121}}{\partial x^2} + \Gamma_{11i}\Gamma^i_{12} - \Gamma_{11i}\Gamma^i_{22}$$

(modern notation). This is the first example of a differential invariant; it only depends on the g_{ik} and their first and second derivatives. Gauss emphasized its invariant character by means of his main theorem, which he called "*theorema egregium*": "If a curved surface is developed upon any other surface whatever, the measure of curvature in each point remains unchanged" (Gauss 1828, section 12). In the original Latin version one reads the word "invariata" instead of "unchanged," which fits much better.

In the following chapter (section 13) Gauss defined surfaces from a new point of view, i.e., a two-dimensional manifold. Gauss again used the term *invariant* as well as the expression *absolute* when he spoke of properties that do not depend on the form into which the surfaces can be bent without tension ("qualitates superficiei . . . absolutae sunt, atque invariatae manent, in quamcunque formam illa flectatur"). Gauss had used the term *absolute* even earlier in some notes from the time between 1822 and 1825; he spoke of absolute curvature regarding the geodesic curvature (Gauss 1900, p. 387). The Gauss curvature was the first example of a differential invariant. Further examples were the so-called differential operators that Eugenio Beltrami had introduced (Beltrami 1864/65). Beltrami distinguished between the relative properties and the absolute properties of surfaces. His differential operators were absolute, because they were not dependent on the special form of the surface.

Gauss's ideas were further developed by Bernhard Riemann (1826–1866). In his lecture of 1854, "On the Hypotheses Which Lie at the Foundation of Geometry," he extended two-dimensional surfaces, i.e., the manifolds of Gauss, to n-dimensional manifolds and, accordingly, the Gauss curvature to what was later called Riemann curvature:

$$R_{klij} = \frac{\partial \Gamma_{lik}}{\partial x^j} - \frac{\partial \Gamma_{ljk}}{\partial x^i} + \Gamma^s_{jl}\Gamma_{iks} - \Gamma^s_{il}\Gamma_{jks}.$$

Riemann regarded his curvature as a geometric magnitude; he did not use the terms absolute or invariant, and his point of view was merely geometrical. Neither Elwin Bruno Christoffel (1829–1900) nor Ricci shared Riemann's geometrical standpoint. They referred to the ideas and vocabulary of algebraic invariant theory when describing differential forms (quantics), covariants and invariants, and the calculus of these newly developed magnitudes. In his paper "Ueber die Transformation der homogenen Differential ausdrücke zweiten Grades," Christoffel had algebraic invariant theory in mind when he determined R_{klij} as a result of integrability conditions; there is no hint of curvature. In consequence, Christoffel introduced a process that allowed the creation of a series of differential covariants, which was his aim:

$$a_{i_1\cdots i_r,i} = \frac{\partial a_{i_1\cdots i_r}}{\partial x^i} - \sum_{m=1}^{r} a_{i_1\cdots i_{m-1}\,h\,i_{m+1}\cdots i_r}\Gamma^h_{i_m i}$$

(Christoffel 1869, p. 57). He called the $a_{i_1\cdots i_r}$ a "system of transformation relations," i.e., tensors.

In 1884, Ricci began a series of papers referring especially to Christoffel and Beltrami, that is, to his differential operators and their significance as absolute functions. In 1887 Ricci introduced the term *covariant derivation* to describe Christoffel's process and derived the following very important theorem: The second covariant derivation enjoys the property of commutativity only in the case of the plane, i.e., Euclidean manifolds; this implies a connection between the covariant derivation and the Riemann tensor:

$$a_{m_1\cdots m_r,ij} - a_{m_1\cdots m_r,ji} = R^s_{m_n ij}a_{m_1\cdots m_{n-1}\,s\,m_{n+1}\cdots m_r}$$

(Ricci 1887, p. 203). In 1893, Ricci used the term *absolute differential calculus* for the first time (Ricci 1893). A summary of these results was first published outside of Italy in the French language as early as 1892. An extensive presentation followed in 1901. Felix Klein (1849–1925) had asked Ricci for this full-length paper.

The term *differential invariant* was created by Sophus Lie (1842–1899). In 1884, the same year in which Ricci started with his papers on differential forms, Lie published his memoir "Ueber Differential-invarianten," where he quoted Lie, of course (group concept), but also Beltrami and Christoffel on the one hand (differential forms) and Arthur Cayley, James Joseph Sylvester, Siegfried Aronhold, Alfred Clebsch, and Paul Gordan on the other hand (invariant theory). He considered the quantities $x_1 \cdots x_n z_1 \cdots z_q$, which are connected with the new variables $x'_1 \cdots x'_n z'_1 \cdots z'_q$ by special transformation equations forming a group. The function

$$\Omega\left(x_1 \cdots x_n z_1 \cdots z_q \frac{\partial z_1}{\partial x_1} \cdots \frac{\partial^2 z_1}{\partial x_k \partial x_m} \cdots\right)$$

is called a differential invariant of the regarded group in the case that

$$\Omega\left(x_1 \cdots z_q \frac{\partial z_1}{\partial x_1} \cdots \frac{\partial^2 z_1}{\partial x_k \partial x_m} \cdots\right) = \Omega\left(x_1' \cdots z_q' \frac{\partial z_1'}{\partial x_1'} \cdots \frac{\partial^2 z_1'}{\partial x_k' \partial x_m'} \cdots\right).$$

Lie's definition also included differential invariants that were dependent on even higher derivatives. His fundamental theorem stated: Each finite or infinite continuous group defines an infinite series of differential invariants that can be interpreted as solutions of complete systems (Lie 1884, p. 539).

In the following years, the group concept and the idea of differential invariants became very important. One of the first mathematicians who reacted to Lie's newly introduced concept was Kasimir Zorawski (1866–1953), a Polish scientist who worked in Warsaw. As examples of differential invariants, Zorawski mentioned Gauss curvature, Beltrami operators, and Minding's geodesic curvature. Similar to Lie, he asked for the number of invariants of different orders (Zorawski 1892/93). From then on, two concepts played a major role within the theory of differential invariants: the concept of the absolute differential calculus, initiated by Ricci, and the group concept that was introduced by Lie and improved by Zorawski. Levi-Civita, for example, made use of both concepts in his very important paper "On Absolute Invariants" (Levi-Civita 1893/4).

For a long time there were practically no links between differential invariants and geometry, and there were no allusions to vector calculus. The theory of differential invariants became what its contributors intended it to be—a formal theory that developed as a discipline within algebra, that is, invariant theory. The papers on differential invariants were immediately and regularly reviewed in the *Jahrbuch über die Fortschritte der Mathematik* under the heading "Algebra: Theory of Forms (Invariant Theory): Theory of Differential Forms (Differential Invariants)." The theory of differential forms soon became a well-established discipline, contributions came from all over the world: from Austria, Emil Waelsch; from Belgium, Théophile de Donder; from Germany, Gerhard Hessenberg, Johannes Knoblauch, Hermann Kühne, and Rudolf Rothe; from Great Britain, Andrew Russell Forsyth; from Italy, Tullio Levi-Civita, Ernesto Pascal, Gregorio Ricci, Luigi Sinigallia, Carlo Somigliana, and Guido Tognoli; from Norway, Sophus Lie; from Poland, Kasimir Zorawski; from the USA, Charles N. Haskins, Louis Ingold, Edward Kasner, Gilbert N. Lewis, Heinrich Maschke,

Clarence L.E. Moore, James B. Shaw, Edwin B. Wilson, and Joseph E. Wright. Most of these authors cited the absolute differential calculus or used its ideas indirectly. This means that Ricci's tensor calculus was well recognized not only in Italy but internationally, as was mentioned before.

3. The American Contribution

Mathematicians involved in the theory of differential invariants followed in the steps of Ricci and Lie; both sides of the Atlantic adopted, combined, and extended their theories. Two mathematicians invented a new symbolism. The first attempt was made by a German, Gerhard Hessenberg, who was not very successful with his presentation. The other attempt was due to Heinrich Maschke, who was American. Several mathematicians succeeded Maschke and followed his approach. Some mathematicians tried to generalize the methods and results of their predecessors. There were mainly three types of generalization:

(1) the question of differential parameters of higher than second degree;

(2) the question of differential invariants on the base of a not only quadratic linear element, i.e., the general differential form

$$F = g_{ikl\ldots n}\, \mathrm{d}x^i\, \mathrm{d}x^k\, \mathrm{d}x^l \cdots \mathrm{d}x^n .$$

Several results could be achieved in the case $n = 3$.

(3) the quadratic linear element was generalized by neglecting its condition of symmetry, $g_{ik} \neq g_{ki}$.

These generalizations were mostly the domain of European mathematicians, but there were two new directions favored especially by American mathematicians. Later, after Einstein's general relativity theory, the following directions became the main paths of development:

(1) The combination of the theory of differential invariants with vector calculus. Though Ricci and Levi-Civita had given first applications (Ricci and Levi-Civita 1901, pp. 135–137), these ideas did not become fruitful immediately.

(2) The combination of tensor calculus with differential geometry. Ricci had mentioned differential geometry as a possible field of application. He also had written a textbook on the subject (Ricci 1898), but not only was he not successful, he was rejected by the traditional differential geometers (Reich 1989, pp. 282–285, 295). The first positive ideas concerning differential geometry on the basis of tensor calculus were presented on the occasion of the St. Louis Congress in 1904.

3.1 The Combination of the Absolute Differential Calculus and the Group Concept: Haskins

Charles N. Haskins, born in New Bedford, Massachusetts, in 1874, attended MIT, where he graduated in 1897. He continued his studies at Harvard and received the degrees of Master of Science, in 1899, Master of Arts, in 1900, and Doctor of Philosophy, in 1901. Before he became an assistant professor at the University of Illinois in 1906, he held a teaching position at Harvard. In 1909 he left for Dartmouth College, where he was promoted to professor of mathematics in 1916. In 1920 he was nominated vice-president of the American Mathematical Society, where he had been a council member from 1914 to 1916. In 1928 he was awarded the honorary degree of Doctor of Science (Haskins 1928). He was 68 years old when he died in 1942 (Haskins 1943).

For his Ph.D., Haskins had written a 109-page memoir, *On the Invariants of Quadratic Differential Forms* (Haskins 1901). This was also the title of two articles published later in the *Transactions* (Haskins 1902; 1904). Lie's continuous groups were the background; Haskins quoted Zorawski but also Ricci and Levi-Civita. He determined the number of invariants of the general quadratic differential form in n variables

$$\phi = \sum_{i,k=1}^{n}\sum a_{ik}(x_1, \ldots, x_n)\, \mathrm{d}x_i\, \mathrm{d}x_k .$$

In the second part, Haskins asked for the number of differential parameters of quadratic differential forms. At about the same time he examined forms of a degree higher than two (Haskins 1903). In his article "On the Differential Invariants of a Plane" (Haskins 1906), he summarized the results of Andrew R. Forsyth, which had only been published recently, compared them with the corresponding results of Ricci in 1885; this was of special interest because Forsyth and Ricci had achieved different results for the same problem, Forsyth in considering the complete system of linear partial equations, Ricci by his method of absolute differential calculus.

3.2 The Symbolic Method: Maschke

Heinrich Maschke was born in 1853 in Breslau (Wroclaw), Germany. He studied in Heidelberg and in Berlin, mainly under Leo Königsberger, Leopold Kronecker, Ernst Kummer, and Karl Weierstrass. In 1878 Maschke finished in Berlin after an examination qualifying him as a teacher. Lecturing in a Berlin high school, Maschke wrote a thesis entitled *On Triply Orthogonal Systems of Surfaces*, which was accepted by the University of

Göttingen (Maschke 1880). This work is closely connected with the work of Gaston Darboux, the author of many papers and even a textbook on triple orthogonal systems of surfaces (Darboux 1897). In 1886, Maschke left Berlin for one year, which he spent in Göttingen. There, he continued his studies under the auspices of Felix Klein, according to whom Maschke's main field became group theory. Because it was hopeless to obtain a position at a German university at that time, Maschke emigrated to the United States in 1891, where he finally got a position as an assistant professor at the newly founded University of Chicago in 1892. Later, he became an associate and then a full professor (Bolza 1908).

Maschke gave lectures on nearly all fields. During a lecture on differential geometry in 1899–1900, he invented a new symbolic method for treating quadratic differential forms. This new symbolism soon became his main field of investigation. He published his first results in his paper "A New Method of Determining the Differential Parameters and Invariants of Quadratic Differential Quantics" (Maschke 1900). He quoted Luigi Bianchi as the leading mathematician in differential geometry, as well as Gerhard Hessenberg and Johannes Knoblauch, whom he knew from his time in Berlin. Maschke started his article with the following remarks:

> I propose to exhibit in a preliminary way a symbolic method, in close analogy with the symbolism used in the algebraic theory of invariants, for the construction and investigation of invariants of quadratic differential quantics. The method proves to be fully as successful as in algebra, the chief advantage lying in the fact that after the establishment of the fundamental principles of the method further reference to the formulas of transformation becomes unnecessary. (Maschke 1900, p. 197)

In the following, Maschke considered $A = \sum_{i,k=1}^{2} a_{ik}\, \mathrm{d}x_i\, \mathrm{d}x_k$ and denoted the derivatives of the function f by $f_i f_k = a_{ik}$; this meant $\partial a_{ik}/\partial x_l = f_i f_{kl} + f_k f_{il}$. In this new symbolic system, the Christoffel symbols $\Gamma_{ikl} = \left[{kl \atop i}\right]$ were presented by

$$f_i f_{kl} = \tfrac{1}{2}\Big(\frac{\partial a_{ik}}{\partial x_l} + \frac{\partial a_{il}}{\partial x_k} - \frac{\partial a_{kl}}{\partial x_i}\Big).$$

Maschke showed that his method led easily to the formation of expressions remaining invariant with respect to the transformation of quadratic differential forms.

In his first paper Maschke had only treated the case $n = 2$; in his next paper, "Invariants and Covariants of Quadratic Differential Quantics of n Variables" (Maschke 1903/4), he considered the general case and also showed the usefulness of his new method for Riemann's curvature, which he called "quadrilinear covariant." These investigations were continued in "A Symbolic Treatment of Invariants of Quadratic Differential Quantics of *n* Variables" (Maschke 1903). Here, Maschke expressed in words the advantages of his denotation in form of the following theorems:

> The value of an invariant expression in symbolic form is not changed if two equivalent symbols are interchanged. An invariant expression in symbolic form vanishes if by the interchange of two equivalent symbols its sign is changed. (Maschke 1903, p. 449)

In quoting Ricci's first publication in French (Ricci 1892), Maschke also treated covariant differentiation and its significance for the quadrilinear covariant, i.e., Riemann's curvature. After also considering higher covariants, Maschke discussed all possible kinds of invariants. This paper was reviewed in detail (two pages!) in the *Jahrbuch* by Franz Meyer, who lectured in Königsberg (Kaliningrad). Meyer was an outstanding specialist of invariant theory and also the author of the exhaustive "Report on the Actual Situation of Invariant Theory" (Meyer 1890/91). Meyer concluded his review with the remark, "The whole makes it possible to realize the fertility of the symbolic method" (Meyer 1903).

Later, Maschke also generalized the theory of differential parameters by means of his new symbolic method (Maschke 1906). Maschke belonged to the small, elected group of invited speakers at the St. Louis Congress in 1904 (see Section 3.4). Unfortunately, Maschke died young, suddenly, in 1908.

3.3 On the Way to the First Textbook: Wright

Born in Liverpool in 1878, Joseph Edmund Wright entered Trinity College in Cambridge in 1897. He was senior wrangler in 1900 and elected fellow of Trinity in 1903. In the same year, he emigrated to the U.S. He got a position as associate professor at Bryn Mawr College where he stayed for seven years; it was his period of highest productivity (Scott 1910). In the monograph series of his college, Wright published 11 works, most of them exhaustive textbooks on invariant theory, differential invariants, group theory, differential equations, and differential geometry (Wright 1904–1908). In connection with these works, Wright also wrote several papers that appeared in common mathematical journals and were reviewed. Wright died in 1910, at only 31½ years old.

Wright was familiar with the actual situation in invariant theory; this is proved by his paper "Covariants of Power Series" (Wright 1905b), where he tried to treat forms of arbitrary order *n* by means of the Aronhold symbolism. From invariant theory Wright moved to differential invariants. In his article "On Differential Invariants" (Wright 1905c) he summarized the results of Lie, Zorawski, and Forsyth before he determined all first-order invariants in the case of contact transformations. In his paper "The Differential Invariants of Space," Wright at first defined his understanding of an invariant as including "the whole class of Gaussian invariants, parameters and covariants" and continued with the determination of invariants, considering the problem to be solved "when a method is given for determining a complete functionally independent set of invariants by direct processes" (Wright 1905a, p. 323). He also applied the theory of differential invariants to triply orthogonal systems of surfaces. Darboux had proved that such a family of surfaces must satisfy a third-order differential equation given in the form of a determinant; Wright showed that this determinant could be considered to be a differential invariant expressible as an algebraic invariant of certain forms (Wright 1906).

Wright's most outstanding contribution was his 90-page monograph *Invariants of Quadratic Forms* (Wright 1908), the first textbook on this subject. In the preface, Wright pointed out that "the aim of this tract is to give, as far as possible in so short a book, an account of the invariant theory connected with a single quadratic differential form." In the first chapter, he outlined the history of differential invariants. Invariance was necessarily connected with the idea of transformation, and it was therefore clear that every invariant was invariant under a group of transformations. According to Wright, there were "three main methods of attack." The first was due to Christoffel and owed its further development to Ricci and Levi-Civita; Wright referred to it as "the method of Christoffel." In the preface he characterized it as "the most successful method." The second method had its roots in Lie's group theory. The third was that of Heinrich Maschke, who had introduced a symbolism similar to that of algebraic invariants. In the text that followed, Wright devoted a whole chapter to the presentation of each of these methods. The exhaustive chapter 5, dealing with geometrical and dynamical applications, was very appealing. In the introduction, Wright pointed out that

> the geometry of the manifold thus breaks up into two parts:
>
> (1) the determination of all invariants and all relations connecting them;
>
> (2) the geometrical interpretation of all these invariants in the manifold.
>
> (Wright 1908, p. 4)

It was an exceptional point of view at that time, to think of a geometrical interpretation of the invariants, a challenge that was accepted only much later. In this fifth chapter Wright emphasized once again the close relationship between differential geometry and differential invariants: "we have now... to interpret geometrical magnitudes in terms of invariants" (Wright 1908, p. 52). In his later paper "Corresponding Dynamical Systems," Wright focused again on the dynamical applications (Wright 1909).

Franz Meyer reviewed Wright's works in the *Jahrbuch*, among them Wright's textbook. Meyer's review was very detailed; he concluded with the remark that a monograph like this was desirable because it was easy to handle, cheap, and easy to read. According to Meyer the reader was offered such a rich amount of material in an elegant way; it also should be possible to publish something equivalent in German (Meyer 1908).

Luther Pfahler Eisenhart's main field of investigation was differential geometry (Lefschetz 1969, pp. 72–77). In 1900 Eisenhart had become an instructor at Princeton University. He became familiar with absolute differential calculus during his visit to Padua in 1905, where he had met Levi-Civita. From 1909 onward (he was promoted full professor in that year), he also gave lectures on Ricci's calculus (Ruse 1953). Eisenhart wrote an extensive review, 11 pages long, on Wright's book (Eisenhart 1911). In complaining about the too numerous mathematical publications, Eisenhart remarked, "one is delighted to find here and there a digest of the work in a particular field." But Eisenhart had to conclude his review with the sad announcement of Wright's death:

> It is impossible to close this review without remarking the loss to American mathematics by the death of Mr. Wright. His brilliant record at Cambridge and his subsequent career in this country had won for him a high place in his field.

It is surprising and remarkable that several physicists knew Wright's textbook. Harry Bateman quoted it in his paper "The Transformations of Coordinates which Can Be Used to Transform One Physical Problem into Another" (Bateman 1910, p. 472). Friedrich Kottler, an Austrian physicist, who among others worked on relativity theory, also knew it. He mentioned Wright's book seven times in his paper "On the Space-Time Lines in the Minkowski World" (Kottler 1912, pp. 1666–1689).

3.4 The St. Louis Congress, 1904

In 1904, the World Fair took place in St. Louis, between September 19 and 24; it was accompanied by the "International Congress of Arts and Sciences." The sciences were arranged in seven sections; mathematics was a part of the so-called normative sciences. Mathematics had three parts:

(1) Algebra and Analysis, with invited speakers E. Picard from Paris and H. Maschke from Chicago University;

(2) Geometry, with invited speakers G. Darboux from Paris and E. Kasner from Columbia University;

(3) Applied Mathematics, with speakers L. Boltzmann from Vienna and H. Poincaré from Paris.

Part (3) as well as Picard's talk "On the Development of Mathematical Analysis and Its Relation to Other Sciences" may be omitted here. H. Maschke delivered the address "On Present Problems of Algebra and Analysis"; he immediately specialized in the theory of invariants and quadratic differential quantics:

> Invariants suggest at once algebra, differential quantics: analysis. At the same time the subject also leads into geometry—it contains, for instance, a great part of differential geometry and of geometry of hyperspace. But is there, indeed, any algebraic or analytic problem which does not allow geometrical interpretation in some way or other? And when it comes to geometry of hyperspace,—it is then only geometrical language that we are using,—what we are actually considering are analytic or algebraic forms. (Maschke 1905, p. 518)

He emphasized that the presentation of the theory of differential invariants is "in strict analogy with the algebraic theory of invariants" and summarized its development. He started with the differential quadratic quantics as they occur in geometry, he quoted Gauss, Beltrami, Codazzi, Riemann, Darboux, and others and continued with the purely analytic representation according to Christoffel, Ricci, Levi-Civita, Lie, and Haskins. Maschke distinguished between the three methods: the absolute differential calculus, his symbolic method, and Lie's theory of continuous groups. He suggested that in the future a combination of two or all three methods would be favored, a forecast that turned out to be wrong. Maschke finished with the statement: "But here, as always, it is the man, not the method, that solves the problem" (Maschke 1905, p. 530).

To clarify the connection between the differential invariants and differential geometry was unusual among European mathematicians. Maschke's address was very different from Gaston Darboux's (1842–1917). Darboux was a typical European differential geometer. He was totally convinced of the effectiveness and usefulness of geometry; his standpoint was a purely geometrical one. He began his address, "A Survey of the Development of Geometric Methods," with a detailed history. According to Darboux, Lagrange became tired of research in analysis and mechanics and therefore turned to chemistry. The reason was that the program of investigations opened up by the discovery of the calculus was nearly exhausted. Only at the end of the 18th century was geometry able to celebrate a triumphant comeback; Darboux called Monge "the regenerator of modern geometry" (Darboux 1905, pp. 518–519). Monge's successors, for example, Dupin, Chasles, and Poncelet, directed the field and generated a movement toward geometry. Only in passing did Darboux mention the theory of quantics and Lie's group concept. This homage to geometry was underlined by reproaches for analysis. Darboux regretted that, although the number of mathematical contributions grew exponentially on all sides, the number of those in which pure geometry was cultivated was very limited:

> This is a danger against which it is of some importance to guard.... It was in the school of geometry that we have learned, and there our successors will have to learn it, never blindly to trust to too general methods.... Therefore, let us cultivate geometry, which has its own advantages, and this without wishing to make it equal in all points to its rival. (Darboux 1905, pp. 542–543)

And Darboux compared geometry with Mother Earth; other mathematical branches, symbolized by the giant Antaeus, regained their strength only in touching their Mother Earth, geometry.

Like Darboux, Edward Kasner delivered an address within the division of geometry: "On the Present Problems of Geometry" (Kasner 1905), but he did not share Darboux's narrow idea of pure geometry, being closer to the view of Maschke. Born in 1878 in New York City, Kasner began his studies at City College; as a graduate student, he went to Columbia, where he received an M.A. in 1897 and a Ph.D. in 1899. His doctoral dissertation was *The Invariant Theory of the Inversion Group: Geometry upon a Quadric Surface* (Kasner 1900). As was usual at that time, he went abroad for one year, completing his studies in Göttingen, where he attended lectures of Klein and Hilbert. Returning to Columbia in 1900, Kasner became a tutor in mathematics, then an instructor in 1905, an adjunct professor in 1906, and a full professor in 1910. Differential geometry became his main field of study; he even founded a seminar on differential geometry. Kasner was on good terms with Eduard Study and Levi-Civita, who sent him reprints of their publications (Douglas 1958).

It was a great honor for Kasner to be selected as a principal speaker for the St. Louis congress. He was the youngest speaker, and even Poincaré was among his audience. Kasner's address was a complete success, arousing interest even abroad, where it was also published in a Polish translation (Douglas 1958; p. 190). For Kasner, geometry included more than Darboux would have allowed; he considered the domain of geometry to be intermediate between analysis on the one hand and mathematical physics on the other. As examples, Kasner mentioned the concepts of transformation and invariant, the space of n dimensions, etc., which owe their origin to analysis, and the theory of vector fields, questions of applicability and deformation of surfaces, which began in mechanics. Kasner spoke further about the foundations of geometry as well as the geometry of multiple forms and on natural and intrinsic geometry (Kasner 1905, pp. 287, 297, 300). For him, the basis of natural geometry was essentially the theory of differential invariants, and he quoted Bianchi as well as Ricci, Maschke, and Lie. Kasner also mentioned vector fields together with tensor fields (*sic*), which arose in Maxwell's theory of electromagnetism (Kasner 1905, pp. 312f). The phrase *tensor field* is remarkable. The term *tensor* had been introduced in 1898 by Woldemar Voigt, a professor of crystallography in Göttingen. It is probable that Kasner had learned about tensors during his stay in Göttingen. He belonged to the very small group of non-German scientists who were familiar with Voigt's tensors (Reich 1993, table VII).

Maschke and Kasner pursued the correct course, and subsequent developments justified their attitudes. Darboux, on the contrary, hoped for a new Monge, who did not arrive.

3.5 Differential Invariants in Connection with Vectors

At the beginning of the 20th century, mathematicians, especially European mathematicians, had a major problem with vector calculus; some accepted it, and some only regarded it as a kind of notation, not able to offer new results. There was no question that nevertheless vectors had become quite common in physics, mostly in electrodynamics. Kasner had added a new aspect when he brought the theory of differential invariants into direct contact with vector and tensor fields. This aspect bore within it the germ of future development. In Europe, only Emil Waelsch, an Austrian mathematician, had made an attempt to combine his so-called binar analysis (derived from binar forms) with vector calculus. In America, however, several important papers were published on this subject.

Louis Ingold, born in 1872 in Luray, Missouri, had received his B.A. in 1901 and his M.A. in 1902 from the University of Missouri, where he worked as an instructor from 1905 to 1906 and as an assistant from 1906 to 1910. He got his Ph.D. from the University of Chicago in 1910, at the same time that he became a professor in Missouri. Ingold had worked under the auspices of Heinrich Maschke. His paper, "Vector Theory, in Terms of Symbolic Differential Parameters," was presented on March 30, 1907, to the Chicago section of the American Mathematical Society; this paper was published in somewhat different form in 1910 under the title "Vector Interpretation of Symbolic Differential Parameters." Ingold intended

> to establish a relation between the symbolic theory of invariants of differential forms, due to... Maschke, and the theory of extensive quantity (vectors), due to Grassmann. It will be shown that those symbolic expressions used by Maschke which lacked an interpretation in his theory, may be represented as vectors of the Grassmann type, and that all of Maschke's expressions, including his actual differential parameters, are expressible in the vector system. The theory of such vectors will be extended, new formulas in the symbolic theory will be obtained and applications to geometry will be made. (Ingold 1910, p. 449)

James Byrnie Shaw (1866–1948) had studied at Purdue University, Indiana, where he had also written his thesis on "I. Algebra. II. Mathematics. The Science of Algorithms" (Shaw 1893). In early 1890 he became a professor of mathematics at the Central University in Pella. After several changes he became an assistant professor at the University of Urbana (Illinois) in 1910, associate professor in 1915, and professor in 1918. Besides the philosophy of mathematics, his main fields were algebra, associative algebras, and quaternions.

In 1913 Shaw published his memoir "On Differential Invariants," which was read in part before the Chicago section of the American Mathematical Society, as Ingold's paper had been (Shaw 1913, p. 395). Shaw wanted to present the expressions of certain differential operators and differential parameters in vector form. As predecessors he mentioned Ricci, Levi-Civita, Maschke, and Ingold. Shaw started with a chapter on vector algebra of n dimensions (Shaw 1913, pp. 393–399). After this, he treated, among others, the Codazzi equations, the so-called Christoffel expression (i.e., the Riemannian curvature), covariant differentiation, and the differential operator Δ. The last chapter was devoted to symbolic invariants. As Ingold had, Shaw proved that Maschke's symbolic method could easily be transferred into vector notation.

3.6 Differential Geometry on the Basis of Vector and Tensor Calculus: Wilson and His Collaborators

Edwin Bidwell Wilson (1879–1964) had been the last student of Josiah Willard Gibbs (Hunsaker and MacLane 1973, p. 287). In 1899 he graduated from Harvard; in 1901 he got his Ph.D. at Yale, where he had worked as an instructor since 1900. He spent the following year, 1902–1903, in Paris to complete his studies at the Ecole Normale Supérieure. Back at Yale, he became an assistant professor in 1907. A year later he left Yale for MIT, where he started his career as an associate professor, becoming full professor in 1911. He had only held positions as a mathematician when in

1917 he also became a professor of mathematical physics and head of the department of physics.

In his textbook, *Elements of Vector Analysis*, Gibbs presented three-dimensional vectors and a vector calculus equivalent to modern vector calculus (Gibbs 1881/84). He had also introduced linear vector functions, which he called dyadics. A *dyadic* was a linear mapping of a vector with another vector, its components being transformed like tensor components. Based on Gibbs' papers, Wilson produced a new edition, which appeared in a revised form in 1913 (Wilson 1901). In this book, which became a standard text, Wilson extended Gibbs' idea of dyadics in creating triadics, tetradics, and polyadics, but he did not develop a calculus of these higher vector functions, i.e., tensors. He provided a sophisticated theory in handling the properties of linear transformations in invariant form. According to Wilson, vector calculus was as important a tool for physics as invariant theory was for geometry (Schlegel 1902).

In the same year (1901), a review of Ricci's textbook *Lezioni sulla teoria delle superficie* was published in *Bulletin of the American Mathematical Society* (James 1901). The author gave an impression of the essentials of Ricci's absolute differential calculus:

> The method leads to formulae and equations always presenting themselves under the same form for any system of independent variables, and the difficulties which are incidental and formal rather than intrinsic are thus to some extent done away with, and the research assumes a uniformity absent in other methods. The entire discussion is based on the properties of differential quadratic forms.

James also stated that Ricci's calculus had a wider applicability than only differential geometry; it allowed many questions of pure mathematics and of mathematical physics to be treated advantageously.

In 1912, Wilson and Gilbert N. Lewis (1875–1946) published a detailed paper on special relativity (Wilson and Lewis 1912; Hildebrand 1958, p. 211) that included a four-dimensional vector analysis. Wilson had written many reviews for the *Bulletin*, more than 30 during the years 1911–1914 on various subjects. In 1914 he reviewed Einstein and Grossmann's "Entwurf einer verallgeminerten Relativitätstheorie und einer Theorie der Gravitation" (Wilson 1914). Wilson thought the mathematical part of their theory especially interesting for those familiar with quadratic differential forms and Ricci's absolute calculus. More than a year later, during the twenty-second annual meeting of the American Mathematical Society in New York on December 27–28, 1915, Wilson read a paper entitled "Ricci's Absolute Calculus and Its Applications to the Theory of Surfaces" (Wilson 1916), which obviously was not published. With that paper Wilson tried to call attention to "Ricci's generally neglected absolute calculus and to its suggestiveness as an implement of research in developing the theory of surfaces of two dimensions in Euclidean space of *n* dimensions" (Wilson 1916). During this meeting, Clarence Lemuel Elisha Moore (1876–1931) was also present; in the following year Wilson and Moore published a summary (Wilson and Moore 1916b) and an exhaustive paper on differential geometry on the basis of Ricci's calculus (Wilson and Moore 1916a). The authors complained that Ricci's *Lezioni sulla teoria delle superficie* (Ricci 1898) was only available in very few American libraries and therefore they thought it to be necessary to give a detailed presentation of Ricci's absolute differential calculus.

Within differential geometry on the basis of Ricci's calculus, Wilson and Moore also integrated vector calculus. For them, Ricci's absolute differential calculus was nothing more than a generalized vector analysis. They mentioned that other differential geometers like Johannes Knoblauch (1855–1915), for example, did not accept Ricci's calculus because geometric magnitudes were described only as "systems of coefficients." Wilson and Moore, however, hoped to obviate this difficulty by using the notations of multiple algebra, i.e., vectors (Wilson and Moore 1916a, p. 294). At this time vector notation was not generally used in differential geometry. Therefore it is quite remarkable that Wilson and Moore presented this combination: surface theory, vector calculus, and absolute differential calculus. With it they delivered the mathematical background of general relativity.

In the same year that Wilson and Moore published their memoir, Einstein published his theory of general relativity (Einstein 1916). It is known that Einstein first developed the new physics without being able to transfer it into mathematical language. To fill this gap he looked for a special calculus; he had in mind a generalized vector calculus. With the help of Marcel Grossmann, Einstein became acquainted with Ricci's differential calculus through Ricci's and Levi-Civita's paper of 1901. Einstein and Grossmann transformed this calculus into a generalized vector calculus and created a new form: tensor calculus; they denoted Ricci's systems as tensors. Einstein also tried to give tensors a more geometric interpretation but this goal was fully achieved only by Hermann Weyl (1855–1955), who gave an excellent presentation of general relativity on the basis of differential geometry in his textbook *Space, Time, Matter* (Weyl 1918). Wilson and Moore had pursued a similar direction but, since it was chronologically not possible, their work did not include general relativity and did not influence Einstein.

It is quite remarkable that, in 1916, Wilson practically stopped his mathematical work and changed to physics, his interest shifting away from geometry toward mechanical problems. Several weeks before his death, Wilson mentioned this shift in a letter to Saunders MacLane (November 4, 1964):

> C.L.E. Moore and I did a Differential Geometry of Two-Dimensional Surfaces in Hyperspace... (November 1916) which was all new original stuff and in a subsequent review of the literature many years later was cited as the most important contribution in the field. It was about the last thing I did in pure mathematics. (Hunsaker and MacLane 1973, p. 291)

4. Conclusion

The development of the theory of differential invariants was promoted internationally, although European mathematicians pursued different directions than Americans. There were almost no examples or even hints of how to combine vector calculus, tensor calculus, and differential geometry. It was typical that Ricci's absolute differential calculus was not acknowledged among geometers; it was not allowed to be brought into connection with differential geometry. The most famous differential geometers of the time, Luigi Bianchi and Gaston Darboux, did not include tensor calculus in their work; they also did not recognize vector calculus as an important mathematical tool. American mathematicians, however, were prepared and willing to accept the above-mentioned combination of vector calculus, tensor calculus, and differential geometry. As early as 1904 Maschke and Kasner closed the first links between the theory of differential invariants and geometry. This was an important step toward the geometrization of tensor calculus.

The University of Chicago played a main role in the connection of the theory of differential invariants with vector calculus. Though Maschke had died in 1908, his student Louis Ingold and also James B. Shaw presented major papers on the subject there. In Europe vector calculus was favored mainly by physicists and not primarily by mathematicians. Physicists applied it to elasticity theory and, especially, electromagnetism. In physics, vector analysis was comparatively widespread, but at the same time, physicists were generally not acquainted with Ricci's absolute differential calculus. The first physicists to mention it before Einstein were Max Abraham in 1901, Orazio Tedone in 1906, Harry Bateman in 1910, and Friedrich Kottler in 1912.

For further development, it became crucial to emphasize the relationship between vector calculus, tensor calculus, and differential geometry. Einstein and Grossmann achieved the combination of vector calculus and

Ricci's absolute differential calculus, and Hermann Weyl took care of its geometrization, i.e., he presented general relativity within differential geometry and vice versa. Only looking at mathematics, however, Wilson and his collaborators reached the same goal. Unfortunately, Wilson and Moore published their paper at the wrong time, too late to be mentioned by Einstein in his theory of general relativity and too early to itself include general relativity. Otherwise, perhaps, Wilson and Moore's contribution could have played the role of Hermann Weyl's *Space, Time, Matter*. Nevertheless, Wilson and Moore were the first who presented differential geometry, the foundation of which was vector and tensor calculus.

References

Bateman, Harry (1910). "The Transformation of the Coordinates which Can Be Used to Tranform One Physical Problem into Another." *Proceedings of the London Mathematical Society* 8(2): 469–488.

Beltrami, Eugenio (1864/65). "Ricerche di analisi applicata alla geometria." *Giornale di matematica* 2 and 3. Also in *Opere matematiche* Vol. 1. Milano: Ulrico Hoepli, pp.107–198.

Bolza, Oskar (1908). "Zur Erinnerung an Heinriche Maschke." *Jahresbericht der Deutschen Mathematikervereiningung* 17: 345–355.

Christoffel, Elwin Bruno (1869). "Ueber die Transformation der homogenen Differentialausdrücke zweiten Grades." *Journal für die reine und angewandte Mathmatik* 70: 46–70.

Darboux, Gaston (1897). *Leçons sur les systèmes orthogonaux et les coordonnées curvilignes*. Paris: Gauthier-Villars. (Second edition 1910.)

—— (1905). "A Survey of the Development of Geometric Methods." *Bulletin of the American Mathematical Society* 11: 517–543. Also in *Congress of Arts and Science St. Louis 1904*, Vol. 1, pp. 535–558.

Douglas, Jesse (1958). "Edward Kasner." *Biographical Memoirs*. National Academy of Sciences of the USA 31: 180–209.

Einstein, Albert (1916). "Die Grundlage der allgemeinen Relativitätstheorie." *Annalen der Physik*. 49(4):769–822.

Eisenhart, Luther Pfahler (1911). Review of Wright (1908). *Bulletin of the American Mathematical Society* 17: 140–150.

Gauss, Carl Friedrich (1828). "Disquisitiones generales circa superficies curvas." *Commentationes Societatis Gottingensis, Classis math.* 6: 99–146. Also in *General Investigations of Curved Surfaces of 1827 and 1825*. J.C. Morehead and A.M. Hiltebeitel, eds. Princeton: Princeton University Library, 1902.

—— (1900). *Werke*, Vol. 8. Göttingen: B.G. Teubner, pp. 386–396.

Gibbs, Josiah Willard (1881/84). *Elements of Vector Analysis*. New Haven: Tuttle Morhouse and Taylor. Also in *The Scientific Papers*, Vol. 2, pp. 17–90.

Goodstein, Judith R. (1982/3). "The Italian Mathematicians of Relativity." *Centaurus* 26: 241–261.

Guth, Eugene (1970). "Contribution to the History of Einstein's Geometry as a Branch of Physics." In *Relativity, Proceedings of the Relativity Conference in the Midwest*, held at Cincinnati, Ohio, June 2–6, 1969. M. Carmeli, S.I. Fickler, L. Witten, eds. New York: Plenum Press, pp. 161–207.

Haskins, Charles Nelson (1901). *On the Invariants of Quadratic Differential Forms*. Harvard University, Ph.D. thesis, 109 p.

—— (1902). "On the Invariants of Quadratic Differential Forms." *Transactions of the American Mathematical Society* 3: 71–91.

—— (1903). "On the Invariants of Differential Forms of Degree Higher than Two." *Transactions of the American Mathematical Society* 4: 38–43.

—— (1904). "On the Invariants of Quadratic Differential Forms." *Transactions of the American Mathematical Society* 5: 167–192.

—— (1906). "On the Differential Invariants of a Plane." *Transactions of the American Mathematical Society* 7: 588–590.

—— (1928). "Charles Nelson Haskins." *The Dartmouth Alumni Magazine*, August: 818.

—— (1943). "Charles Nelson Haskins." *The Dartmouth Alumni Magazine*, January: 112.

Hildebrand, Joel H. (1958). "Gilbert Newton Lewis." *Biographical Memoirs*. National Academy of the Sciences of the USA 31: 209–235.

Hunsaker, Jerome and MacLane, Saunders (1973). "Edwin Bidwell Wilson." *Biographical Memoirs*. National Academy of the Sciences of the USA 43: 285–320.

Ingold, Louis (1910). "Vector Interpretation of Symbolic Differential Parameters." *Transactions of the American Mathematical Society* 11: 449–474.

James, George Oscar (1901). Review of Ricci (1898). *Bulletin of the American Mathematical Society* 7: 359–360.

Kasner, Edward (1900). "The Invariant Theory of the Inversion Group: Geometry upon a Quadric Surface." *Transactions of the American Mathematical Society* 1: 430–498.

—— (1905). "The Present Problems of Geometry." *Bulletin of the American Mathematical Society* 11: 283–314. Also in *Congress of Arts and Science St. Louis 1904*, Vol. 1, pp. 559–586.

Kottler, Friederich (1912). "Über die Raumzeitlinien der Minkowski' schen Welt." *Sitzungsberichte der Akademie der Wissenschaften Wien*, math.-nat. Klasse, Part IIa, 121: 1659–1759.

Lefschetz, Solomon (1969). "Luther Pfahler Eisenhart." *Biographical Memoirs*. National Academy of Sciences of the USA 40: 68–90.

Levi-Civita, Tullio (1893/4). "Sugli invarianti assoluti." *Atti Ist. Veneto = Opera matematiche* Vol. 1. Bologna: Zanichelli, pp. 41–100.

Lie, Sophus (1884). "Ueber Differentialinvarianten." *Mathematische Annalen* 24: 537–578.

Maschke, Heinrich (1880). "Ueber ein dreifach orthogonales Flächensystem, gebildet aus Flächen dritter Ordnung." Dissertation Göttingen, Halle: Otto Hendel.

——— (1900). "A New Method of Determining the Differential Parameters and Invariants of Quadratic Differential Quantities." *Transactions of the American Mathematical Society* 1: 197–204.

——— (1903). "A Symbolic Treatment of the Theory of Invariants of Quadratic Differential Quantics of *n* Variables." *Transactions of the American Mathematical Society* 4: 445–469.

——— (1903/04). "Invariants and Covariants of Quadratic Differential Quantics of *n* Variables." *Decennial Publications of the University of Chicago* 9(1): 125–138.

——— (1905). "On Present Problems of Algebra and Analysis." *Congress of Arts and Science St. Louis* 1904, Vol. 1, pp. 518–530.

——— (1906). "Differential Parameters of the First Order." *Transactions of the American Mathematical Society* 7: 69–80.

Meyer, Franz (1890/91). "Bericht über den gegenwärtigen Stand der Invariantentheorie." *Jahresbericht der Deutschen Mathematikervereinigung* 1: 79–292.

——— (1903). Review of Maschke (1903). *Jahrbuch über die Fortschritte der Mathematik* 34: 141–142.

——— (1908). Review of Wright (1908). *Jahrbuch über die Fortschritte der Mathematik* 39: 180–184.

Reich, Karin (1989). "Das Eindringen des Vektorkalküls in die Differentialgeometrie." *Archive for History of Exact Sciences* 40: 275–303.

——— (1993): *Die Entwicklung des Tensorkalküls. Vom absoluten Differentialkalkül zur Relativitätstheorie.* Basel: Birkhäuser Verlag.

Ricci, Gregorio (1887). "Sulla derivazione covariante ad una forma quadratica differenziale." *Rendiconti Accademia dei Lincei, Opere*, Vol. 1. Rome: Edizioni Cremonese, pp. 199–203.

——— (1892). "Résumé de quelques travaux sur les systémes variables de fonctions associés à une forme differéntielle quadratique." *Bulletin des sciences mathématiques* 16(2): 167–189. Also in *Opere*, Vol. 1. Rome: Edizioni Cremonese, pp. 288–310.

——— (1893). "Di alcune applicazioni del calcolo differenziale assoluto alla teoria delle forme differenziali quadratiche binarie e dei sistemi a due variablili." *Atti Ist. Veneto, Opere*, Vol. 1. Rome: Edizioni Cremonese, pp. 311–335.

——— (1898). *Lezioni sulla teoria delle superficie*. Verona: Fratelli Drucker.

Ricci, Gregorio and Levi-Civita, Tullio (1901). "Méthodes du calcul différentiel absolu et leurs applications." *Mathematische Annalen* 54: 125–201.

Roth, L. (1942). Obituaries. Prof. T. Levi-Civita. *Nature* 149: 266.

Ruse, H.S. (1953). "The Ricci Calculus." *Nature* 171: 61–62.

Schlegel, Viktor (1902). Review of Wilson (1901). *Jahrbuch über die Fortschritte der Mathematik* 33: 96–97.

Scott, Charlotte Angas (1910). In memoriam. J. Edmund Wright. Born at Liverpool, England, 1878, died at Bryn Mawr, 1910. *The Bryn Mawr Alumnae Quarterly* 33–34.

Shaw, James Byrnie (1893). *I. Algebra. II. Mathematics. The Science of Algorithms.* Ph.D. Purdue University, Jacksonville, Illinois: Henderson and Depew.

——— (1913). "On Differential Invariants." *American Journal of Mathematics* 35: 395–406.

Struik, Dirk Jan (1932). "C.L.E. Moore." *Journal of Mathematics and Physics* 11: 1–11.

Sylvester, James Joseph (1852). "On the Principles of the Calculus of Forms." *Cambridge and Dublin Mathematical Journal* 7: 52–97, 179–217.

——— (1853). "On the Calculus of Forms, Otherwise the Theory of Invariants." *Cambridge and Dublin Mathematical Journal* 8: 256–269; 9: 85–103.

Weyl, Hermann (1918). *Raum-Zeit-Materie.* Berlin: Springer. *Space, Time, Matter.* English translation by Henry L. Brose. London: Methuen, 1922. Reprint, New York: Dover Publications, 1951, 1952.

Wilson, Edwin Bidwell, ed. (1901). *Vector Analysis.* New York: Scribner's Sons. Third edition, New Haven: Yale University Press, 1913.

——— (1914). Review of A. Einstein and M. Grossmann: "Entwurf einer verallgemeinerten Relativitätstheorie und einer Theorie der Gravitation." *Bulletin of the American Mathematical Society* 20: 273.

——— (1916). "Ricci's Absolute Calculus and Its Application to the Theory of Surfaces." *Bulletin of the American Mathematical Society* 22: 265, 271.

Wilson, Edwin Bidwell and Lewis, Gilbert Newton (1912). "The Space-Time Manifold of Relativity. The Non-Euclidean Geometry of Mechanics and Electromagnetics." *Proceedings of the American Academy of Arts and Sciences* 48: 389–507.

Wilson, Edwin Bidwell and Moore, Clarence Lemuel Elisha (1916a). "Differential Geometry of Two-Dimensional Surfaces in Hyperspace." *Proceedings of the American Academy of Arts and Sciences* 52: 267–368.

——— (1916b). "A General Theory of Surfaces." *Proceedings of the National Academy of Sciences* 2: 273–278.

Wrede, Robert C. (1972). *Introduction into Vector and Tensor Analysis.* New York: Dover.

Wright, Joseph Edmund (1904–1908). Bryn Mawr College, PA, Monographs Reprint Series.

(a) *On Differential Invariants.* 315 p., Vol. 8, No. 2, 1904.

(b) *Differential Invariants of Space.* 342 p., Vol. 8, No. 11, 1905.

(c) *On Lamé's Six Equations Connected with Triply Orthogonal Systems of Surfaces.* 163 p., Vol. 8, No. 7, 1905.

(d) *Lines of Curvature of a Surface.* 304 p., Vol. 8, No. 12, 1906.

(e) *Application of the Theory of Differential Invariants to Triply Orthogonal Systems of Surfaces.* 382 p., Vol. 8, No. 4, 1906.

(f) *Correspondences and the Theory of Groups*. 400 p., Vol. 8, No. 6, 1906.
(g) *Ovals of the Plane Sextic Curve*. 308 p., Vol. 8, No. 11, 1907.
(h) *Double Prints of Unicursal Curves*. 391 p., Vol. 8, No. 10, 1907.
(i) *Corresponding Dynamical Systems*. 26 p., Vol. 8, No. 13, 1908.
(j) *Differential Equations Admitting a Given Group*. 302 p., Vol. 8, No. 16, 1908.
(k) *Differential Equations Satisfied by Abelian Theta Functions of Genus Three*. 298 p., Vol. 8, No. 15, no date.

——— (1905a). "The Differential Invariants of Space." *American Journal of Mathematics* 27: 323–342.

——— (1905b). "Covariants of Power Series." *Proceedings of the London Mathematical Society* 2(2): 470–477.

——— (1905c). "On Differential Invariants." *Transactions of the American Mathematical Society* 6: 286–315.

——— (1906). "An Application of the Theory of Differential Invariants to Triply Orthogonal Systems of Surfaces." *Bulletin of the American Mathematical Society* 12: 379–382.

——— (1908). *Invariants of Quadratic Differential Forms*. Cambridge Tracts in Mathematics and Mathematical Physics No. 9, Cambridge: Cambridge University Press.

——— (1909). "Corresponding Dynamical Systems." *Annali di matematica pura ed applicata* 16(3): 1–26.

Zorawski, Kasimir (1892/3). "Ueber Biegungsinvarianten. Eine Anwendung der Lie'schen Gruppentheorie." *Acta mathematica* 16: 1–64.

The Reaction to Relativity Theory in Germany, III: "A Hundred Authors against Einstein"

Hubert Goenner

1. Introduction

It is known that the theory of relativity was quickly accepted in Germany by theoretical physicists and most of their experimental colleagues. Notable opponents were the experimental physicists Philipp Lenard, winner of the Nobel Prize of 1905, and Ernst Gehrcke. In the politically and economically unstable Berlin of 1920 a public attack on Einstein and his theory was mounted by the right-wing political agitator and anti-Semite Paul Weyland, founder of the Association of German Scientists for the Preservation of Pure Science.[1] In the same year, a scientific discussion between Einstein and Lenard during the 86th meeting of German Researchers in the Exact Sciences and Physicians in Bad Nauheim resulted in the weakening of the anti-relativistic campaign and induced a steady decline of the opposition to relativity theory as time went by. The term *anti-relativists* refers to all those who opposed special or general relativity as a whole, not just a particular feature of it or some technical detail.

In 1922, Lenard, Gehrcke, and 17 others, mostly physicists, mathematicians, astronomers, and philosophers, signed a public protest against the theory of relativity on the occasion of the centenary celebration of the Association of German Researchers in the Exact Sciences and Physicians, in Leipzig. The protest was distributed at the meeting and printed in the press.[2] A formal reason for this protest was the fact that the organizers had allowed a talk on the theory of relativity during the main plenary session—much to the discomfort of Lenard and Gehrcke.[3]

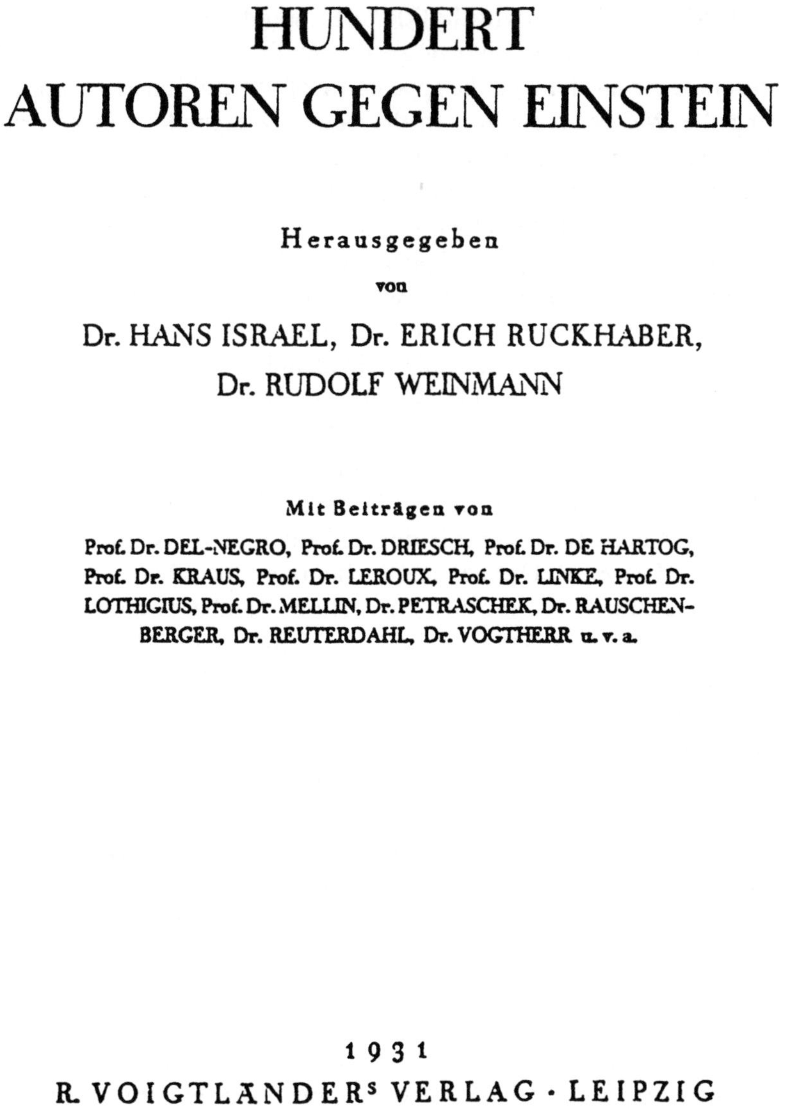

HUNDERT
AUTOREN GEGEN EINSTEIN

Herausgegeben
von

Dr. HANS ISRAEL, Dr. ERICH RUCKHABER,
Dr. RUDOLF WEINMANN

Mit Beiträgen von

Prof. Dr. DEL-NEGRO, Prof. Dr. DRIESCH, Prof. Dr. DE HARTOG, Prof. Dr. KRAUS, Prof. Dr. LEROUX, Prof. Dr. LINKE, Prof. Dr. LOTHIGIUS, Prof. Dr. MELLIN, Dr. PETRASCHEK, Dr. RAUSCHENBERGER, Dr. REUTERDAHL, Dr. VOGTHERR u. v. a.

1931
R. VOIGTLÄNDERS VERLAG · LEIPZIG

Figure 1. Title page of "A Hundred Authors against Einstein."

In the following years, public interest in the theory of relativity waned. Also, a majority of theoretical physicists in Germany moved away from a theory with little potential for experiments and testable consequences. Instead, they took part in the development of quantum theory and its application to atomic and nuclear physics.

In 1931, two years before the Nazi takeover, a booklet (see Figure 1) of roughly a hundred pages with the title "A Hundred Authors against Einstein" (abbreviated HAE in the following) was published (Israel et al. 1931). It represented the last joint public appearance of some of Einstein's opponents during the Weimar republic. In the book, 28 signed, short contributed statements against relativity theory (special or general) were collected. A list of 92 further authors declared by the editors to be opponents [of Einstein] is given, 20 of whom had already died. (A list of all names appearing in HAE is given in this chapter's Appendix.) Of these 92 authors, another 19 were selected and presented with excerpts opposing the theory of relativity. The book definitely does *not* reflect an *intraphysics* dispute but rather the reaction of part of the academically trained middle class of German society.

A. Pais suggests (Pais 1982) that HAE is related to Weyland's infamous association referred to above. An indication in this direction may be seen in a reference, in the book's preface, to the public protest of 1922 in Leipzig by the group including Lenard and Gehrcke. In fact, five of the protest's signers are among those contributing actively to HAE. With the exception of three (L. Glaser, R. Orthner, and J. Riem), all the others are included in the list of 92 authors assembled by the editors. Neither Lenard nor Gehrcke were active contributors to HAE, however. Lenard, since 1924, had openly backed Hitler and his National Socialist Movement but did not engage any further in public anti-Einstein activities.[4] From the correspondence of Ernst Gehrcke with Nobel prize winner Johannes Stark[5] during the years 1924–1931, we know that Gehrcke permanently watched and complained about what he called "Einstein-billing" (*Einstein-Reklame*). He tried to rally opposition against Einstein and the theory of relativity, openly, by writing a book (Gehrcke 1924) and, covertly, by trying to pull strings to influence newspapers through the right-wing magnate Stinnes,[6] and get Lenard or Stark on the Board of the Physikalisch-Technische Reichsanstalt in Berlin. Gehrcke wanted a counterweight for the then acting president Paschen, whom he disqualified as "a philo-Semite and democrat."[7] In the correspondence, Stark suggested that Gehrcke write another brochure against Einstein for the general public and offered to recommend the manuscript to the publisher of the Nazi movement.[8] According to Gehrcke's working schedule he felt unable to finish such a book before 1932.[9] HAE is not mentioned in the correspondence of 1931 despite its earlier appearance and the prior publication of two book reviews. It appears unlikely to me that Gehrcke or Lenard stood behind HAE. The same can be said of Weyland, who had no further political use for the theory of relativity after 1925 (cf. Goenner 1993). Stark had claimed earlier that Gehrcke could not become a full professor, in Germany, because of his fight against the theory of relativity (Stark 1922).

The referencing journal of the German Physical Society, *Physikalische Berichte*, printed a brief notice on HAE signed by its junior editor. He noncommittally reprinted two sentences from the book's preface describing its rationale: "It is the aim of this publication to confront the terror of the Einsteinians with an overview of the quality and quantity of the opponents [of the theory of relativity] and opposing arguments. It is also our goal to enlighten the general public and to assist in the solution of the problems under debate."

Hans Reichenbach, in his usual role as intellectual bodyguard for Einstein and his theories, wrote a report on HAE for the entertainment section of the Berlin daily *Vossiche Zeitung* (Reichenbach 1931). In it, he ridiculed the book as "a magnificent collection of naive mistakes" and as "unintended droll literature." His technique was to quote and comment on some of the most bizarre statements in the book without disclosing the particular authors. He also wondered why the well-known publisher Voigtländer in Leipzig had bothered to offer HAE to the public.

A more detailed report was given by the astronomer A. von Brunn (von Brunn 1931), an unpaid collaborator in the Einstein observatory in Potsdam and co-worker of Freundlich and van Klüber. Von Brunn earlier had successfully challenged a note of Einstein in the *Sitzungsberichte* of the Prussian Academy of Sciences concerning irregular fluctuations in the moon's position (Einstein 1919; von Brunn 1919). Von Brunn characterized HAE as a pamphlet "of such deplorable impotence as occurring elsewhere only in politics" and "a fallback into the 16th and 17th centuries." He asked whether, perhaps, *weltanschauliche* antipathies were the unique motive of the book. Under *weltanschauliche* we must not understand the Kantian philosophy ascribed by him to most of the contributors, but rather an allusion to political and ethical views. His scathing report ends with the remark: "It can only be hoped that German science will not again be embarrassed by such sad scribblings."

Although both reviewers are perfectly right in discrediting HAE in terms of scientific and scholarly value, I think that neither of them managed to leave the narrow angle of his discipline. To me, the book is an example of a committed but inadequate reaction by the educated middle class in Germany (*Bildungsbürgertum*) to a topic in the exact sciences without immediate economical or technological consequences: relativity theory. There are not very many examples of such an interaction; in the 19th century, Darwin's evolution theory would be an example; in our times, maybe, the crisis in the foundation of mathematics as felt by its ripples through society in the form of the "new math." In the following, I will try to convince you that the inclusion of sociological and psychological aspects in the history of physics can be helpful for a better understanding of the motives of past actions. We can see clearly that three themes are hidden behind much of the anti-relativistic thinking in HAE: the dispute about the relationship between "objective reality" and physical theory, the lack of knowledge about the manner by which public consent among scientists required for the validation of a scientific theory is established, and the question of what possible role nonscientists can play in this process. Intertwined with these intrascience problems are questions of *scientific style*, e.g., personality cult and the intrusion of advertising into science; *social status*—in Germany during the first decades of this century, teachers at the higher schools preparing students for university studies were losing both prestige and contact with university research; *political and moral conservatism*, and finally, *racial prejudice* (anti-Semitism).

2. Einstein's Opponents

In order to substantiate my thesis I will ask some obvious questions, i.e., what made the editors band together, who were the contributors to HAE, and what were their problems in understanding special and general relativity?

My first surprise came when I found out that none of the editors of HAE was a practicing scientist and, moreover, none was working in a profession close to physics or to any other of the exact sciences, although all three of them had obtained Ph.D. degrees. The senior editor (*Schriftführer*), Rudolf Weinmann (1870–?) wrote his thesis in Munich, in 1895, under the guidance of the philosopher and psychologist Stumpf on "the concept of specific sense energies" (Weinmann 1895). In it, he investigated a hypothesis made in the first third of the 19th century by Johannes Müller, i.e., that particular energies mediating perception exist in each sensory organ. Weinmann concluded that the concept of "specific sense energies" is of no epistemological significance. Four years later, Weinmann became an actor in Heidelberg and, subsequently, in Graz, Köln, Vienna, Dresden, and Berlin. In 1913 he published an adaptation for the stage of Friedrich Schiller's drama, *Die Räuber*. Weinmann became a member of the Deutsches Theater in Berlin and is listed as an actor in the Berlin address book as late as 1928.[10] In 1935, he described his interests as epistemology and relativity theory.[11] Before 1931, in addition to his Ph.D. thesis, Weinmann had published five booklets with four of them directed against relativity theory (Weinmann 1922a, 1922b, 1923, 1926a). He also published five articles criticizing relativity theory in philosophical and general culture journals (Weinmann 1926b, 1927, 1929a, 1929b, 1930). Weinmann solicited contributions for HAE, sometimes unsuccessfully as in the case of the philosopher Hugo Dingler.

Dingler, in 1930, refused to participate because he had the impression that "something is put to a majority vote which can only be decided upon by scientific arguments."[12] Already in April of 1930, Weinmann, at the end of his paper on "the fallacy of the special theory of relativity" (Weinmann 1930), listed 21 anti-relativists, 10 of whom were to contribute actively to HAE.

Erich Ruckhaber (1876–1956) worked as an interpreter, first private, then official (*amtlich*), and translator in Berlin.[13] He had a foible for philosophy as well as for poetry and drama. While he expressed his literary ambitions in a volume on the awakening of the world spirit (Ruckhaber 1919), his philosophical interests led to books on the mechanism of human thinking, memory and thinking as a function of the system of muscles (Ruckhaber 1910, 1911, 1915) and to a doctorate degree in 1927, i.e., when he was 51 years old. His thesis was concerned with logic (Ruckhaber 1927). Relativity theory seems to have been only one of his numerous other interests reaching from writings on the cause of aging (Ruckhaber 1938) to the discovery of a coding machine for military use[14] (Ruckhaber 1924) and the teaching of shorthand writing (Ruckhaber 1924) to presenting "world wisdom in 100 theses" (Ruckhaber 1939). On relativity, he wrote a satirical essay and a brochure aimed at refuting the special theory of relativity by a logical argument (Ruckhaber 1928, 1929a). Ruckhaber also ventured into microphysics by unearthing Descartes' theory of ether eddies. He transformed it by postulating that the attraction and repulsion of atoms are not explained by electromagnetism but by a sort of spin–spin interaction between vortices in the ether (Ruckhaber 1953). His pamphlet concerning planetary science, i.e., the ether mechanics of the solar system (Ruckhaber 1941) drew a devastating critique in the journal of the German Nazi student movement (Korn 1941). Worse still, Ruckhaber suggested a proof that imaginary numbers do not exist (Ruckhaber 1929b, 1930).

The youngest of the editors, Hans Israel (1881–?), first obtained a degree in engineering (*Dipl. Ing.*) and then, in 1905, wrote his Ph.D. thesis in the chemistry laboratory of the University of Rostock on the theory of flow time (of a fluid from a vessel). In it he considered the time needed by a fluid in a container to stream out through a hole and studied its dependence on the goemetry of the container (Israel 1905). Years later, Israel returned to his native Berlin, where he made his living as a drugstore owner and chemist.[15] He must have felt, though, that his vocation was philosophy, because he published two books concerned with Kant's critique of pure reason (Israel 1911, 1925). In them, he tried to "link successively the systems of physics, logic, and ethics. It also will prove advantageous to join esthetics...." A closer look reveals him as a genuine crackpot, who set out to show "that the truth of the world integral withstands all Kantian arguments against a solution of the deepest metaphysical questions."[16] His "world integral" derived from the limit equations $0 \cdot \infty = \text{const}$, an equation to be handled with extreme mathematical care for it to make sense. According to his report, he had come upon and been puzzled by this equation in the course of his Ph.D. work. I consider his philosophical writing to be sheer fantasy, sometimes bordering insanity.

Obviously, these three men were united not only by their common interest in philosophy and opposition to relativity theory but also by their incompetence in the fields of mathematics and physics. Israel, the only one of the editors with a background in engineering, chemistry, and physics, was not an exception. A brief reading of his brochure against special relativity theory shows this (Israel 1929). In discussing Michelson's interferometric experiment, Israel claims to have derived, for the reflected beam parallel to the velocity of the earth, a factor in the Lorentz transformation of the form $\left(1 + (v/c)^2\right)^{-1/2}$ in place of the usual $\left(1 - (v/c)^2\right)^{-1/2}$. His derivation contains at least two obvious errors. The well-known astronomer E.F. Freundlich wrote a report on Israel's brochure for the Prussian Minister of Science, the Arts, and General Education.[17] He concluded that "the publication belongs to the kind of anti-relativistic writings in which pathological parts and contributions resulting from misunderstandings can be discerned only with difficulty.... In any case, the objections against the theory of relativity [in Israel (1929)] cannot be taken seriously."

Which respected academic would join forces with such editors? Unbelievably, two of the 28 contributors to HAE receive international recognition even today: the German philosopher Hans Driesch[18] (1867–1941) and the Finnish mathematician Hjalmar Mellin[19] (1854–1933). We also should not forget the other university professors who joined the lot, i.e., the philosophers O. Kraus and P. Linke, and the theoretical physicist J. LeRoux.

H. Mellin is known for the Mellin transform, an integral transformation equivalent to the Laplace transformation. In his exchange with Hans Reichenbach, Mellin described himself as a conventionalist in the spirit of H. Poincaré. For him, the objects of mathematics are purely mental: "There is no bridge between mathematics and reality" (Mellin 1931). Therefore, asking whether space is Euclidean or non-Euclidean does not take make sense to him. Moreover, if this question could be given a meaning, according to his opinion empirical experience cannot form a truth criterion for geometry. Mellin also blamed Gauss as well as other workers in the field of non-Euclidean geometry (Bolyai, Riemann, Helmholtz) for having made the mistake of an incomplete logical disjunction: space could be neither Euclidean nor non-Euclidean.

Also, Mellin does not accept Reichenbach's definition of time-ordering through causality. Possibly he has a point here, because we cannot discern between cause and effect without an arrow of time. By pointing out that *time* has to be defined before we speak of *clocks*, Mellin shows that he did not distinguish between the undefined terms, in an axiomatic approach, from the definitions given with their help (Mellin 1933).

Hans Driesch, professor of philosophy at the University of Leipzig, is considered the outstanding representative of neovitalism. Nevertheless, Driesch was also a critical realist and an "inductive" metaphysician along the lines of Aristotle and Leibniz. It is thus no wonder that he characterized Einstein as "a contemporaneous physicist in the grip of a functional-mathematical view of the world" in contrast to his own view, which he called "causal, of natural logic" (Driesch 1924). Nevertheless, Driesch protected Einstein from the vulgarizers of his theory of relativity by commenting that "Einstein and Spengler have as much in common as the critique of pure reason and the main railway station of Leipzig."[20]

Driesch had studied zoology with Weisman and Haeckel and experimented intensively with the embryonic development of sea urchins. By this experience, he broke away from Haeckel's mechanistic interpretation of the organism and replaced it with the assumption of some nonmechanical holistic factor in nature. The organism is not just the sum of its parts but the sum of some holistic causality. In dealing with the mind and the unconscious, Driesch believed that we may find, in parapsychological phenomena, traces of a supraindividual wholeness (Driesch 1932). Such studies of occult phenomena did not make him friends among physicists. In fact, von Brunn, in his book review of HAE, without explicitly using Driesch's name, referred to him as "one who does not even know the borderlines between exact science and an idle pastime." This is an arrogant statement indeed, in view of Driesch's extensive experimental work in biology. Ironically, the Nazis did not let him preside at a meeting of the International Society for Psychic Research in Oslo, in 1935.

Driesch states that he does accept the special theory of relativity as mathematically consistent physical theory but disputes its value for an interpretation of the world. He could not swallow the principle of the constancy of the velocity of light, i.e., its independence of the light source. Accordingly, for him, the Michelson experiment finds its explanation in the model of an ether carried along by the earth. Special relativity just deals with constraints on measurements that are logically thinkable but almost never realized in practice. From the point of view of ontology, Driesch rejected the assumption of different times for different observers. As to general relativity, he confessed "that I do know only a few things as securely as I know of the absolute validity of Euclidean geometry for physics" (Driesch 1924).

Occasionally, though, Driesch had insights that were physically well founded. At the time, there was a debate on whether the Michelson experiment should be carried through with a light source not moving with the earth; e.g., with light from a star. Driesch concluded that there could be no such external light source. This amounts to saying, in modern language, that the extinction length for visible light in air is smaller than the thickness of the earth's atmosphere.

After this brief encounter with the two most prominent contributors to HAE, we come back to the question of what induced the editors and contributing authors to work together on this book. It seems not unlikely that the editors got to know each other through their common publisher, O. Hillmann, Leipzig, or possibly through joint membership in the Berlin section of one of the philosophical clubs popular at the time, i.e., the *Kantgesellschaft* or *Schopenhauergesellschaft*, and then decided to battle jointly with Einstein and his theory of relativity. It is also not excluded that one of the philosophers involved in the fight with relativists, notably Professor Oskar Kraus of the German University in Prague, brought forth the idea for the book. I have not yet located correspondence that might clear up this point. Kraus, one of the contributors to HAE, published in the journals *Kantstudien* and *Annalen der Philosophie*, the organ of the fictionalists—as did the editors. Kraus referred to Weinmann in his brochure "Open Letters to A. Einstein and M. von Laue," in 1925 (Kraus 1925). With the exception of two (Nachreiner and Strehl) all German and Austrian contributors are linked through their publishers. One of them, G. Wendel, retired principal of a *gymnasium* in Frankfurt/Oder was also president of a German writers union.[21] Until further evidence appears, I am inclined to conclude that the philosophical outlook and the journalistic interests of the majority of the 28 contributing authors of HAE form their common denominator.

This judgment does not exclude further motives for bringing together all these men. It is interesting to note that, even before A. von Brunn's alleged *weltanschauliche* antipathies of anti-relativists, the philosophers O. Kraus and H. Driesch had expressly excluded nonscientific motivations. During the Nazi rule, Driesch, who felt akin to Einstein "as a human being and in ethical–political outlook,"[22] was characterized as inclined to "political liberalism, cultural cosmopolitanism, and a restrained pacifism" in a popular encyclopedia.[23] He was prematurely placed in emeritus status by the Nazi government (Oppenheimer 1970). Kraus, in trying to fend off the criticism he expected concerning political antipathies, referred to his book dealing with the law of nations and world peace that had appeared during World

War I (Kraus 1915). I suspect that most of the contributors had similar conservative political opinions (with the exception of Driesch and de Hartog), but I cannot now give proper evidence.[24]

We briefly look at some of the data concerning the group of people actively involved in HAE and also those listed there as alleged opponents to the theory of relativity by the editors. The group of 28 contributors is academically well trained, with 24 of them having obtained Ph.D. degrees. However, only slightly more than a third of them received this training in physics (4), mathematics (4), or other exact sciences (2). The majority came from the humanities (7), law (4), theology (1), and medicine (1). A minority made its living as university professors (6) or as teachers in schools leading to university (5). At the time of publication of HAE, only two were working in physics.

One of them was Jean LeRoux (1863–1949), professor at the faculty of sciences of Rennes, in Brittany, France, whose specialty seems first to have been mathematical methods applied to physics and, then, the theories of special and general relativity. In 1922, he received the Grand Prix des Sciences Mathématiques of the Academy of Science in Paris. He also wrote a Gaelic novel and was a member in the council of the national as well as local offices for the orphans of World War I.[25] The other physicist, Karl W. Strehl (1864–?), a Bavarian, specialized in optics, had retired from teaching physics and mathematics at the *Gymnasium* in 1923, and worked on as a private scholar.[26]

Four contributors to HAE were freelance writers. One of them wrote a book that I encountered first as a student in the library of the Göttingen mathematical institute. Emanuel Lasker (1868–1941) was the mathematician and chess world champion from 1894 to 1921. (The book in the library was about chess.) Lasker had received his Ph.D. at the University of Erlangen and taught mathematics at the University of Manchester.[27] Later, he lived in Berlin and wrote many books on chess and other games (Lasker 1907, 1919, 1925, 1926, 1929, 1931a, 1931b, 1931c). In philosophical thinking, he was a Bergsonian (Lasker 1913, 1916). Lasker was honored by a postal stamp in the former GDR. To HAE he contributed a mere eight lines, criticizing Einstein for giving a finite value to the velocity of light.

The other writer whose thread I found without knowing his anti-relativity stand was Salomo Friedländer (1871–1946), better known under his pen name *Mynona* (which is "anonym," i.e., anonymous, reversed): I bought his book entitled *Kant for Children* (Friedländer 1924). Friedländer received his Ph.D. at the University of Jena in 1902 with the thesis "A critique of Schopenhauer's position with regard to the epistemological foundation of the critique of pure reason" (Friedländer 1902). In 1932, he published a booklet against the theory of relativity (Friedländer 1932). Although he also wrote a biography of Julius Robert Mayer, the pioneer of energy conservation (Friedländer 1905), Mynona was better known by his widely read novels, short stories, essays, and droll stories (Friedländer 1914, 1919, 1921, 1922, 1924, 1931, 1935). He also published a book related to a well-known antiwar novel of Remarque (Remarque 1929; Mynona 1929) and an "Anti-Freud" paper (Mynona 1925), thus placing himself near the conservative sector of the political spectrum. Friedländer is remembered even today (Jäger 1991).

Friedländer disputed Einstein's claim that the velocity of light in vacuum does not depend on the observer measuring it: "His claim that the motion of light is independent and does not play a special role relative to other motions is unfounded and cannot be understood; therefore the whole theory is untenable." In this regard he follows an esteemed friend and fellow Kantian judge, E. Marcus, who had criticized Einstein on logical grounds (Marcus 1926; Friedländer 1930).

Two-thirds of the contributing authors were from Germany, with the majority living in Berlin and North Germany (Prussia). I do not know who established contacts with the authors coming from France, the Netherlands, Sweden, Finland, and the U.S.A. Possibly Hugo Dingler was one of the middlemen. In 1921, he had corresponded with A. Reuterdahl, then dean of the Department of Engineering and Architecture at the College of St. Thomas in St. Paul, Minnesota. Reuterdahl had invited Dingler to contribute to his planned book, *Fallacies of Einstein.* He was also in contact with another contributor to HAE, Dr. Sten Lothigius of Stockholm, who backed a mechanical theory of light (Lothigius 1920). There exist traceable connections to Austria and Czechoslovakia.

The age distribution of those contributing is revealing: of the $\simeq$ 90% whose birth year I found, less than a third fall into the same age bracket (i.e., ±3 years) as Einstein, while 57% are older (up to 26 years). Only two men were considerably younger (by 6 and 19 years, respectively).

Life is not as one-dimensional and streamlined as to facilitate the writing of history. Four of the 28 contributing authors were Jewish; the three of them still alive when power was handed over to the Nazis left Berlin and Prague (Friedländer, Lasker, Kraus).[28] Two group members were or became outspoken anti-Semites, interestingly, an Austrian, L. von Mitis, (von Mitis 1936), and a U.S. citizen of Swedish origin, Reuterdahl.[29] There are indications that four others collaborated with the Nazi movement, Del-Negro, Petraschek, Rauschenberger, and Vogtherr. Coeditor Ruckhaber also might have had more than just a philosophical resentment when writing: "The fanatics of relativity theory have been cured, more or less, by the number of logically thinking opponents, from their delusion that the great messiah of philosophy has come."[30]

Del-Negro's affinity to the Nazi movement seems to follow from his biographical data.[31] Born in 1898, he advanced in 1940 from a teacher at a *Gymnasium* to a university lecturer (*Universitätsdozent*) in philosophy at the University of Innsbruck; he held this post until 1945 (Del-Negro 1926, 1942). It was only 20 years later, however, that he again obtained a similar position at the University of Salzburg. In between, he worked as a geologist (Del-Negro 1949).

W. Rauschenberger (1880–?), director of the Senckenberg library in Frankfurt, received his Ph.D. degree in law (Rauschenberger 1906). After some publications concerning questions of federal jurisdiction, he devoted himself to the study of heredity and race. He had an illustrious ancestor himself: Andreas Osiander, a Lutheran professor of divinity in the 16th century involved in the dispute between the reformers Luther and Zwingli.[32] It was this A. Osiander who wrote the anonymous preface to "De revolutionibus orbium celestium" of N. Copernicus, in which he reduced to mere hypotheses the revolutionary facts Copernicus tried to establish. Rauschenberger is the author of such writings as *The Philosophical Genius and His Racial Descent* (Rauschenberger 1922a), *The Characterological and Racial Meaning of the Eagle Nose* (Rauschenberger 1922b), and *Hereditary and Racial Psychology of Creative Personalities* (Rauschenberger 1942). In this last book, within a chapter on Nietzsche, he confesses that "all democratical and socialistic teachings and systems derive in the end from Christian thinking and thus from the middle-Eastern and oriental race. . . ." Karl Vogtherr (1882–?), a physician for railway personnel, wrote articles on the measurement of time, simultaneity, and the theory of relativity in philosophy and physics journals, and also a book on simultaneity (Vogtherr 1933). Two of his articles appeared in the journal of the Nazi student movement (Vogtherr 1937/38, 1944). He used a supernova explosion that was given different dates by observers in relative motion as an example of the alleged inconsistency of Einstein's theory of special relativity because "one real unique event cannot happen at two different times in one and the same real world."

Considering now the list of 92 authors referred to as "further opponents or authors of opposing publications," we notice immediately that it must have been assembled without sufficient knowledge of the scientific community and without the consent of those grouped together. Twenty people on the list had died prior to the publication of HAE. The list given includes the names Max Abraham, Friedrich Adler, Henri Bergson, Paul Ehrenfest, Ernst Mach, Paul Painlevé, Henri Poincaré, Wilhelm Wien, Otto Wiener, Emil Wiechert, Erich Kretschmann, and Ernst Reichenbächer. From these and other names on the list we infer that the editors did not distinguish between a scientific debate with, possibly, disagreement on particular technical questions and outspoken disbelief and opposition. It may be that, in the editors' minds, the list is meant to be an appeal to authority: 55% of those listed were university professors, lecturers, and other scientific workers. About one-half worked in physics, astronomy, and mathematics; a sizable fraction were philosophers, but we also find colleagues from chemistry, forestry, pedagogy, anatomy, physiology, biology, and engineering. Not more that 10% of the names collected belong to outright dilettantes. The age distribution corresponds closely to that of the actual contributors to HAE.

Among the names quoted, we find Philipp Lenard and some of his colleagues and collaborators at the Physics Institute of the University of Heidelberg, such as A. Becker, R. Tomaschek, and E. Rupp. Rupp later was involved in a scientific fraud that he tried to cover up by presenting a medical referee report concerning a nervous disease.[33] While J. Stark is not included in the list, we find the mathematician G. Hamel (1877–1954), who did not hide his sympathy for the Nazi rule (Hamel 1938; Lindner 1980). I also traced three outspoken anti-Semites (Gartelmann, Stickers, Ziegler), none of whom belonged to the group of academics. Gartelmann and Stickers were teachers,[34] Ziegler a successful inventor in dye-chemistry turned private scholar (Saager 1930). Stickers and Ziegler were Swiss citizens. All three wrote pamphlets against the theory of relativity and Einstein (Gartelmann 1920, 1934; Stickers 1922; Ziegler 1923). Another name to be added here is that of the industrialist, Nazi politician, and economics expert A. Pfaff, a descendant of the famous mathematician Pfaff.[35] Further research may well reveal anti-Semitism among the academic colleagues of Einstein. Dingler's opportunistic turn from methodological to racial arguments against the theory of relativity is well documented (Wolters 1992). Wolters also produced, from Dingler's diary, anti-Semitic remarks by the astronomer Hugo von Seeliger, who was included in the list of 92 authors.

3. On the Thinking of the Anti-Relativists

We now again address the question of what the opponents of relativity theory contributing to HAE thought. As a guiding principle, I use W. Wien's list of three possible ways for disproving the theory of relativity (Wien 1921):

(1) the uncovering of contradictions within the mathematical formulation;

(2) the demonstration of consequences that are falsified by empirical evidence;

(3) proof that the theory is unsuitable for representing the laws of nature because simple conceptual foundations have been given up.

We combine the first two lines of argument under the label *intrascience* arguments from mathematics or physics. Some of the contributors to HAE must have believed that they had satisfied some or all of these criteria.

The claim of mathematical inconsistency of Einstein's theory was made by H. Israel, while a physical inconsistency, i.e., a conflict between the basic postulates of special relativity—the relativity principle and the principle of the constancy of the velocity of light—was seen by O. Kraus and G. Wendel. Theoretical physicist J. LeRoux thought that Einstein's general relativity predicted the wrong value for the perihelions shift because he forgot to subtract the shift due to the Newtonian many-body interactions. It is interesting that H. Keller anticipated an observable consequence of relativity theory: trips around the world in easterly or westerly directions take different times. This effect was measured by Hafele and Keating carrying atom clocks on transworld flights in the 1960s. For Keller, such a consequence was mere fantasy and enough reason to declare the theory of relativity to be contradictory. Reuterdahl, as well as Walte, put into question whether any of the three effects in the planetary system had been measured. Wendel joined them with regard to red shift and light deflection. This argument concerned only the empirical status of the theory at the time and is much weaker than the argument Wien asked for. A similarly weak point of view is represented by Keller, who criticized the special theory of relativity for not being helpful in atomic physics. All contributing authors claiming that Michelson's experiment could be explained without the special theory of relativity (Kraus, Rauschenberger, Weinmann, Walte) did not satisfy Wien's demand. Wien's third line of argument possibly was meant to be an intrascience argument as well, but I will bring it together with philosophical arguments against relativity theory. As seen through the eyes of the physicist S. Valentiner (Valentiner 1921), philosophers "in the majority of cases do not try to contradict the foundations of relativity theory from its foundations, but aim at proving the theory as unimaginable (*unvorstellbar*) and, consequently, impossible." In particular, he mentions four points occurring regularly (and in HAE as well), i.e.:

(1) Time dilation and length contraction are construed as conflicting with the necessity of cause and effect;

(2) Absolute simultaneity must be preserved. This view is represented by Gimmerthal, Mellin, Rauschenberger, Vogtherr, and Wendel;[36]

(3) Non-Euclidean space is unacceptable.

Valentiner's last point concerns an argument that makes an appeal to authority:

(4) Einstein's theory contradicts Kant's *a priori* judgments concerning space and time, and hence cannot be accepted. Friedländer and Goldschmidt, both devout Kantians, as well as Nachreiner, use this kind of reasoning.

Perhaps we should not give too much weight to this classification of philosophical arguments by a physicist. A philosopher, P. Linke (1876–1955), discussed the epistemological foundations of special relativity, in particular the operational fixing of simultaneity, and found them unacceptable. He considered problematic the concept of simultaneity of events at the same place because it depends on the concept of time and "there exist various times on an equal footing according to the theory of relativity. In the sense of which of these times do we introduce simultaneity?" Others, like Driesch and Del-Negro or Weinmann saw ontological blunders as well as, for example, an infinity of "realities," referring to the infinity of observers in place of the unique objective reality. Often, philosophers (trained or untrained) endeavored to prove the logical inconsistency of the theory of relativity. Not being equipped with sufficient mathematical knowledge, however, they had to confine themselves to verbal fencing without any chance of reaching the body of the theory attacked. I do not need to go into details here; Klaus Hentschel, in his voluminous book, broadly discussed the various misinterpretations of the theory of relativity by philosophers (Hentschel 1990).

What neither Wien nor Valentiner cared to mention were the arguments advanced against relativity theory following from the ignorance (and the lack of mathematical competence) of some of the opponents of relativity theory. One of these self-made problems concerns semantical misunderstandings. As an example, we note Minkowski's famous statement on the union of space and time which was misconstrued (e.g., by Driesch) to signify full equivalence of space and time and thus rejected as foolish.

Here, a subtle point should be discussed, concerning the obvious discrepancy between the often stated claim that the theory of relativity requires for its understanding the highest intellectual abilities and the endeavor of Einstein (and his continued support for others with the same intention) to bring his theories to the attention of a wider public with only preuniversity knowledge of physics and mathematics. Oskar Kraus[37] was an example of a professor of philosophy trapped between such an invitation to the educated layman (i.e., someone trying to understand relativity theory) and a cold warning sign held up when he raised questions: "No entrance without knowledge of mathematics." In addition, Kraus may have been irritated by Einstein's attack on philosophers (Einstein 1960). Kraus considered himself "as someone convinced by *a priori* truths." He saw a conflict between the principle of relativity and the principle of the constancy of the velocity of light. He complained bitterly about being prevented from answering von Laue or Reichenbach in physics and philosophy journals of good standing (*Zeitschrift für Physik, Logos*) (Kraus 1925). The incommensurability of Kraus' philosophical thinking with the approach of physicists such as, for example, Philipp Frank, is shown clearly by Hentschel (Hentschel 1990).

For him, "the breakdown of relativity cannot be avoided. The enemies of German culture will then wish to triumph in view of this scientific embarrassment." He goes on to say "that then this shame will be reduced by the fact that a few critical people on the German side [like himself] did discover immediately the erroneous foundation of the theory of relativity."

Another group of arguments may be called *extrascience*. It includes the well-known debate, convincingly documented by Hentschel, about whether relativity theory conforms to common sense. Again, within this group of extrascience arguments, semantical misinterpretations lead to opposition against the theory of relativity. For example, relativity theory was made responsible for the extrapolation of the concept "relative" to ethics ("all is relative") and then condemned as "the sick product of a sick time."[38] The Dutch theologian de Hartog expressly stated that his only motivation for contributing to HAE was the objection to this sort of moral relativity—not the wish to dispute Einstein's great gifts and importance in science.

Another item on the list of extrascience arguments against Einstein and his relativity theories is the discussion about priorities, or worse, the plagiarism charge made against Einstein. As to priorities, Geissler, Richter, and Reuterdahl each claimed to have been first. Geissler in HAE: "As early as 1900, I published a comprehensive, general "possible" theory of relativity in space, time, etc., while Einstein published some particulars on relativity from 1905 on without referring to my book" (Geissler 1921). The book referred to is Geissler 1900. Geissler thought that he was first in suggesting the relativity of time. Richter flatly stated: "Without having heard of Einstein, already 10 years ago (i.e., 1911), I reached the same final results as he did through pure philosophical reasoning without needing the relativity of time" (Richter 1921). And finally, Reuterdahl (in HAE): "The writer of the present contribution sketched, in 1902, the thought of unitary field comprising all kinds of forces.... In 1913 I coined the hyphenated expression 'space-time' which received the copyright in 1915.... Einstein's molluscous reference system has been built after the plan of my system of potential zones...." Reuterdahl also suggests Palágyi, von Soldner, and Gerber as Einstein's precursors.

Another line of argumentation against relativity theory and Einstein foreign to science stems from the presentation of the theory in public, i.e., the advertisement it received, the style of some publications (including Einstein's frontispieces), and the appraisal of the theory as one of the finest achievements in science of mankind well before its empirical basis was secure enough. Einstein himself sometimes was declared free of guilt in this context (Driesch 1924). Reuterdahl did not shy away from invectives and, in HAE, once called him the "Barnum of science." It is hard to imagine the extent to which, in the 1920s, Einstein's theory of relativity must have been the talk of the town. To some, the popular acclaim the theory received just was too much: "He who does not understand the least of his theory professes Einstein, professes the theory of relativity. He owns a *Weltanschauung* of relativity, a positive belief in relative nothingness; he is mor Catholic than a Roman Catholic; he is a follower of Einstein: *credo quia absurdum*" (Lewin 1932).

4. Conclusion

The opponents of relativity theory and Einstein gathered in HAE form a group of ambitions, extroverted people with broad intellectual interests. It is surprising to me that the majority of them made contact with the theory of relativity. Their professional flexibility and experience was a poor match for the philosophical and ethical rigorism expressed; it also contrasted with their technical incompetence in matters of physics and mathematics. Within the politically and economically highly unstable environment of the Weimar Republic they could not adapt to the changes brought to our world view due to Einstein's special and general theory of relativity. In this sense, it was both nonacceptance and lack of understanding that made these men anti-relativists.

In their interaction with the theory of relativity, these people, scientists or not, lacked the delicate balance between a healthy feeling of self-assurance and the well-developed sense of self-criticism necessary for creative work. In about half of the contributors to HAE, the discrepancy between technical incompetence in physics and mathematics is compensated by either hubris (Einstein and relativists are making trivial mistakes) or by the feeling of being repressed and censored (relativists form a gang terrorizing those with differing opinions). HAE is a classic piece of evidence for the fact that the validation of a theory in the exact sciences can be achieved only from within the body of experts. What is required is a sufficient knowledge of the methods for gaining empirical data, for relating these data by self-consistent mathematical models, and for reaching public consent—always revisable—on the explicatory and predictive value of the resulting theory. By "public" here the *public of scientists* is meant. Acceptance or nonacceptance of the theory by the general public is irrelevant, including, up to a point, the criticism by and the evaluations through other scientific disciplines, such as, for example, the philosophy of science. Although HAE does not exactly present an encouraging case of the interaction of theoretical physics and the educated layman, it nevertheless indicates clearly that the history of science cannot be separated from the general cultural and social environment of science.

ACKNOWLEDGMENTS. I would like to thank John Earman and John Norton for the opportunity to present a first version of this paper at the 3rd International Conference in the History and Philosophy of General Relativity, and Lorraine Daston for inviting me to give a talk at her seminar on the history of science at the University of Göttingen. John Norton's editing improved my English greatly. The reference to Lewin 1932 and to the report of Freundlich (Note 17) I owe to Klaus Hentschel, Göttingen. Gereon Wolters, Konstanz, kindly informed me about Dingler's correspondence with Reuterdahl. I also profited from the critical remarks of Skuli Sigurdsson, Göttingen.

NOTES

[1] Cf. Goenner (1993), "The Reaction to Relativity Theory I: The Anti-Einstein Campaign in Germany in 1920."

[2] *Der Berliner Westen* Nr. 212, September 20, 1922, p. 6, under the heading "A Learned Protest."

[3] Cf. my paper in preparation: "The Reaction to Relativity Theory II—The Years 1921–1926."

[4] Cf. the article by Lenard in *Großdeutsche Zeitung*, May 8, 1924, and *Beyerchen*, 1977. Nevertheless, in the preface to a reissue of his World War I pamphlet against the English, Lenard combined his anti-Semitic stance with a criticism of "the permanent homage to Einstein's thinking" (Lenard 1940).

[5] The correspondence is kept in the Staatsbibliothek zu Berlin–Preußischer Kulturbesitz (Nachlass J. Stark), to which I give thanks for the permission to quote from it.

[6] Gehrcke to Stark, April 5, 1924.

[7] Gehrcke to Stark, April 5, 1924.

[8] Stark to Gehrcke, May 13, 1931; June 9, 1931.

[9] Gehrcke to Stark, May 17, 1931.

[10] Berliner Adressbuch 1919–1931.

[11] Kürschners Deutscher Gelehrtenkalender 1935, 5. Ausgabe, G. Lüdtke, ed. Berlin: W. De Gruyter, Spalte 1508.

[12] Quoted after Wolters 1992, p. 281.

[13] Berliner Adressbuch 1919–1931 (cf. Note 15).

[14] From the dust jacket of Ruckhaber 1953.

[15] Berliner Adressbuch 1919–1931, entries 1925–1931, K. Umlauf, ed. (On microfiche in the Bibliothek Preußischer Staatsbesitz, Berlin.)

[16] Israel 1911, p. 5.

[17] Letter from Freundlich to Prussian Minister for Science, the Arts and General Education, December 10, 1929, Zentralarchiv der DDR, Astrophysikalisches Observatorium Nr. 148.

[18] Cf. *Encyclopedia of Philosophy* (P. Edwards, ed.), New York, 1967. Cf. also Oppenheimer 1970.

[19] Cf. Suomen Elämä-Kerrasto, Helsinki, 1955.

[20] Driesch 1924, p. 45. A reference to Spengler 1927.

[21] Kürschners Deutscher Gelehrtenkalender 1928/29. 3. Ausgabe, Berlin: W. De Gruyter. Cf. also the 4th edition, 1931.

[22] Driesch 1924, p. 41.

[23] Meyers Lexikon, 8. Aufl., 3 Bd., Leipzig (1937), p. 265.

[24] Leonore Ripke-Kühn (1878–1955), the only woman anti-relativist in the list of opponents, was active in the women's section of the right-wing political party DNVP (Deutsche Nationale Volkspartei) in Berlin and Danzig. Cf. Deutschbaltisches Biographisches Lexikon 1710–1960, W. Lenz, ed. Köln, Vienna 1970.

[25] *Qui êtes-vous?* Paris (1924), p. 476, and Poggendorff VII, Teil 5 (1976), p. 2830.

[26] Poggendorff's biographisch-literarisches Handwörterbuch V (1926), p. 1221.

[27] Meyers Lexikon, 8. Aufl., Band 7, Leipzig, 1939.

[28] L. Goldschmidt, an insurance mathematician and teacher, died in 1931. Cf. Heuer, Renate (1981–1988). *Bibliographia Judaica*, Three Volumes. Frankfurt/New York: Campus.

[29] Arvid Reuterdahl (1876–1933) taught mathematics, theoretical and applied mechanics and received, in 1923, a doctorate in science by an Academy of Nations. He published books on reinforced concrete arches and on theism versus materialism and disputed the value of Einstein's theories in public. Cf. *Who's Who in America* 14: 1601 (1926/27) and *Who Was Who in America* 1: 1022 (1962).

[30] Ruckhaber 1941, p. 10.

[31] Kürschners Deutscher Gelehrtenkalender 1940/41, 6. Ausgabe Nachträge and Kürschners Deutscher Gelehrtenkalender 1976, 12. Ausgabe. Berlin: W. De Gruyter.

[32] Reichshandbuch der Deutschen Gesellschaft Bd. 1/2, Berlin 1930–1931. Deutscher Wirtschaftsverlag AG, Berlin.

[33] Cf. two notes in *Zeitschrift f. Physik* (1935)95: 801; 96: 278; and Gerlach 1979.

[34] For Gartelmann, cf. Kürschners Gelehrtenkalender 1935. Concerning Stickers, this is an indirect conclusion drawn from Stickers 1922.

[35] Cf. G. Wolters 1987, p. 348.

[36] Armin Gimmerthal was a playwright (Gimmerthal 1901, 1902), who also wrote a brochure against Einstein's theory of relativity (Gimmerthal 1926).

[37] Cf. Wien 1921.

[38] Von Mitis, in HAE, p. 35.

References

Beyerchen, Alan D. (1977). *Scientists under Hitler.* New Haven and London: Yale University Press.

Del-Negro, Walter (1926). *Der Sinn des Erkennens.* Munich: Reinhardt.

——— (1942). *Die Philosophie der Gegenwart in Deutschland.* Leipzig: Meiner.

——— (1949). *Geologie von Salzburg.* Innsbruck: Wagner.

Driesch, Hans (1924). *Relativitätstheorie und Weltanschauung.* Leipzig: Quelle und Meyer.

——— (1932). *Parapsychologie.* Die Wissenschaft von den okkulten Erscheinungen. Munich: Bruckmann.

Ebert, Hermann (1931). "Buchanzeige." *Physikalische Berichte* 12: 821.

Einstein, Albert (1919). "Bemerkung über periodische Schwankungen der Mondlänge, welche bisher nach der Newtonschen Mechanik nicht erklärbar schienen." *Sitzungsberichte Preussische Akademie der Wissenschaften,* (Berlin): 433–436.

——— (1920). *Über die spezielle und die allgemeine Relativitätstheorie.* Braunschweig: Vieweg.

——— (1960). *Grundzüge der Relativitätstheorie.* Braunschweig: Vieweg.

Freundlich, Erwin, von Klüber, H., and von Brunn, Albert (1931). "Über die Ablenkung des Lichts im Schwerefeld der Sonne. Dritte Mitteilung der Postdamer Sonnenfinsterniss-Expedition 1929, Sumatra." *Abhandlungen Preussische Akademie der Wissenschaften Phys. Math. Kl.* Nr. 1.

Friedländer, Salomo [Mynona] (1902): *Versuch einer Kritik der Stellung Schopenhauers zu den erkenntnis-theoretischen Grundlagen der Kritik der reinen Vernunft.* Berlin: Selbstverlag.

——— (1905). *Julius Robert Mayer.* Leipzig: Thomas.

——— (1914). *Für Hunde und andere Menschen.* Berlin: Der Sturm.

——— (1919). *Die Bank der Spötter.* Ein Unroman. Munich and Leipzig: Wolff.

——— (1921). *Das widerspenstige Brautbett und andere Grotesken.* Munich: Wolff.

——— (1922). *Graue Magie.* Berliner Nachschlüsselroman. Berlin: Kaemmerer.

——— (1924). *Kant für Kinder.* Hannover: Steegemann.

——— (1925). *Anti-Freud.* Heitere Geschichten. Berlin: Gottschalk.

——— (1930). *Der Philosoph Ernst Marcus als Nachfolger Kants.* Essen: Baedecker.

——— (1931). *Der Holzweg zurück, oder Knackers Umgang mit Flöhen.* Berlin: Steegemann.

——— (1932). *Kant gegen Einstein.* Berlin: Der neue Geist.

——— (1935). *Der lachende Hiob und andere Grotesken.* Paris: Edition du Phénix.

Gartelmann, Henri (1920). *Zur Relativitätslehre.* Eine kritische Betrachtung. Berlin: Verlag der Neuen Weltanschauung.

——— (1934). *Der Fall Einstein.* Dresden: Pierson.

Gehrcke, Ernst (1924). *Die Massensuggestion der Relativitätstheorie*. Berlin: Meuser.

Geissler, Kurt F.J. (1900). *Eine mögliche Wesenserklärung (Relativität) von Raum, Zeit, dem Unendlichen und der Kausalität*. Berlin: Gutenberg.

—— (1921). *Gemeinverständliche Widerlegung des formalen Relativismus (von Einstein und Verwandten)* etc. Leipzig: Hillmann.

Gerlach, Walter (1979). "Erinnerungen an A. Einstein 1908–1930." In *Albert Einstein*, Peter C. Aichelburg and Roman U. Sexl, eds. Braunschweig and Wiesbaden: 199–210.

Gimmerthal, Armin (1901). *Hinter der Maske*. Sudermann und Hauptmann in den Dramen Johannes, Die drei Reiherfedern, Schluck und Jau. Berlin: Schwetschke und Sohn.

—— (1902). *Aschenbachs*. Schauspiel in 4 Aufzügen. Berlin: Schwetschke und Sohn.

—— (1926). *Die Irrtümer und Trugschlüsse in Einsteins Relativitätstheorie*. Langendreer: Gimmerthal.

Goenner, Hubert (1993). "The Reaction to Relativity Theory I: The Anti-Einstein Campaign in Germany in 1920." Göttingen: *Science in Context* 6, No. 1, 107–136.

Hamel, Georg (1938). Buchbesprechung K.A. Doxiadis: Raumordnung im Grichischen Städtebau. *Deutsche Mathematik* 3: 345–346.

Hentschel, Klaus (1990). *Interpretationen und Fehlinterpretationen der speziellen und der allgemeinen Relativitätstheorie durch Zeitgenossen Albert Einsteins*. Basel: Birkhäuser.

Israel, Hans (1905). *Theorie der Ausflußzeiten einer Flüssigkeit*. Rostock: Selbstverlag (Ph.D. thesis).

—— (1911). *Auflösung der Widerspruchslehre Kants. Erster Teil*. Der Kritik der reinen Vernunft. Analytik der Begriffe. Berlin: Mayer und Müller.

—— (1925). *Auflösung der Widerspruchslehre Kants. Zweiter Teil*. Der Kritik der reinen Vernunft. Grundsätze–Antinomie. Berlin: Schwetschke und Sohn.

—— (1929). *Beweise, weshalb die Einsteinsche Relativitätstheorie ad acta zu legen ist*. Leipzig: Hillmann.

Israel, Hans, Ruckmann, Erich, and Weinmann, Rudolf, eds. (1931). *Hundert Autoren gegen Einstein*. Leipzig: Voigtländer.

Jäger, Gabriele (1991). *Wilmersdorfer Portraits*. Berlin: Stapp.

Korn, J. (1941). "Buchbesprechung E. Ruckhaber. Die Ätherdynamik des Sonnensystems." *Zeitschrift für die gesamten Naturwissenschaften* 7: 366.

Kraus, Oskar (1915). *Jeremy Benthams Grundzüge für ein künftiges Völkerrecht und einen dauernden Frieden*. Halle: Niemeyer.

—— (1925). *Offene Briefe an Albert Einstein und Max von Laue über die gedanklichen Grundlagen der speziellen und allgemeinen Relativitätstheorie*. Vienna and Leipzig: Braumüller.

Lasker, Emanuel (1907). *Das Pokerspiel*. Berlin: Steinitz.

——— (1913). *Das Begreifen der Welt*. Berlin: Josef.

——— (1916). *Die Philosophie des Unvollendeten*. Leipzig and Berlin: Vereinigung wissenschaftlicher Verleger.

——— (1919). *Die Anfangsgründe des Schachspiels*. Berlin: Kagern.

——— (1925). *Gesunder menschenverstand im Schach*. Berlin: Siedentop.

——— (1926). *Lehrbuch des Schachspiels*. Berlin: Wertbuchhandlung.

——— (1929). *Das verständige Kartenspiel*. Berlin: Scherl.

——— (1931a). *Brettspiele der Völker*. Berlin: Scherl.

——— (1931b). *Das Bridgespiel*. Berlin: Scherl.

——— (1931c). *Das Skatspiel*. Berlin: Scherl.

Lenard, Philipp (1940). *Ideelle Kontintentalsperre*. Kriegschriften der Reichsstudentführung. Munich: Eher.

Lewin, Robert K. (1932). "Hundert Autoren und Einer gegen Einstein." *Allgemeine Rundschau*, Nr. 14 vom 2. April 1932: 156–159.

Lindner, Helmut (1980). Deutsche und gegentypische Mathematik. In: Mehrtens, Herbert and Richter, Steffen: *Naturwissenschaft, Technik und NS-Ideologie*. Frankfurt: Suhrkamp.

Lothigius, Sten (1920). *Esquisse d'une thèorie nouvelle de la lumière*. Stockholm: Dahlström.

Marcus, Ernst (1926). *Kritik des Aufbaus (Syllogismus) der speziellen Relativitätstheorie und Kritik der herrschenden Hypothese der Lichtausbreitung*. Berlin: Der Sturm.

Mellin, Hjalmar (1931). "Das Zeit-Raum Problem." *Annales Academiae Scientarum Fennicae* 34(A), Heft 3: 1–23.

——— (1933). "Die Widersprüche in der Relativitätstheorie." *Annales Academiae Scientarum Fennicae* 37, Heft 8: 1–13.

Mynona [Friedländer, Salomo] (1925). *Das Eisenbahnunglück oder der Anti-Freud*. Berlin: Gottschalk.

——— (1929). *Hat Erich Maria Remarque wirklich gelebt?* Berlin: Steegemann.

Oppenheimer, Jane (1970). "Hans Adolf Eduard Driesch." In *Dictionary of Scientific Biography*, Charles C. Gillespie, editor-in-chief. New York: Scribner, 1970–1980, Volume 4.

Pais, Abraham (1982). *Subtle is the Lord*. Oxford and New York: Oxford University Press.

Petraschek, Karl O. (1922). *Der Grundwiderspruch in der Speziellen Relativitätstheorie und seine Folgen*. Leipzig: Hillmann.

——— (1926). *Die Logik des Unbewussten*. Munich: Reinhardt.

——— (1938). *System der Philosophie des Staates und des Völkerrechtes*. Zurich and Leipzig: Verlag für Recht und Gesellschaft.

Rauschenberger, Walther (1906). *Der Anteil des Bundesrates an der Reichsgesetzgebung*. Borna and Leipzig: Noske.

——— (1922a). *Das philosophische Genie und seine Rassenabstammung*. Frankfurt am Main: Selbstverlag.

——— (1922b). *Die charakteriologische und Rasse-Bedeutung der Adlenase.* Frankfurt am Main: Selbstverlag.

——— (1942). *Erb- und Rassenpsychologie schöpferischer Persönlichkeiten.* Jena: Fischer.

Reichenbach, Hans (1931). "Hundert gegen Einstein." *Vossische Zeitung Berlin*, Nr. 47 of 24.2.

Remarque, Erich-Maria (1929). *Im Westen nichts Neues.* Berlin: Propyläen.

Richter, Gustav (1921). *Kritik der Relativitätstheorie Einsteins.* Leipzig: Hillmann.

Ruckhaber, Erich (1910). *Des Daseins und Denkens Mechanik und Metamechanik.* Hirschberg: Springer.

——— (1911). *Der Mechanismus des menschlichen Denkens.* Brackwede: Breitenbach.

——— (1915). *Die Steigerung des Gedächtnisses und der Denkfähigkeit.* Berlin: Psychol.-Soziolog. Verlag.

——— (1919). *Des Weltgeists Erwachen.* Eine Dichtung. Berlin: Neue Weltanschauung.

——— (1924). *Natürliche Kurzschrift.* Berlin: Selbstverlag.

——— (1927). *Untersuchung über das Prinzip des Widerspruchs.* Berlin: Ebering.

——— (1928). *Die Relativitätstheorie widerlegt durch das Widerspruchsprinzip und die natürliche Erklřung des Michelsonversuchs.* Leipzig: Hillmann.

——— (1929a). *Relativia.* Der Roman eines Propheten, eine philosophische Humoreske. Berlin-Spandau: Kuntz.

——— (1929b). "Logische und sprachliche Verneinung. Die neutralen, positiven und negativen Zahlen." *Annalen der Philosophie* 8: 348–352.

——— (1930). *Annalen der Philosophie* 10, Heft 8/10.

——— (1938). *Biomechanik*: Das gelöste Lebensproblem. Die alleinige Bekämpfung des Alterns und sämtlicher Alterserscheinungen, die einzige Hilfe für Gelähmte. Berlin: Pape.

——— (1939). *Weltweisheit in 100 Thesen.* Berlin: AGV-Verlag.

——— (1941): *Die Ätherdynamik des Sonnensystems.* Berlin: AGV-Verlag.

——— (1953). *Die Ätherwirbeltheorie vielfach bewiesen.* Die Folgen für die Relativitätstheorie. Berlin: Schikowski.

Saager, Adolf (1930). *Der Winterthurer Naturphilosoph Johann Heinrich Ziegler.* Winterthur. Zurich: Welformelverlag.

Spengler, Oswald (1927). *Der Untergang des Abendlandes.* Munich: C.H. Beck.

Stark, Johannes (1922). *Die gegenwärtige Krise in der Deutschen Physik.* Leipzig: Barth.

Stickers, Joe (1922). *Die wahre Relativitätstheorie der Physik und die Mißgriffe Einsteins.* Bielefeld: Breitenbach.

Valentiner, Siegfried (1921). *Wie stellt sich die Philosophie zum Prinzip der Relativität?* Rede gehalten bei der Rektoratsübergabe in Clausthal am 7.11.1921. Clausthal: Pieper.

Vogtherr, Karl (1933). *Das Problem der Gleichzeitigkeit*. Munich: Reinhardt.

——— (1937/38). "Über die Erkenntnis von Raum und Zeit." *Zeitschrift für die gesamten Naturwissenschaften* 3: 145–146, 201–219.

——— (1944). "Das Dilemma der Relativitätstheorie." *Zeitschrift für die gesamten Naturwissenschaften* 10: 41–67.

von Brunn, Albert F.J.W.L. (1919). "Zu Herrn Einsteins Bemerkung über die unregelmäßigen Schwankungen der Mondläge von der genäherten Periode." *Sitzungsberichte Preussische Akademie der Wissenschaften*, (Berlin): 710–711.

——— 1931). "Buchbesprechung Hundert Autoren gegen Einstein." *Naturwissenschaften* 19: 254–256.

von Mitis, Lothar (1936). *Einsteins Grundirrtum*. Leipzig: Hillmann.

——— (1936). *Beaconsfield, Benjamin Disraeli: Die jüdische Weltherrschaft*. Übersetzt und erläutert von L. von Mitis. Leipzig: Hillmann.

Weinmann, Rudolf (1895). *Die Lehre von den spezifischen Sinneswahrnehmungen*. Hamburg and Leipzig: Voss.

——— (1913). *F. v. Schiller, Die Räuber*. Für die Bühne bearbeitet. Leipzig: Reclam.

——— (1922a). *Philosophie Welt und Wirklichkeit*. Eine erkenntnistheoretische Skizze. Munich and Berlin: Oldenburg.

——— (1922b). *Gegen Einsteins Relativierung von Zeit und Raum*. Gemeinverständlich. Munich and Berlin: Oldenburg.

——— (1923). *Anti-Einstein*. Leipzig: Hillmann.

——— (1925). *Widersprüche und Selbstwidersprüche der Relativitätstheorie*. Leipzig: Hillmann.

——— (1926a). *Versuch einer endgültigen Widerlegung der speziellen Relativitätstheorie*. Leipzig: Hillmann.

——— (1926b). "Kommt der Relativitätstheorie philosophische Bedeutung zu?" *Philosophie und Leben* 2, Heft 1: 154–159.

——— (1927). "Anti-Einstein Quintessenz." *Archiv für Systematische Philosophie und Soziologie* 30: 263–270.

——— (1929a). "Der Widersinn und die Überflüssigkeit der speziellen Relativitätstheorie." *Annalen der Philosophie* 8: 46–57.

——— (1929b). "Über einige philosophische Argumente gegen die Relativitätstheorie (zum Aufsatz von Hugo Bergmann)." *Kantstudien* 34: 254–255.

——— (1930). "Die Unhaltbarkeit der speziellen Relativitätstheorie." *Natur und Kultur* 27: 121–125.

Wien, Wilhelm (1921). *Die Relativitätstheorie vom Standpunkt der Physik und Erkenntnislehre*. (Vortrag bei Fa. Siemens am 18.3.1921.) Leipzig: Barth.

Wolters, Gereon (1987). *Mach I, Mach II, Einstein und die Relativitätstheorie*. Berlin: de Gruyter.

——— (1992). Opportunismus als Naturanlage: Hugo Dingler und das ⟨Dritte Reich⟩. In: *Entwicklungen der methodischen Philosophie*, Peter Janich, Hrsg. Frankfurt: Suhrkamp.

Ziegler, Johann H. (1923). *Das Ding an sich und das Ende der sog. Relativitätstheorie*. Zurich: Weltformel-Verlag.

Appendix

Contributors and Authors Listed in the Name Index of 'A Hundred Authors against Einstein'

The volume's name index was divided into a list of contributors to the volume, a list of further authors who were identified as opponents of Einstein or as having published writings opposing him and a list of excerpts from publications opposing Einstein. These lists are reproduced below without associated page numbers.

Contributors:

Del-Negro, Walter
Driesch, Hans
Friedländer, S.
Geißler, J.K.
Gimmerthal, Armin
Goldschmidt, Ludwig
Hartog, A.H. de
Israel, Hans
Keller, Hugo
Kraus, O.
Kuntz, W.
Lasker, Emanuel
LeRoux, J.
Linke, P.F.
Lothigius, Sten
Mellin, H.
Mitis, Lothar
Nachreiner, Vincenz
Petraschek, K.O.
Rauschenberger, Walther
Reuterdahl, Arvid
Richter, Gustav
Ruckhaber, Erich
Strehl, Karl W.
Vogtherr, Karl
Walte, W.
Weinmann, Rudolf
Wendel, Georg

Excerpts from Opposing Publications:

Fricke, H.
Friedländer, S.
Frischeisen-Köhler, M.
Gehrcke, E.
Gilbert, L.
Kirschmann, A.
Kraus, O.
Kremer, J.
Lenard, P.
Linke, P.F.
Lipsius, F.
Mohorovičič, S.
Nyman, A.
Palágyi, M.
Ripke-Kühn, L.
Thedinga, E.
Weinstein, B.
Wittig, H.
Ziehen, T.

FURTHER OPPONENTS OR AUTHORS OF OPPOSING PUBLICATONS:

Abraham, M.
Adler, F.
Alliata, G.
Anderson, W.
Balster, W.
Becher, E.
Becker, A.
Benedicks, K.
Bergson, H.
Bottlinger, K.F.
Bucherer, A.H.
Budde, E.
Dennert, E.
Dingler, H.
Drechsler, J.
Ehrenfest, P.
Fricke, H.
Friedrichs, G.
Frischeisen-Köhler, M.
Gartelmann, H.
Gawronsky, D.
Gehrcke, E.
Geppert, H.
Gilbert, L.
Gleich, G. von
Großmann, E.
Häring, T.
Hamel, G.
Hartwig, Ernst
Hirzel, J.E.G.
Höfler, A.
Isenkrahe, C.
Jovičič, Milorad
Karollus, F.
Kirschmann, A.
Klages, L.
Krauße, A.
Kremer, J.
Kretschmann, E.
Kries, J. von
Lauer, H.E.
Lecher, E.
Lenard, P.
Leopold, C.
Lipsius, F.
Mach, E.
Maier, H.
Mauthner, Fritz
Mohorovičič, S.
Nyman, A.
Painlevé, P.
Palâgyi, M.
Péczi, G.
Pfaff, A.
Podeck
Poincaré, H.
Prey, A.
Raschevsky, N. von
Rehmke, J.
Reichenbächer, E.
Riedinger, Franz
Ripke-Kühn, L.
Rothe, R.
Rupp, E.
Sagnac, G.
Schultz, J.
Schwinge, O.
See, T.A.
Seeliger, H. von
Selety, F.
Sittig
Stickers, I.
Strasser, H.
Thedinga, Eddo
Thiry, R.
Tomaschek, R.
Triebel, H.
Tummers, J.H.
Del Vecchio, Giorgio
Wächter, F.
Weinstein, M.B.
Westin, O.E.
Wiechert, J.E.
Wien, W.
Wiener, O.H.
Wittig, H.
Wodetzky, I.
Wolf, M.
Zboril, I.
Ziegler, J.H.
Ziehen, T.
Zlamal, H.

Attempts at Unified Field Theories (1919–1955). Alleged Failure and Intrinsic Validation/Refutation Criteria

Silvio Bergia

1. Introduction

In the 1970s and 1980s, the success of gauge theories has prompted many authors[1] to point out that the idea of gauge invariance, although originally with a different meaning, was first expressed in 1918 by Hermann Weyl in his attempt to formulate a unified geometrical theory of gravitation and electromagnetism along the lines opened up by general relativity. On the other hand, the revival of attempts at formulating multidimensional unified theories à la Kaluza–Klein has had the effect that the formalism first proposed by these authors was rediscovered and developed and that the original papers have received many citations.[2]

However, despite the acknowledgment of some of these investigations as the historical antecedents of modern attempts at unified field theories, the general feeling is that one is dealing with a set of premature approaches, doomed to failure essentially because in those days no adequate view of the interactions operating in nature had been acquired.[3] Most of the authors involved can be criticized for actually refusing to deal with interactions other than gravitation and electromagnetism though the time seemed ripe to do so[4] (the only point that occasionally received some attention was how to include matter fields). Early attempts at unification have also been criticized for their purely classical approach. In particular, the link between the electromagnetic field and complex fields describing the particles established by the gauge viewpoint could not be perceived from a purely geometrical viewpoint.[5]

No one today, however, would dream of criticizing Maxwell's unification of electricity and magnetism for ignoring the electroweak unification. Moreover, it can be speculated that unification at the classical level would in any case be necessary as a preliminary to the formulation of a unified quantum theory. It therefore seems fair to subject these theories to more intrinsic tests, such as their ability to reproduce the Einstein–Maxwell theory in first approximation, to predict new effects, etc. A whole set of increasingly stringent criteria of this kind was formulated by, among others, Weyl, Pauli, and Lichnerowicz. Einstein himself was keenly aware of the necessity of high epistemological standards in this field.[6]

In this paper, I first summarize these criteria (Section 2); I then briefly review the attempts at unified field theories (UFTs) based on extensions of Riemannian geometry, from Weyl's first "attempt at a final synthesis" of 1918 to the late efforts of Einstein and Schrödinger, discussing the extent to which they conform to the criteria (Section 3). Theories such as Nordström's (1914) and Hilbert's (1915), therefore, will not be dealt with. (For an account of these theories and for a general presentation of the genesis of the unified field program the reader is referred to Vizgin 1989; see also Sigurdsson 1991, especially pp. 109–114 for an account of Mie's matter theory.) A short section is devoted to a discussion of the so-called Blackett effect, which has implications for some of the proposed theories (Section 4); finally (Section 5), the attempts previously examined are synoptically compared to each other and evaluated in the light of the criteria.

An alternative view on unification was introduced by Rainich[7] (1925; see also Rainich 1950), and elaborated upon by Misner and Wheeler (1957). It emerges from the following approach: consider the set consisting of Maxwell's and Einstein's equations (with the electromagnetic field as the only source term); then solve the latter for the electromagnetic field in terms of the contracted curvature tensor, and substitute this solution into Maxwell's equations. The result is what Misner and Wheeler call "an *already* unified field theory," in which "electric and magnetic fields are not signals to *invent* a unified field theory or to introduce one or another new kind of geometry." Rather, the theory "describes electric and magnetic fields in terms of the rate of change of curvature of pure Riemannian geometry, and nothing more" (Misner and Wheeler 1957, p. 530).

This is by no means intended as a comprehensive review of the UFTs worked out in the period considered. A general presentation aimed at providing the background to Einstein's own efforts is given in chapter 17 of Pais's biography (Pais 1982; see also Vizgin 1989). The rich geometrical structure of UFTs from Eddington's theory to the theories of the present time is thoroughly analyzed in Goenner 1984. An analysis of the early attempts at UFTs, which pinpoints their highest achievements from the point of view of mathematical physics, has been carried out by Ferraris and Francaviglia (1986).

2. Validation/Refutation Criteria

What should be asked of a purely classical theory aiming at giving a unified field picture of gravitation and electromagnetism? The following list of requirements is an elaboration of the criteria indicated by various authors.

(1) Let us define, as is customary, an Einstein–Maxwell theory as a theory that describes the electromagnetic field in terms of the Maxwell equations and the gravitational field in terms of the Einstein equations, the latter having the energy-momentum tensor of the electromagnetic field as the source term in the absence of matter; then, as the first requirement, one should ask that the theory contain the Einstein–Maxwell theory as a limiting case.

(2) The mere reproduction of the Einstein–Maxwell theory, however, is not sufficient, as stressed, for instance, by Lichnerowicz, Pauli, Weyl, and Einstein.

According to Lichnerowicz, one can stipulate to call a theory *unified in a broad sense*

> if, in the representation of the fields and in the formation of the equations, it attributes symmetrical roles to the gravitational and the electromagnetic fields; in particular, since the gravitational field, in the conceptions of general relativity, is linked to the geometrical structure of the universe, it will be desirable to choose a structure such that the two fields emanate from the same geometry.

One can, on the other hand, stipulate to call a theory *unified in a strict sense*

> to the extent that the exact equations govern a non-decomposable hyperfield,[8] and that they can only approximately be decomposed into propagation equations of the gravitational and the electromagnetic fields when the physical conditions are such that one of the fields dominates the other. (Lichnerowicz 1955, p. 152)

The sense in which the nondecomposability of the hyperfield should be intended was formalized by Pauli. He expressed the viewpoint that a unified theory should conform to the "*principle that only irreducible quantities should be used in field theories.*" Not only is this principle "satisfactory from a formal point of view," it has also "been verified empirically without exception in physics until now." "Therefore," Pauli continues,

> I believe that *cogent mathematical reasons* (for instance invariance postulates of a wider group of transformation) *have to be given why a decomposition of the reducible quantities used in the theory (for instance R_{ik}, g_{ik}, and Γ^{l}_{ik}) does not occur.* (Pauli 1958, p. 226)

As is well known, there are theories in which the gravitational and electromagnetic fields form the symmetric and antisymmetric part of a tensor respectively, a tensor which, therefore, is not irreducible; it is to such theories that Pauli's remark applies. It should be noted that the invariance under a symmetry group reducing to a direct or semidirect product cannot represent a solution, being a purely formal expedient. Pauli often made such harsh comments aimed at attempts to formulate unified theories. Weyl recalls that he used to express his skepticism with the aphorism: "Men shall not join what God has torn asunder" (Weyl 1950 [1968, p. 431]; translation as in Pais 1982, p. 350). Weyl expressed himself in terms similar to Pauli's.[9] Einstein formulated the criterion of irreducibility quoting the covariant formulation of electromagnetism as an example:

> The unification here [is] that the entire field considered is described as a skew-symmetric tensor. The basic group of Lorentz transformations does not enable us to split this field independently of the system of coordinates, into an electric and a magnetic one. (Einstein 1954, p. 578)

As a final specification concerning this second criterion, one may add that the equations ruling the hyperfield should be derived from a variational principle, formulated in terms of a Lagrangian made plausible on a physical and/or geometrical basis. The Lagrangian or the Hamiltonian function should not be the sum of several invariant parts (Einstein 1945, p. 578). This criterion can be met even if the requirement of irreducibility of the field is not. A theory of this kind would be considered "unified only in a limited sense" (Einstein 1945, p. 578).

(3) A theory satisfying the second criterion also satisifes, as a consequence, another requirement of an epistemological character: that of not representing a mere recodifying of the existent. The third requirement that must be formulated is that some explicit predictive power should correspond to this feature. That is, the theory must predict new physical effects, such as the electromagnetic waves predicted by Maxwell's theory, capable of refuting or corroborating it.

(4) Finally, as is obvious, the observed effects should agree with the predictions.

3. Attempts at Unified Field Theories

3.1 Weyl's "Attempt at a Final Synthesis"[10]

In his basic paper on the subject (Weyl 1918), Weyl began with the statement that it is the infinitesimal displacement of a vector that should be taken as the basic concept for the natural construction of Riemannian geometry.[11] Since the displacement is non-integrable, i.e., different paths between two points lead to different orientations of the final vector, Riemannian geometry is a "geometry of the nearby" (*Nahe-Geometrie*). However, a residual element of a "geometry of the faraway" (*Fern-Geometrie*) has survived in this geometry—on no objective ground, as far as he could see—viz. that the length of a vector is integrable under parallel displacement. In his view, there is no reason "why the problem of the displacement of a length from one point to another at a finite distance should be assumed to be integrable more than the problem of the displacement of the direction" (Weyl 1918 [1968, p. 148]). How should a geometry of this kind be characterized? Since the comparison of lengths is still possible locally, the manifold should be endowed with a metric tensor. However, one should not require conservation of lengths and scalar products under parallel displacement, i.e., the metricity condition. This does not mean that the connection is purely affine. Actually, one assumes a (path-dependent) rescaling determined by a covariant vector field ϕ_μ, proportional to the initial length. The result is a connection (Weyl's connection), determined by the fields $\mathbf{g}$ and ϕ_μ, that may be called semimetric.[12]

The connection is invariant under the simultaneous tranformations

$$g_{\mu\nu} \longrightarrow \lambda g_{\mu\nu} \tag{1}$$

$$\phi_\mu \longrightarrow \phi_\mu - \frac{1}{\lambda}\frac{\partial\lambda}{\partial x^\mu}, \tag{2}$$

where λ is a positive function of the point on the manifold.

In such a theory, therefore, the fields $\mathbf{g}$ and ϕ_μ are not uniquely defined in each point: they are subject to simultaneous rescaling. Weyl referred to this property as *Eichinvarianz* or *Maßstabinvarianz*, which were later rendered as gauge invariance. Gauge invariance, as expressed by Eqs. (1) and (2), must be considered to be on the same footing with the general transformations of coordinates; the arbitrariness in the choice of the gauge factor corresponds to the arbitrariness in the choice of the coordinate system.

The enlarged version of Riemannian geometry considered by Weyl allows the introduction of new physical quantities (the potentials ϕ_μ), side by side with the gravitational potentials expressed by the components of $\mathbf{g}$.

The quantities ϕ_μ are the natural candidates for expressing the electromagnetic potentials: one immediately observes that Eq. (2) has the form of a gauge transformation for the potentials. The form

$$F_{\mu\nu} = \partial_\mu \phi_\nu - \partial_\nu \phi_\mu$$

is therefore gauge invariant and represents the electromagnetic field. (For more details on Weyl's theory, see Vizgin 1989.)

Weyl's theory satisfies Lichnerowicz's definition of a unified theory in the broad sense, since both the gravitational and the electromagnetic field emanate from the same geometry; the two fields, however, do not form a nondecomposable hyperfield, so that the theory cannot be considered unified in the strict sense. Moreover, they do not form an irreducible tensor (actually, they do not form a tensor at all), and do not satisfy Pauli's criterion. The field equations can be derived from a variational principle, simultaneously invariant under the group of general relativity (general coordinate transformations) and the group of the transformations (1). As pointed out by Vizgin, the latter demand led to Lagrange functions quadratic in the curvature scalar or bilinear in the Riemann tensor (Vizgin 1989). Pauli was able to show (Pauli 1921 [1958]) that one still obtained the Schwarzschild solution, so that the theory gave the usual general relativistic results for the well-known classical effects. Quadratic Lagrangians, however, produce various difficulties (Vizgin 1989). As far as the criteria about predictivity and experimental controls are concerned, the situation is very interesting. The nonintegrability of the norm implies the nonintegrability of proper time intervals; this, in turn, implies that the rate of *relativistic*, or *standard*, clocks should depend on their world lines. If it is assumed, as is customary, that the clocks provided by the atomic spectral lines provide an example of standard clocks, very stringent bounds are imposed on this path dependence by the precision with which the spectral lines are known. Hence, it turns out that Weyl's theory satisfies the purely epistemological requirement of predicting new effects, but unfortunately not the requirement that its predictions be in agreement with experiment. The weak point of the theory was immediately spotted by Einstein. His first reaction to Weyl's paper was enthusiastic ("It is a stroke of genius of the highest order"[13]), and he accepted the responsibility of communicating Weyl's paper to the Prussian Academy. "Your ideas show a wonderful cohesion," he wrote, "apart from the agreement with reality, it is at any rate a grandiose achievement of the mind."[14] Agreement with reality, however, was impossible to achieve according to Einstein. "However beautiful your thought is, I must admit frankly that, according to my opinion, it is impossible that this theory corresponds to nature,"[15] he wrote to Weyl in a subsequent letter, in which he gave the argument outlined above, to conclude: "if agreement with ruler and clock measurements is dropped, the theory of relativity loses its empirical meaning in general."[16] Weyl's paper was published immediately followed by a critical note from Einstein (Einstein 1918). In reply, Weyl stated that Einstein's argument "was not an objection to the theory since the latter [was] not concerned with the behavior of real rulers, clocks, and atoms" (Weyl 1920 [1968, p. 141]). Pauli's comment on the way out proposed by Weyl was: "this relinquishment seems to have very serious consequences. While there now no longer exists a direct contradiction with the experiment, the theory appears nevertheless to have been robbed of its inherent convincing power from a physical point of view" (Pauli 1958, p. 196).

It should perhaps be mentioned that Weyl's way out has subsequently been proposed by various other authors. Ehlers, Pirani, and Schild, for instance, after stating that Weyl's theory must be abandoned "if equality of gravitational time . . . and atomic time is assumed,"[17] ask: "How compelling is the time-equality postulate?" (Ehlers et al. 1972, p. 82). A similar way out was suggested by Dirac in the framework of his "large number hypothesis" and the time variation of the gravitational "constant" he derived from such a hypothesis. Einstein's theory of gravitation must then be modified. Dirac proposed a modification in the direction of Weyl's geometry; Einstein's argument against it could be overcome, in his view, by disassociating the line element dealt with by the theory from the line element dealt with by the theory from the line element measured by some "atomic apparatus" (Dirac 1973).

Another way out of the difficulty encountered by Weyl's theory was discussed by D.K. Sen, who traced its origin back to a 1951 paper by Lyra (see Sen 1968 and references therein). This approach is based on the introduction of a gauge function on an otherwise structureless manifold. Field equations identical to Weyl's, apart from a cosmological term and constant factors, are thus obtained, but no difficulty connected with nonintegrability of length transfer arises (Sen 1968, p. 85).

The fate of Weyl's idea has been tracked by F.W. Hehl, J.D. McCrea, and E.W. Mielke in a recent paper (Hehl et al. 1988). The authors recall how the Bjorken scaling law verified in the deep inelastic electron-nucleon scattering brought new life to the idea of scale or recalibration invariance as a broken symmetry of the physical world (Hehl et al. 1988, p. 244). On the other hand, the classical Hilbert–Einstein action is not scale invariant and, as a consequence, general relativity does not exhibit approximate Bjorken-type scaling, a property which is believed to be indispensable for renormalizability. A clue for a possible way out of the difficulty encountered by the theory from a modern point of view might be the local extension of space-time symmetries, such as Lorentz's and, in fact, scale invariance (Hehl et al. 1989, p. 1075). The Noether current associated to this local symmetry requires an additional current, i.e., a material dilation current; it is to this current, and not to the electric current of Maxwell's theory that the Weyl vector is coupled. The essential idea behind Weyl's theory would thus survive, although with a completely different outlook compared to his original theory (Hehl et al. 1988, p. 244).

As is well known, modern gauge theories owe a lot to Weyl's idea of *Eichinvarianz*. After the formulation of wave mechanics, Fritz London (London 1927), V. Fock, and others realized that the rescaling of the metric tensor had to be replaced by a phase transformation of Schrödinger's wave function for the electron:

$$\psi(x) \longrightarrow \psi'(x) = e^{-i\epsilon\alpha(x)}\psi(x).$$

As is universally known, the fundamental step toward the non-Abelian case was taken by Yang and Mills (1954).

3.2 Eddington's First Affine Theory

Weyl had shown how one could disassociate the connection theory from the metricity condition, even though his unified theory was, as we said, still semimetric. The first to try to exploit this freedom in formulating a unified theory on a new basis was Eddington. In Eddington's space-time only a (symmetric) linear connection is initially defined. In Weyl's theory, he observed, a "particular standard of length should only be used at the time and place where it is." Nevertheless, "we *do* compare lengths on the sun and the earth." This means that there exists a "natural gauge system" determined by physics. And in fact it is the introduction of the natural gauge system that "marks the transition from pure geometry to physics" (Eddington 1921, p. 105). According to Eddington, the gauge system can only be determined by the "gauging equation"

$$G_{\mu\nu} = \lambda g_{\mu\nu},$$

which coincides with the equation governing De Sitter's universe. The possibility of formulating a unified theory in this framework arises from the fact that in Eddington's theory the Ricci tensor contains an antisymmetric part, which Eddington thought could be identified with the electromagnetic field. It is therefore clear from the start that the theory does not satisfy Lichnerowicz's second criterion. In its original formulation, it is also unsatisfactory because it does not specify the equations governing the dynamics of the connection considered as a field variable. Such equations were given by Einstein in three papers of 1923 (see, for instance, Einstein 1923), whose content was further elaborated in Eddington's book of 1924, *The Mathematical Theory of Relativity* (Eddington 1924). Einstein was the first to point out clearly the necessity of starting from a Lagrangian depending only on a linear connection and its first derivatives, taking the variation directly with respect to the connection. This approach was further developed by Schrödinger several years later, as will be discussed below.

3.3 Five-Dimensional Theories

The starting point of Kaluza's[18] attempt was the speculation that the electromagnetic field tensor might be a truncated Christoffel symbol. Since, in a four-dimensional world, these symbols are saturated by the components of the gravitational field, one is led "to the extremely odd decision to ask for help from a new, fifth dimension of the world" (Kaluza 1921, p. 967; translation by Muta, in Lee 1984).

In order to understand Kaluza's point, one may compare the expressions for the electromagnetic field tensor

$$F_{\mu\nu} = \phi_{\mu,\nu} - \phi_{\nu,\mu}$$

and for a Christoffel symbol

$$\Gamma^{\Lambda}_{\mu\nu} = g^{\Lambda\sigma}(g_{\nu\sigma,\mu} + g_{\sigma\mu,\nu} - g_{\mu\nu,\sigma})$$

for fixed Λ.

The outlined possibility could be implemented by assuming proportionality of the electromagnetic potential to the mixed ($\mu 5$) components of the metric. Kaluza's article contained almost all essential points of what we now call a Kaluza–Klein theory, namely the field equations of the Einstein–Maxwell theory (plus a scalar equation for the (55) component of the metric), the particle world lines resulting from the projection of the five-dimensional geodesics, the interpretation of the electric charge as the fifth component of the momentum ("a further fusion of two formerly heterogeneous basic concepts..." Kaluza 1921, p. 969), and the charge conservation along the world lines.

From a formal point of view, however, Kaluza did not quite work out the theory such as we know it. First of all, he did not introduce the constraint referred to as Klein's constraint,

$$\partial_M \gamma_{55} = 0, \quad M = 1, \ldots, 5, \tag{3}$$

but only the condition

$$\partial_5 \gamma_{MN} = 0 \tag{4}$$

(which he called the "cylinder condition"). Moreover, and this is a point that has not received due consideration, he did not have the correct metric for the ordinary four-dimensional space-time manifold. In his own words "the fundamental metric tensor... becomes the gravitational tensor potential framed by the electromagnetic four-potential" (Kaluza 1921, p. 968). This means that, denoting the metric of R_5 as γ and of R_4 as g, Kaluza had $g_{\mu\nu} = \gamma_{\mu\nu}$. Apart from this detail, it must be observed that Kaluza's theory satisfies Lichnerowicz's definition of a unified theory in the broad sense, since the gravitational and electromagnetic fields emanate from the same geometry. It also satisfies (though only formally, as I will argue) Pauli's criterion of irreducibility, since both fields appear in the theory as components of a symmetric metric tensor. However, it does not satisfy the criterion required for a unified theory in the strict sense, since it gives nothing more than the Einstein–Maxwell theory.[19] (Note, by the way, that this result appeared important enough to Salam to speak of the "Kaluza–Klein miracle" [Salam 1980, p. 20]). Actually, any trace of the fifth dimension disappears from the final field equations, so that it may be said that it is a four-dimensional theory in disguise. This is what induced Pauli to state that "Kaluza's geometric form of the generally covariant laws of the electromagnetic field... *is in no way a 'unification' of the electromagnetic and gravitational field.* On the contrary, every theory which is generally covariant and gauge invariant can also be formulated in Kaluza's form" (Pauli 1958, p. 230).

The story of the fifth dimension was to have a further important development in 1926 thanks to O. Klein. His work was not motivated by Kaluza's, which he apparently did not see until completion of his first paper on the subject (according to his own testimony [see Klein 1969; see also Kragh 1984]). Rather, Klein was striving, in a way not dissimilar from the path de Broglie was independently following in those years, to find support for the idea that the propagation of a wave could be associated with the motion of particles, or in his words, "to find a wave background to the quantization rules" (Klein 1969, p. 63). So, he

> played with the idea that waves representing the motion of a free particle had to be propagated with a constant velocity, in analogy with light waves—but in a space of four dimensions—so that the motion we observe is a projection on our ordinary three-dimensional space of what is really taking place in four-dimensional space. (Klein 1969, p. 63)

Klein's starting point (Klein 1926a) was the observation that in a five-dimensional space-time, as a consequence of the independence of the metric's components of the fifth coordinate, allowed coordinate transformations are those of the product group of the group of transformations of the usual four-dimensional theory G_4:

$$x^\mu \longrightarrow x'^\mu = x'^\mu(x^\nu) \tag{5a}$$

and the group S_1 of the transformations

$$x^5 \longrightarrow x'^5 = bx^5 + \psi(x^\nu) \tag{5b}$$

(where b is a positive constant, and where a change of scale for the time parameter is allowed for). Since it can easily be verified that γ_{55} is not altered under the transformations (5), Klein observed that it can be assumed to be constant; he then observed that the quantities

$$\mathrm{d}\theta = \mathrm{d}x^5 + \frac{\gamma_{5\nu}}{\gamma_{55}}\mathrm{d}x^\nu$$

and

$$\left(\gamma_{\mu\nu} - \frac{\gamma_{5\mu}\gamma_{5\nu}}{\gamma_{55}}\right)\mathrm{d}x^\mu\,\mathrm{d}x^\nu$$

are invariant under the transformations (5). The quantities $(\gamma_{\mu\nu} - \frac{\gamma_{5\mu}\gamma_{5\nu}}{\gamma_{55}})$ could thus be identified with the components of the four-dimensional metric (in this way Klein introduced for the first time the correct projected metric). From the invariance of $\mathrm{d}\theta$ it follows that, at fixed x^5, the four quantities $\gamma_{5\nu}$ transform like a four-vector. They can then, as in Kaluza's 1921 paper, be taken as proportional to the components of the electromagnetic potential. In modern notation, one would set $\gamma_{5\nu} = \sqrt{2\kappa}\,\phi_\nu$, with $\kappa = 8\pi G$ (G = gravitational constant). Their transformation properties under S_1 show that this group may be considered as the geometrized version of the local electromagnetic gauge group. One may note that, since the form of the five-dimensional metric is not preserved under a general coordinate transformation in five dimensions but only under the transformations of the product group, Eqs. (5a, 5b), Pauli's irreducibility criterion is only formally satisfied.

As is well known, Klein obtained what we would today call a stationary "Klein–Gordon" equation as the projection of a five-dimensional wave equation, thus achieving his original goal. The details of the derivation are not of interest to us here; what is of interest here, though, is that Klein's particle quantum theory is characterized by a natural mass unit, $e^2/2\kappa \simeq 10^{18}$ Gev. Subsequent analyses showed that considering Klein's fields leads naturally to masses of charged particles (which cannot be massless) that are multiples of this fundamental mass. This can to some extent be considered as a prediction of the five-dimensional theories, and would appear to refute them altogether. During the revival of multidimensional theories in the early 1980s, the problem of the mass spectrum was also addressed. The author finds it difficult to say what the conclusions were (if any).[20]

In a subsequent short paper (Klein 1926b), Klein stressed that in his first paper on the subject, the classical equations of motion of the charged particles had been obtained as projections of the geodesics of the five-dimensional space-time, by choosing

$$P_5 = \frac{Ne}{\sqrt{2\kappa}\, c}, \tag{6}$$

where Ne is the particle's charge and $\kappa = 8\pi G$. To begin with, this suggests that the fifth component of the momentum be associated with the charge. But there is more to it: the formula suggests, in the author's words, "that the atomicity of electricity may be interpreted as a quantum theory law" (Klein 1926b, p. 516). In fact, if the five-dimensional space is assumed to be closed along x^5 with period l, and the usual quantization rule due to periodicity along the fifth dimension is applied, the possible values of P_5 are given by

$$P_5 = N\frac{h}{l} \tag{7}$$

where N is now a quantum number accounting for the structure of Eq. (6). From Eqs. (6) and (7) it follows that the "fiber's length" l is given by

$$l = \frac{hc\sqrt{2\kappa}}{e} = 0.8 \cdot 10^{-30}\,\text{cm}.$$

Klein was thus led to the conclusion: "The small value of this length together with the periodicity in the fifth dimension may perhaps be taken as a support of the theory of Kaluza in the sense that they may explain the nonappearance of the fifth dimension in ordinary experiments as the result of averaging over the fifth dimension" (ibid.). The "reality," or "observability," of extra spatial dimensions is philosophically interesting. Klein's hypothesis may be seen as an *ad hoc* removal of the problem set by their actual inobservability. It is not, therefore, a prediction; it is rather to be considered as a reformulation of the theory under the pressure of a prediction refuted by the observations. Compactification of the extra space dimensions into some compact "small" manifold, like Klein's fiber, does not eliminate in principle their observability; however, it does so in practice, since they could be explored only by means of probes with energies in the order of 10^{18} Gev.

3.4 Einstein and the Fifth Dimension

Several years after he first became involved with five-dimensional theories, Einstein wrote a paper with P. Bergmann that shows his concern about the possible "reality" of the fifth dimension and its actual inobservability. Section 2 of the paper, in fact, begins with the following passage:

> If Kaluza's attempt is a real step forward, then it is because of the introduction of the five-dimensional space. There have been many attempts to retain the essential formal results obtained by Kaluza without sacrificing the four-dimensional character of physical space. This shows distinctly how vividly our physical intuition resists the introduction of the fifth dimension. But by considering and comparing all these attempts one must come to the conclusion that all these endeavors did not improve the situation. It seems impossible to formulate Kaluza's ideas in a simple way without introducing the fifth dimension. (Einstein and Bergmann 1938, p. 688)

Einstein's interest in the five-dimensional approach dates from the very beginning, since it was Einstein to whom Kaluza turned when, in 1919, he thought of publishing his paper in the proceedings of the Prussian Academy. The story of the exchange between Einstein and Kaluza and of the delayed publication has been told elsewhere (see Lee 1984; Middleton 1991). Einstein produced two short notes "on Kaluza's theory" after the appearance of Klein's paper; the few additional elements set out in these papers are not of interest to us here. He came back to Kaluza's theory in 1931, in collaboration with W. Mayer (Einstein and Mayer 1931, 1932). The authors presented a new formalism that, however "psychologically related" to it, "avoids the extension of the physical continuum to five dimensions" (Einstein and Mayer 1931, p. 541). The basic idea is to associate with the four-dimensional continuum a five-dimensional *vector* space M_5. The implementation of the theory is achieved by prescribing the embedding of the Minkowskian local approximation to the five-dimensional space-time in M_5, and the way tensors of M_5 decompose with respect to M_4. The connection must now specify how the M_5 vector spaces associated with different points of R_4 relate to each other. It therefore becomes, so to speak, five-dimensional; the appearance of an antisymmetric part of its four-dimensional projection makes it possible, in principle, to identify this part with the electromagnetic field **F**. It must, however, be *assumed* that **F** is a rotational. This theory is hardly relevant for our purpose, the main point of interest arising from the fact that it cannot be derived from a variational principle.[21]

The so-called projective relativity, which was developed from 1930 onward by several authors,[22] is to some extent related to the Einstein–

Mayer theory. The basic idea is to treat the continuum as four-dimensional, but to introduce, as in projective geometry, five homogeneous coordinates: values of the coordinates differing by a common factor belong to the same point of the continuum. For the purposes of this paper, it seems we do not have the look into these theories.

The paper by Einstein and Bergmann, mentioned above, is slightly more interesting from our point of view. Its apparatus is close to that considered by Klein: space curls and closes up along the fifth dimension; as in Klein, the apparent four-dimensionality of space-time is recovered if the length of the fiber is very short; as in Klein, γ_{55} is assumed to be constant. However, a dependence of the other metric components on the fifth coordinate is considered. One must then consider metric fluctuations around the background value obtained by averaging over the fifth dimension, to which full physical existence is thus attributed. Metric components have a Fourier expansion with respect to the fifth coordinate. According to Bargmann and Bergmann,[23] Einstein thought the higher Fourier components could in some way be related to quantum fields, which, by the way, is the modern view. In this respect, it can be said that the theory may predict new physical effects, although at the time it was not elaborated to the level of making *explicit* predictions. The situation is discussed in a fairly recent paper by Appelquist and Chodos (1983).

3.5 The Jordan–Thiry Theory

Another extension of the Kaluza–Klein scheme that has had a notable impact on recent developments is associated with the names of Jordan and Thiry. The Jordan–Thiry development was , in fact, implicit in Kaluza's original formulation, in which, as we have pointed out, constancy of the 55 metric component was not required. After imposing Kaluza's constraint, that makes it independent of the fifth coordinate, the 55 component becomes a scalar field on the space-time manifold. This idea was further and independently explored by Jordan and Thiry.

The starting point of the German author, well known for his contributions to the formulation of quantum mechanics, was Dirac's idea[24] that the gravitational "constant" may actually not be a constant, but subject to variations on the cosmological scale, thus becoming a scalar field on the space-time manifold (Jordan 1946). To Jordan, a five-dimensional scheme, with a variable 55 metric component, seemed to be a natural framework for implementing Dirac's idea. Jordan favored, in particular, the projective version of the theory.[25]

Thiry's motives were completely independent. He gave a formal analysis, using techniques previously developed by Lichnerowicz and Cartan (his note was presented by the latter), of a Kaluza–Klein scheme, in which the Klein constraint was not implemented, to analyze the structure of the theory's field equations (Thiry 1948).

The Jordan–Thiry field has, first of all, a gravitational role. In this sense, the Jordan–Thiry theory anticipates the Brans–Dicke scalar–tensor theory of gravitation (Brans and Dicke 1961), which, however, had an independent motivation in the author's desire to incorporate Mach's principle in a metric theory of gravitation.

As a unified theory, the Jordan–Thiry theory from the very beginning appears to be more satisfactory from an epistemological point of view than the Kaluza–Klein theory, in that it predicts new effects, event though they may be hard to detect. These new effects are (Tonnelat 1955, p. 8)

(1) the variability of the gravitational "constant" χ;

(2) the presence of extra terms in the equations for the gravitational and electromagnetic fields; such terms are linked to the variability of χ; if χ is constant, the classical equations are recovered (Einstein–Maxwell's theory);

(3) a fifteenth field equation is added to the fourteen equations ruling the gravitational and electromagnetic fields. It implies that a magnetic field may be created by (moving) matter, even in the absence of electric charges. One is thus led to the prediction of a mechano-magnetic effect, which may arise, in particular, in the presence of a rotating body. This effect is, of course, completely foreign to the body of familiar physical effects, to the point that one would be tempted to discard the theory right away on the basis of this prediction. A moment's reflection, however, will convince us that any attempt at a not merely formal reunification of gravitation and electromagnetism necessarily implies a link, whatever its nature, between gravitation and electromagnetism, and hence between matter and the electromagnetic field. Since a similar prediction also obtains in the asymmetric theories, which will be discussed in the next sections, I postpone its discussion until after the presentation of these theories.

Jordan's approach was developed and extended by various authors (see Just 1954 and references therein[26]). Just was particularly concerned with the problems set by the corrections to general relativity in connection with the solar system effects. In a series of papers (see Just 1956 and references therein), he used in particular the conditions set by the agreement of general relativity with the data on Mercury's perihelion shift to discriminate between possible versions of the theory.

In France, attention was focused on a difficulty of the theory related to the identification of the space-time manifold with the four-dimensional quotient space identified by the theory's isometry group. The problem was tackled, and a solution was proposed, by F. Hennequin and later by P. Pigeaud (see Pigeaud 1963 and references therein).[27] In the final part of his thorough study, Pigeaud treated the Schwarzschild problem in the generalized theory according to the proposed solution, finding in particular the exact Einstein expression for the perihelion shift. In conclusion, he stated that, in the absence of an electromagnetic field, the theory accounted for the whole set of relativistic phenomena (Pigeaud 1963, p. 216).

3.6 Einstein's Metrico-Affine Theory

I have already mentioned that Einstein contributed several critical remarks or developments to the approaches to unified theories followed by various authors. It was only in 1925, however, after some 10 years of gestation, as pointed out by Pais (1982, p. 382), that he attempted to formulate a unified theory of his own. I will give a fairly detailed outline here of this first attempt (Einstein 1925), since it presents some of the features common to subsequent attempts at formulating asymmetric theories, by Einstein himself and by other authors.

In the theory, it is assumed that a connection and a fundamental tensor $g^{\mu\nu}$, both nonsymmetric, are given as independent variables in space-time. It is, of course, the symmetric part of the fundamental tensor $g_{\mu\nu}$, introduced through $g_{\mu\alpha}g^{\nu\alpha} = g_{\alpha\mu}g^{\alpha\nu} = \delta^{\nu}_{\mu}$, that is identified with the metric, while the antisymmetric part is the natural candidate to represent the electromagnetic field. It is, first of all, the introduction of these two fields as independent quantities, a feature retained in subsequent attempts, that we will examine here. They belong to the class of the metrico-affine theories, even if the term is mostly used in theories of pure gravitation.

It must be stressed that in his first paper Einstein considered only one connection, Γ, while it would be legitimate to consider also its transposed $\tilde{\Gamma}$, defined as[28]

$$(\tilde{\Gamma}^{\mu})_{\alpha\beta} \doteq (\Gamma^{\mu})_{\beta\alpha}.$$

On this basis, the Riemann and Ricci tensors are evaluated, and the scalar density $g^{\mu\nu}R_{\mu\nu}$ is formed. It should be noted that these cases present the possibility of a further independent contraction of the Riemann tensor, which gives rise to a simple expression in the derivatives of the connection. Taking into account the "transposed" connection as well, one has, in principle, a total of four second-order contracted curvature tensors, for which no *a priori* selection criterion exists. In his first paper, however, Einstein considered only the usual Ricci tensor, as mentioned above. He then required the vanishing of the corresponding action integral under independent variations of the fundamental tensor and the connection. This is considered to be "the first rigorous formulation of a metrico-affine variational principle with a nonsymmetric connection."[29]

It should be mentioned that nonsymmetric connections had first been introduced by Cartan (1922); Einstein's work, however, was independent.[30]

The variational procedure leads to $16 + 64$ field equations for as many field variables. Obviously, to recover the purely gravitational limiting case, symmetry of the fundamental tensor is required; the connection then turns out to by symmetric as well, and is determined to be the Levi-Civita connection. However, the symmetry condition is not sufficient; the vanishing of a certain vector field, a function of the metric and the connection, is also required. In the general case, the field equations are unwieldy, and Einstein had to take recourse in a weak field approximation, considering (small) symmetric and antisymmetric ($\phi_{\mu\nu}$) corrections to a unit fundamental tensor. The set of field equations then gives rise to three subsets: the vacuum field equations of gravitation, the Maxwell equations expressing the vanishing of the four-divergence of the field $\phi_{\mu\nu}$, as one should have in the source-free case, and finally the equations

$$\frac{\partial}{\partial x_\alpha}\left(\frac{\partial \phi_{\mu\nu}}{\partial x^\alpha} + \frac{\partial \phi_{\nu\alpha}}{\partial x^\mu} + \frac{\partial \phi_{\alpha\mu}}{\partial x^\nu}\right) = 0.$$

Therefore, the expression

$$\frac{\partial \phi_{\mu\nu}}{\partial x^\alpha} + \frac{\partial \phi_{\nu\alpha}}{\partial x^\mu} + \frac{\partial \phi_{\alpha\mu}}{\partial x^\nu},$$

which is identically zero in Maxwell's theory, does not necessarily vanish in the theory; only its four-divergence vanishes identically. Because of this result, Einstein abandoned this first version of an asymmetric theory soon after its publication.

Einstein resumed his work on asymmetric theories in 1945 and continued studying them until his death. As stressed by Pais,[31] Einstein's efforts in this field in the last decade of his life were much more elaborate than his efforts in 1925; several successive versions were produced. I have chosen not to follow the evolution of Einstein's thoughts; I simply want to single out the common basis of the various versions. This will be enough for my purpose, which is not so much that of giving a full historical account, but rather isolating common points for comparison with epistemological standards. In this presentation, I will follow closely Kaufman (1956) and Tonnelat (1955).

From a formal point of view, all attempts show that Einstein had become aware of the possibility of having two distinct connections and, as a consequence, four contracted second-order curvature tensors. What changed over time was his attitude toward the problem of what the most suitable prescription was for eliminating the ambiguity. In the last version of the theory, developed by Einstein in collaboration with Kaufman (Einstein and Kaufman 1955), it was expressed in terms of the requirement of transposition invariance of the Lagrangian, which Einstein tried to link with the physical requirement that the theory be invariant under the change of sign of the electric charge.[32] A version of this theory appeared in the 1953 edition of Einstein's only technical book on general relativity (Einstein 1953). There he made a remark of epistemological interest. He first observed that each of the two parts of his fundamental tensor

> is by itself a tensor, i.e., under a coordinate transformation, the component of each part transforms independently of the components of the other part, [so that,] considered from the point of view of the relativistic group, the nonsymmetrical g_{ik} is not an irreducible quantity but an arbitrary and unjustified combination of two entities of different nature.... This would seem to be a grave objection.... It should be noted, however, that the group-theoretical point of view is not the only relevant one from which to judge the "uniformity" of the concept of the nonsymmetrical tensor field. [In fact, in Riemann's theory,] the determinant of the g_{ik} makes it possible to correlate a contravariant g^{ik} to the covariant tensor g_{ik} according to the equation
>
> $$g_{is}g^{it} = \delta_s^t = g_{si}g^{ti},$$
>
> where δ_k^i is the Kronecker tensor. This correlation, which plays a fundamental part in the theory of the symmetrical field, can immediately be taken over in the case of the nonsymmetrical field.... [The only difference is that] in the latter case... the order of the indices must be preserved. This represents one argument indicating that, in spite of the objection expressed above, it is indeed natural to introduce the nonsymmetrical tensor field as a generalization of the symmetrical one. [Moreover, the equation given above] makes it possible to raise and lower the tensor indices, [although] in the case of the nonsymmetrical field... this operation is no longer defined *a priori* ($A_s g^{sk}$ and $A_s g^{ks}$ are not equal to each other). (Einstein 1953, p. 135)

I emphasize that the one-to-one correspondence established between vectors and one-forms by a nondegenerate $\binom{0}{2}$ tensor expresses a premetric, and therefore a more general geometrical property of the tensor than the metric properties introduced when the symmetry condition is added. Pauli took notice of the attention Einstein had paid to the objection against reducibility of the fundamental tensor in his theory: "Einstein... was well aware of this objection, which he weighed carefully in his later work" (Pauli 1958, p. 226 [footnote][33]).

Let us now examine in some detail the novel elements introduced in the last version of the theory compared to the Einstein–Maxwell theory (see Tonnelat 1955, p. 10).

(1) In the general case, due to the unwieldiness of the field equations, the occurrence of a magneto-mechanical effect is not immediately evident; the effect, however, is clearly obtained in the particular cases of a field exhibiting cylindrical or spherical symmetry. In Section 3 below, we shall see how this relates to the studies carried out independently by Blackett.

(2) The electromagnetic laws derivable from the theory are nonlinear, thus implying, in principle, light–light scattering at the classical level.

(3) The theory's nonlinearity allows, in principle, for the possiblity of deriving, in accordance with Einstein's view, a field theory in which particles may be reproduced as spherically symmetric singularity-free solutions. The actual possibility appeared doubtful in 1955 (Tonnelat 1955), and as far as I know has never been realized.

In the late 1960s, a detailed exposition of this theory was given by D.K. Sen. Sen reviewed at some length the attempts at obtaining a rigorous solutions in simple special cases and the discussions of weak-field approximations of various orders (see Sen 1968 and references therein). Since the symmetric part of the static, spherically symmetric solution did not coincide with the corresponding solution of the Einstein–Maxwell equations (Sen 1968, p. 102), the question was raised as to whether the symmetric part of the fundamental tensor represented the real metric of the physical space-time (Sen 1968, p. 104). In view of the incompatibility of the result with general relativity, the identification was still considered an open question (Sen 1968, p. 105).

3.7 Einstein and Teleparallelism

Starting in 1928, Einstein tried a different approach, which despite its failure to achieve a satisfactory unified theory is interesting for a number of reasons. In a paper of a purely mathematical character (Einstein 1928), he introduced a new geometry, characterized by the property of distant parallelism, expressed in terms of *n-beins*, i.e., orthogonal tetrads. The latter element is interesting in and of itself, and received attention for various reasons and purposes. N. Wiener and M.S. Vallarta observed that it permitted the introduction of spin (Wiener and Vallarta 1929). Many years later, W.H. McCrea and F.J. Mikhail discussed the possibility of using a tetrad vector field in order to provide field equations admitting the creation of matter according to Hoyle's version of the steady state theory (McCrea and Mikhail 1955). Actually, geometries characterized by distant parallelism had already been introduced by Cartan and developed by other authors. After the publication of Einstein's paper, Cartan pointed out to Einstein that he had mentioned them in a conversation they had in 1922,[34] something Einstein had apparently forgotten. This minor controversy was settled when, on Einstein's invitation, Cartan wrote a historical note on the notion of absolute parallelism.[35] A physical application of this geometry, an attempt at a unified theory, was proposed by Einstein in subsequent papers. He gave a particularly clear exposition of it in a lecture delivered at the Poincaré Institute in November 1929. The contact with physics is essentially established through the antisymmetric part of the connection, Λ, which, Einstein observed, has a tensorial character and $6 \times 4 = 24$ independent components, a promising feature in view of the unification program. In first approximation, the antisymmetric set of the field equations gave Maxwell's equations; the symmetric set turned out to be compatible with the Newtonian theory, but not identical to that obtained within the standard Riemannian approach (Einstein 1930, p. 23). Einstein's new attempt attracted a lot of attention. Eddington, in particular, wrote a short note stressing the aspects in which the new theory differed drastically from existing unified field theories, in particular from his cherished affine theory (AT) (parallel transport is integrable in Einstein's theory, while it is essential that it be nonintegrable in the AT; the connection is symmetric in AT, while it is essential that there be an antisymmetric part in Einstein's theory; finally, the curvature tensor, which provides the field variables in AT, vanishes identically in Einstein's geometry, while it is Einstein's new tensor Λ that vanishes identically in AT). In Eddington's view, the new approach did not offer enough to compensate for the drastic change of viewpoint it implied: "Perhaps one who believes that Weyl's theory and its affine generalization afford considerable enlightenment may be excused for doubting whether the new theory offers sufficient inducement to make an exchange" (Eddington 1929, p. 281). Einstein's theory was developed and defended against criticism, as expressed in particular by Eddington, by R. Zaycoff in a series of papers (Zaycoff 1929a, 1929b, 1929c, 1929d). Zaycoff gave an interpretation of the new geometry maintaining that, although Einstein's world is flat, because of the vanishing of the curvature tensor, "it is not Euclidean in the usual sense, but only as a consequence of the nonvanishing torsion, so to speak, in a nonholonomic sense." In other words, "the Riemannian curvature is compensated by the torsion curvature" (Zaycoff 1929d, p. 724).[36]

From our epistemological viewpoint, the most interesting feature of Einstein's new theory is that, in principle, since the space-time manifold is not endowed with a nonvanishing curvature tensor expressed in terms of the connection, it does not contain general relativity as a limiting case, contrary to all attempts mentioned so far. Pauli wanted to know what had happened to Mercury's perihelion and to the deviation of light rays.[37] Years later Einstein had to admit, "Sie haben also recht gehabt, Sie Spitzbube!"[38]

Notwithstanding its various drawbacks, Einstein's theory continued to receive attention over the years (a generalized field theory based on the use of a tetrad space was elaborated by F.I. Mikhail and M.I. Wanas as late as 1977 [Mikhail and Wanas 1977]). Needless to say, the notion of torsion was to prove central in the so-called Einstein–Cartan theory (for a review, see Hehl et al. 1976), which is, however, an extended theory of gravitation and not an attempt at a unified theory. It acquired a deeper physical meaning[39] through the work of Kondo (1955; see also Kröner 1981; Ross 1989), where the torsional defect produced by spin in the geometry is assumed to be a multiple of the Planck length.

3.8 Schrödinger's Attempts

In the later part of his life, in particular from 1943 to 1951 (Bertotti 1985, p. 87; Moore 1989, p. 385), Schrödinger dedicated most of his time to studies on unified theories.

Schrödinger's first attempt (Schrödinger 1943) was along the lines that had first been indicated by Eddington and Einstein. Like Eddington, he believed the connection to be a more fundamental notion than the metric. Like Einstein, in his development of Eddington's ideas, Schrödinger started from a variational principle with the connection as his only variable. Schrödinger explored first of all what could be derived from the sole assumption that the Lagrangian is an unspecified function of the components of the contracted curvature tensor constructed from a symmetric connection. Direct and inverse metrics were introduced through the symmetric part of the Lagrangian's functional derivative. Despite Schrödinger's claim that this was "already sufficient to produce from pure and straightforward affine geometry the complete system of the differential equations of the combined gravitational and electromagnetic field" (Schrödinger 1943, p. 43), the interpretation of the set of equations is by no means obvious. The specification of a Lagrangian determined a system of equations in which terms of the Born–Infeld nonlinear electromagnetic theory, which had caught Schrödinger's interest as early as 1935, could be identified. Schrödinger expressed the belief that the mesonic field could be accounted for side by side with the electromagnetic field by introducing a second symmetric connection. At the end of the article, he pointed out that "his friend" A.J. Connell had drawn his attention to the fact that, on the one hand, the duplication of the connection appeared fairly strange, while on the other hand the requirement of symmetry was perhaps unnecessary. Schrödinger was to make good use of these observations in his subsequent studies. He returned to the subject the following year, during the course of which he also published a paper with Connell, in which he attempted to find some experimental support for his theory. As mentioned by Moore (1989, p. 417), a finite range μ^{-1} for the electromagnetic interactions (or equivalently, a massive photon) was predicted; the authors were able to extract a lower bound of $\mu^{-1} > 15{,}000$ km, or $\mu < 0.67 \cdot 10^{-7} m^{-1}$ from the data on the earth's magnetic field. In 1968 this limit was improved to $\mu < 1.15 \cdot 10^{-8} m^{-1}$ by Goldhaber and Nieto (Goldhaber and Nieto1968; see also the discussion in Breitenberger 1971). (Today the theory would be immediately discarded: Moore points out that the data sent by the Voyager probe during the 1979 fly-by of Jupiter have reduced the upper bound by several orders of magnitude, "incidentally falsifying the 1943 prediction from Schrödinger's unified field theory" [Moore 1989, p. 453].) From the epistemological point of view, the situation is therefore similar to that of Weyl's theory.

Schrödinger resumed work with new enthusiasm after the end of the war, partly due to the renewal of his correspondence with Einstein (Moore 1989, p. 424). In 1947 he published a paper with the significant title of "The Final Affine Field Laws (I)" (Schrödinger 1947). He must have felt strongly that he had made substantial, even decisive progress. Indeed, in the paper one can read statements like: "Now the correct Lagrangian is found, the fog sinks and everything becomes much simpler" (Schrödinger 1947, p. 163), and: "I am inclined to believe that the field equations (18) are the ultimate word that can be said on the physical fields, short of introducing the quantum aspect" (ibid., p. 169). What was this all about? Schrödinger claimed in the paper he had not deviated "a line's breadth" (ibid., p. 163) from the program he had set himself. And indeed his approach was still formulated in purely affine terms, with a fundamental tensor derived according to the same procedure he followed in 1943. The connection, however, was no longer symmetric, and consequently neither was the fundamental tensor. The Lagrangian is chosen as the square root of the determinant of the Ricci tensor. Eddington's relation,

$$G_{kl} = \lambda g_{kl},$$

though with asymmetric G and g, emerges naturally in this approach.

The field equations have, of course, the correct general relativistic limit. As far as electrodynamics is concerned, the interpretation, once again, is not so immediate. Nevertheless, as Schrödinger writes, "it clearly transpires ... that the 'true' electrodynamics really *is* of the type indicated by Max Born as early as 1934" (ibid., p. 169). Here, however, stress is on the mechano-magnetic effect.

> If these equations embody what they purport to embody, viz. the genuine union of the field of matter and the electromagnetic field, they ought *inter alia* to explain the magnetic field produced, as we know but completely fail to understand, by a rotating mass as the earth or the sun.... There can, I think, be little doubt that the magnetic field is a direct consequence of the mass rotation.... I have no reasonable doubt that the equation (18) will account for the mechano-magnetic phenomenon. (Schrödinger 1947, pp. 169–170)

In the course of the paper, Schrödinger proves "the remarkable fact" that Einstein's recently formulated asymmetric theory could be obtained from his own in the limit $\lambda \to 0$. In the symmetric case, as Schrödinger comments, the term λ "is known to have ... little practical significance, except in the cosmological problem" (ibid., p. 167). It becomes fundamental in the affine theory where it produces "the genuine affine form of the field-equations" (ibid.). As a matter of fact, as stressed by Bertotti (1985), the cosmological term is the only real difference with Einstein's theory. In a letter of February 1947, Einstein commented "... your theory does not really differ from mine, only in the presentation and in the 'cosmological term,' which mine lacks. In mine, in the absence of electromagnetic forces (and matter), space is planar, in yours it is a De Sitter space (due to the cosmological constant)" (quoted in Moore 1989, p. 434).

In the second part of the paper, "The Final Affine Field Laws (II)" (Schrödinger 1948a), Schrödinger, without adding much from the physical point of view, provides a general introduction, didactically very valuable, to the metrical, affine, and metrico-affine theories. Schrödinger has also left us a classic textbook that shows his deep mastery of the geometrical aspects surrounding these theories (Schrödinger 1950). The third part (Schrödinger 1948b) investigates some formal aspects. Written work by Schrödinger on the affine theory ends with two studies on the weak-field approximation (Hittmair and Schrödinger 1951; Schrödinger 1951).

4. The Blackett Effect

As already mentioned several times, a mechano-magnetic effect emerges, albeit with different features, as a prediction of some unified theories (Jordan–Thiry, Einstein, Schrödinger). The year 1947 seems to have been a crucial one for studies on this hypothetical effect. During the year, the effect not only received the attention of some of the authors quoted, but was also the object of an independent investigation by the distinguished British physicist P.M.S. Blackett. In a paper published in *Nature* (Blackett 1947), Blackett recalled the relation

$$P = \beta \frac{G^{1/2}}{c} U$$

in which G and c have the usual meanings and β is a constant of the order of unity), which had long been known to hold between the magnetic moment P and the angular momentum U of both the earth and the sun. From the measurement that year of the magnetic moment of a certain star it could be concluded that the relation, to a good approximation, was valid in that case as well. From the fact that the relation appeared to hold on very different scales, Blackett was led to conclude that it had to be "taken seriously as a possible general law of Nature for all massive rotating bodies" (Blackett 1947, p. 658). "Perhaps," he added, "this relation will provide the long-sought connection between electromagnetic and gravitational phenomena" (ibid.). This hypothesis was supported, in his opinion, by the fact that the formula contained only the constant $G^{1/2}/c$, a circumstance that appeared to exclude specific properties of the rotating body, while suggesting a general role of gravitation.

In his article, Blackett summarized the long history of investigations and discussions on the magnetic fields of the earth and sun, going over the difficulties met in the very natural attempt to explain them in terms of a separation of positive and negative charges within these bodies. It is worth mentioning that, for Blackett, it was "clear that no adequate real charge separation can exist [if] the normal electro-magnetic equations are assumed valid" (ibid., p. 664). Therefore, he wrote, "some alteration in the fundamental equations seems inevitable" (ibid.). Blackett cited an attempt, along similar lines, by H.A. Wilson, but did not make any explicit reference to unified theories.

Blackett's paper caught the attention of those interested in terrestrial magnetism, and did not pass unnoticed between those interested in unified theories, as documented by the reference made to it in Tonnelat's treatise of 1955. No fruitful dialogue seems to have developed among the two communities, however. In this respect, an episode mentioned by Moore is significant: at the 8th Solvay Conference, in 1948, "Blackett gave a lecture [published as Blackett 1949] on the magnetic field of massive rotating bodies; ... surprisingly, Schrödinger did not comment on this problem, although his unified field theory had dealt with it explicitly" (Moore 1989, p. 444).

It should be mentioned that, already in 1946, Elsasser had formulated the "dynamo hypothesis," as an alternative to those being discussed. "A possible explanation in terms of a dynamo produced by lunar tides" today seems "to emerge ineluctably" (Gregori 1990, p. 57). Moreover, Runcorn et al., as early as 1950, showed that the increase of the strength of the geomagnetic field with depth contradicted Blackett's hypothesis (Busse 1980). It does not seem possible to isolate a particular moment at which the predictions of unified theories about the mechano-magnetic effect have been refuted experimentally. It must be admitted, though, that these predictions have never been particularly precise.

5. Conclusions

We may summarize the situation in the following terms. In the first place, as one would have expected (with some exceptions, like Einstein's first attempt of 1925 and his theory based on teleparallelism), nearly all attempts satisfy the first criterion, that of reproducing the Einstein–Maxwell theory in first approximation. This, by the way, in the case of the five-dimensional theories, looked so impressive to Salam as to make him baptize the result "the Kaluza–Klein miracle." Secondly, one can observe that the formal criteria for an effective unification are very often or nearly always met as well. From this point of view, perhaps to our surprise, Einstein and Schrödinger's asymmetric, and Jordan and Thiry's five-dimensional theories are in better shape than the better-known theories by Weyl and by Kaluza and Klein. The situation is far worse from the fundamental point of view of the relation with experiments. Even here, perhaps not without some surprise, we nonetheless discover, or rather rediscover, that some of these theories were not empty, which is in itself a positive feature. They are, however, either refuted by the experiment, as is the case with Weyl's theory (apart from possible, but not very plausible, *escamotages* [sleights of hand]), or do not make sufficiently accurate and stringent predictions, as in the case of the mechano-magnetic effect. In the latter case one is faced with the situation, which is remarkable and worth stressing, that one cannot yet say about some of the proposed unified theories that they have been experimentally refuted.

ACKNOWLEDGMENTS. I wish to express my thanks to the following persons, who have helped me by sending and/or calling my attention to relevant material: B. Bertotti, J. Ehlers, J. Eisenstaedt, H. Goenner, F.W. Hehl,

S. Kichenassamy, and S. Sigurdsson. Permission to reprint letter excerpts was granted by the Albert Einstein Archives, The Hebrew University, Jerusalem.

NOTES

[1] See, for instance, Yang 1975 and Drechsler 1986.

[2] English versions of the original papers by Kaluza and Klein appeared in de Sabbata and Schmutzer 1983 (translated by C. Honselaer), and in Lee 1984 (translated by T. Mura); a collection of the most relevant papers on the Kaluza–Klein theory has been edited by Appelquist et al. 1987; reprints or excerpts appear also in Charap and Okun 1986; comments on work by Kaluza and Klein can be found in Bergia 1987; Bergia et al. 1986; Carazza and Guidetti 1980.

[3] One element of this feeling is certainly "the majority's disbelief in the explanatory power of UFTs [unified field theories] in the realm of elementary particles" (Goenner 1984).

[4] I am indebted to F. Rohrlich for a comment on this point.

[5] "It is... the current opinion that this purely geometrical scheme is not rich enough. For example, electromagnetism is not seen as the skew part of a fundamental tensor, but, stressing its gauge invariance, is considered in conjunction with a complex scalar (or spinor) field; such a field needs a special affinity to be transferred from place to place and the integrability of this transport operation is determined just by the electromagnetic field" (Bertotti 1985). UFTs have a "*geometrically overloaded* but *physically undernourished structure*" (Goenner 1984, p. 192).

[6] Einstein's guidelines are summarized in Goenner 1984. I am indebted to H. Goenner for stressing to me the relevance of Einstein's epistemological contribution. Goenner's paper also gives an up-to-date analysis of the general principles to be satisfied by a unified field theory.

[7] I wish to thank F.W. Hehl for calling my attention to Rainich's approach.

[8] In Einstein's own words, the field must "appear as a unified covariant entity." The classic example is "the unification of the electric and the magnetic fields by the special theory of relativity" (Einstein 1945, p. 578).

[9] I am grateful to J. Ehlers for pointing out to me the relevant passages of Weyl and Lichnerowicz.

[10] The expression appears in Weyl 1921, p. 282.

[11] Weyl 1918, p. 148; Weyl was thus opening a new chapter in the history of the relationship between geometry and relativistic theories. As has been stated by Chern:"It soon became clear that in the applications of Riemannian geometry to relativity, the Levi-Civita parallelism, and not the Riemannian metric itself, plays the crucial role" (Chern 1980). Weyl had in fact improved on Levi-Civita's definition o parallel displacement of a vector (Levi-Civita 1917). The "various factors that conditioned the emergence" of Weyl's theory (like "the Göttingen tradition in mathematical physics," "Weyl's contacts with Einstein," and "his philosophical interests") are analyzed in Vizgin 1989. The influence of similar and other factors on Weyl is also discussed in Sigurdsson 1991. The particular influence of Husserl's thought on the way in which Weyl attacked the crisis of the foundations of mathematics is analyzed in Tonietti 1988.

[12] For this terminology, see Drechsler 1986; Fulton et al. 1962.

[13] Einstein to Hermann Weyl, April 6, 1918. Translation as in Sigurdsson 1991, p. 163.

[14] Einstein to Hermann Weyl, April 8, 1918. Translation as in Pais 1982, p. 341.

[15] Einstein to Hermann Weyl, April 15, 1918. Translation as in Sigurdsson 1991, p. 164.

[16] Einstein to Hermann Weyl, April 15, 1918. Einstein gave the same argument in a letter to Besso (Einstein to Michele Besso, August 20, 1918). Permission for quotations from the letters mentioned in Notes 13 to 16 granted by the Albert Einstein Archives, The Hebrew University, Jerusalem.

[17] See the fine discussion in Moeller 1955. I thank S. Kichenassamy for directing my attention to Moeller's paper.

[18] The story of the delayed acceptance of Kaluza's paper and an analysis of the reasons for the difficulties met by his idea and for the apparent lack of personal recognition was told by E.W. Middleton (1991).

[19] This conclusion can be questioned on the basis of some recent results. J.A. Ferrari (1989) has investigated the static, spherically symmetric solution of the five-dimensional Kaluza–Klein equations (in the case of a spherical charged system), and found that it does not approximate the Reissener–Nordström solution in the limit $r \rightarrow \infty$. J.A. Ferrari, J. Griego, and E.E. Falco claim that "the Kaluza–Klein theory goes beyond merely 'reproducing' classical electrodynamics," insofar as it provides "very simple explanations of the Aharonov–Bohm and Meissner–Ochsenfeld effects" (Ferrari et al. 1989, p. 70), the essential reason being that the prescription $x^5 = \text{const}$ fixes the gauge in which the potential is expressed.

[20] The issue has been recently addressed by D.K. Ross. Ross observes that the Lagrangian density for Klein's fields is customarily assumed to be invariant under general coordinate transformations in a general five-dimensional Riemannian manifold. However, as I pointed out earlier, the theory is only invariant under the transformations of the product group, Eqs. (5a, 5b). If this circumstance is duly taken into account, one "*does not get superheavy masses* and in fact no mass at all," so that "an invariant mass term can be put in by hand, giving the charged particles the mass we like" (Ross 1987, p. 2170).

[21] As stressed by Pais, 1982, p. 334.

[22] Starting with Veblen and Hoffmann 1930; Schouten and van Dantzig, Pauli and others extended the idea; see, in particular, Schouten and van Dantzig 1932.

[23] As reported in Pais 1982, p. 335.

[24] See, for instance, Dirac 1937.

[25] See, for instance, Jordan and Müller 1947.

[26] I thank S. Kichenassamy for drawing my attention to Just's contributions.

[27] I wish to thank S. Kichenassamy for drawing my attention to this work.

[28] See the discussion in Tonnelat 1955, p. 18.

[29] Ferraris and Francaviglia 1986, p. 12; this is what is usually referred to as "Palatini's method."

[30] See Pais 1982, p. 344.

[31] See Pais 1982, p. 348.

[32] Kaufman 1956, p. 228. See, however, Goenner's statement: "I am unaware of any testable or (at least) convincing physical argument for transposition invariance despite ... Einstein's claim, endlessly repeated, ... that it stands for the invariance of the field laws with respect to the sign of electricity" (Goenner 1984, p. 187).

[33] In the footnote, Pauli referred to Einstein 1953 and Einstein and Kaufman 1955.

[34] Elie Cartan to Einstein, May 8, 1929. In Debever 1979.

[35] The exchange between Einstein and Cartan on this subject is discussed in Biezunski 1989.

[36] I wish to thank F.W. Hehl for calling my attention to Zaycoff's papers.

[37] Quoted in Pais 1982, p. 347.

[38] "You were right after all, you rascal!" (translation as in Pais 1982, p. 347).

[39] I wish to thank F.W. Hehl for stressing this point to me.

References

Appelquist, Thomas and Chodos, Alan (1983). "Quantum Dynamics of Kaluza–Klein Theories." *Physical Review* D28: 772–785.

Appelquist, Thomas, Chodos, Alan, and Freund, Peter G.O. (1987). *Modern Kaluza–Klein Theories*. New York: Addison-Wesley.

Bergia, Silvio (1987). "Explication d'une analogie formelle entre la théorie einsteinienne de la gravitation pour des champs stationnaires et la théorie unitaire de Kaluza–Klein." *Annales de la Fondation Louis de Broglie* 12: 349–362.

Bergia, Silvio, D'Angelo, Giuseppe, and Monzoni, Vittorio (1986). "Waves in Five Dimensions and Quantum Physics in O. Klein's Paper of 1926." In *Proceedings of the Fourth Marcel Grossmann Meeting on General Relativity*. Remo Ruffini, ed. Amsterdam: Elsevier, pp. 1795–1803.

Bertotti, Bruno (1985). "The Later Work of E. Schrödinger." *Studies in History and Philosophy of Science* 16: 83–100.

Biezunski, Michel (1989). "Inside the Coconut: The Einstein–Cartan Discussion on Distant Parallelism." In *Einstein and the History of General Relativity*. Don Howard and John Stachel, eds. Boston: Birkhäuser, pp. 315–324.

Blackett, Maynard P.S. (1947). "The Magnetic Field of Massive Rotating Bodies." *Nature* 159: 658–666.

—— (1949). "The Magnetic Fleld of Massive Rotating Bodies." *Philosophical Magazine* Ser. 7, 40: 125–150; the text was prepared by the author for the 8th Solvay Congress, Solvay (Brussels, 1948).

Brans, Carl H. and Dicke, Robert H. (1961). "Mach's Principle and a Relativistic Theory of Gravitation." *Physical Review* 124: 925–935.

Breitenberger, E. (1971). "On the Empirical Foundations of Special Relativity." *Il Nuovo Cimento* 1B: 1–22.

Busse, F.H. (1980). "Motions in the Earth's Core and the Origin of Geomagnetism." In *Proceedings of the International School of Physics 'Enrico Fermi,' Course*

LXXVIII ("Physics of the Earth's Interior"). A.M. Dziewonski, ed. Amsterdam: North Holland, pp. 493–507.

Carazza, Bruno and Guidetti, Gian Paolo (1980). "La nascita dell'equazione di Klein–Gordon." *Archives for the History of Exact Sciences* 22: 373–383.

Cartan, Elie (1922). "Sur une généralization de la notion de courbure de Riemann et les espaces à torsion." *Comptes Rendues* 174: 593–595.

Charap, John M. and Okun, Lev B., eds. (1986). "Special Topics in Gauge Theories." *Surveys in High Energy Physics* Volume 5, Number 3.

Chern, Shiing-shen (1980). "General Relativity and Differential Geometry." In *Some Strangeness in the Proportion*. H. Woolf, ed. New York: Addison-Wesley, pp. 271–280.

Debever, Robert, ed. (1979). *Elie Cartan–Albert Einstein Letters on Absolute Parallelism 1929–1932*. Jules Leroy and Jim Ritter, trans. Princeton: Princeton University Press; Brussels: Académie Royale de Belgique.

De Sabbata, Venzo and Schmutzer, Ernst, eds. (1983). *Unified Field Theories of More than Four Dimensions*. Proceedings of the International School of Cosmology and Gravitation, Erice, Singapore: World Scientific.

Dirac, Paul A.M. (1937). "Cosmological Constants." *Nature* 139: 323.

——— (1973). "Long Range Forces and Broken Symmetries." *Royal Society of London* Proceedings A 333: 403–418.

Drechsler, Wolfgang (1986). "Geometric Formulation of Gauge Theories," paper presented at the conference on the *History of Modern Gauge Theories*, July 20–25, 1987, Logan, Utah, U.S.A.

Eddington, Arthur S. (1921). "A Generalization of Weyl's Theory of the Electromagnetic and Gravitational Fields." *Royal Society of London*. Proceedings A 99: 104–122.

——— (1924). *The Mathematical Theory of Relativity*. Second edition. New York: Chelsea.

——— (1929). "Einstein's Field Theory." *Nature* 123: 280–281.

Ehlers, Juergen, Pirani, Felix A.E., and Schild, Alfred (1972). "The Geometry of Free Fall and Light Propagation." In *General Relativity*. O'Rafertaigh, ed. Oxford, pp. 63–84.

Einstein, Albert (1918). "Nachtrag" to Weyl's "Gravitation und Elektizität." *Königlich Preussische Akademie der Wissenschaften* (Berlin). *Sitzungsberichte*: 465–480.

——— (1923). "The Theory of the Affine Field." *Nature* 112: 448–449.

——— (1925). "Einheitliche Feldtheorie von Gravitation und Elektrizität." *Königlich Preussische Akademie der Wissenschaften* (Berlin). *Sitzungsberichte*: XXII, 414–419.

——— (1928). "Riemann-Geometrie mit Aufrechterhaltung des Begriffes des Fernparallelismus." *Königlich Preussische Akademie der Wissenschaften* (Berlin). *Sitzungsberichte*: XVII, 217–221.

——— (1930). "Théorie unitaire du champ physique." *Annales de l'Institut Henri Poicaré* 1: 1–24.

——— (1945). "A Generalization of the Relativistic Theory of Gravitation." *Annals of Mathematics* 46: 578–584.

——— (1953). *The Meaning of Relativity*. 4th edition. Princeton: Princeton University Press.

Einstein, Albert and Bergmann, Peter G. (1938). "On a Generalization of Kaluza's Theory of Electricity." *Annals of Mathematics* 39: 683–701.

Einstein, Albert and Kaufman, Bruria (1955). "A New Form of the General Relativistic Field Equations." *Annals of Mathematics* 62: 128–138.

Einstein, Albert and Mayer, Walther (1931). "Einheitliche Theorie von Gravitation und Elektrizität." *Königlich Preussische Akademie der Wissenschaften* (Berlin). *Sitzungsberichte*: XXV, 541–557.

——— (1932). "Einheitliche Theorie von Gravitation und Elektrizität. Zweite Abhandlung." *Königlich Preussische Akademie der Wissenschaften* (Berlin). *Sitzungsberichte*: XII, 130–137.

Ferrari, José A. (1989). "On an Approximate Solution for a Charged Object and the Experimental Evidence for the Kaluza–Klein Theory." *General Relativity and Gravitation* 21: 683–695.

Ferrari, José A., Griego, J., and Falco, E.E. (1989). "The Kaluza–Klein Theory and Four-Dimensional Space-Time." *General Relativity and Gravitation* 21: 69–78.

Ferraris, Marco and Francaviglia, Mauro (1986). "Unificazione dei campi classici." Article prepared for the "Dizionario delle Scienze Fisiche" of the Istituto dell'Enciclopedia Italiana, in press.

Fulton, T., Rohrlich, Fritz, and Witten, Louis (1962). "Conformal Invariance in Physics." *Reviews of Modern Physics* 34: 442–457.

Goenner, Hubert F. (1984). "Unified Field Theories: From Eddington and Einstein Up to Now." In Proceedings of the Sir Arthur Eddington Centenary Symposium, Volume 1, *Relativistic Astrophysics and Cosmology*. V. de Sabbata and T.M. Karade, eds. Singapore: World Scientific, pp. 176–196.

Goldhaber, Alfred S. and Nieto, Michael M. (1968). "New Geomagnetic Limit on the Mass of the Photon." *Physical Review Letters* 21: 567–569.

Gregori, Gian P. (1990). "L'origine del campo magnetico della Terra." Invited paper at the LXXXVI Congresso della Società Italiana di Fisica, Trento. Bulletin edited by Editrice Compositori for the Italian Physical Society.

Hehl, Friederich W., von der Heyde, Paul, Kerlick, G. David, and Nester, James M. (1976). "General Relativity with Spin and Torsion: Foundations and Prospects." *Reviews of Modern Physics* 48: 393–416.

Hehl, Friedrich W., McCrea, J. Dermott, and Mielke, Eckehard W. (1988). "Weyl Space-Times, the Dilation Current, and Creation of Gravitating Mass by Symmetry Breaking." In *Exact Sciences and Their Philosophical Foundations—Vorträge des Internationalen Hermann-Weyl-Kongresses, Kiel 1985*. Wolfgang Deppert, Kurk Hübner, Arnold Oberschelp, Volker Weidemann, eds. Frankfurt am Main: P. Lang-Verlag, pp. 241–309.

Hehl, Friedrich W., McCrea, J. Dermott, Mielke, Eckehard W., and Ne'eman, Yuval (1989). "Progress in Metric-Affine Gauge Theories of Gravity with Local Scale Invariance." *Foundations of Physics* 19: 1075–1100.

Hilbert, David (1915). "Die Grundlagen der Physik. (Erste Mitteilung.)" *Königliche Gesellschaft der Wissenschaften zu Göttingen. Matematisch-physikalische Klasse. Nachrichten*: 395–407.

Hittmair, O. and Schrödinger, Erwin (1951). "Studies in the Generalized Theory of Gravitation. II: The Velocity of Light." *Communications of the Dublin Institute for Advanced Studies* Series A, 8: 1–15.

Jordan, Pascual (1946). "Relativistische Gravitationstheorie mit variabler Gravitationskonstante." *Die Naturwissenschaften* Heft 8: 250.

Jordan, Pascual and Müller, Claus (1947). "Über die Feldgleichungen der Gravitation bei variabler 'Gravitationkonstante'." *Zeitschrift für Naturforschung* 2a: 1–2.

Just, Kurt (1954). "Zur Wahl von Feldgleichungen der projektiven Relativitätstheorie." *Zeitschrift für Physik* 139: 498–503.

——— (1956). "Erweiterte Gravitationstheorie und Periheldrehung." *Zeitschrift für Physik* 144: 411–427.

Kaluza, Theodor (1921). "Zum Unitätsproblem der Physik." *Königlich Preussische Akademie der Wissenschaften* (Berlin). *Sitzungsberichte*: 966–972.

Kaufman, Bruria (1956). "Mathematical Structure of the Non-Symmetric Field Theory." *Helvetica Physica Acta* Suppl. IV: 227–238.

Klein, Oskar (1926a). "Quantentheorie un fünfdimensionale Relativitätstheorie." *Zeitschrift für Physik* 37: 895–906.

——— (1926b). "The Atomicity of Electricity as a Quantum Theory Law." *Nature* 118: 516.

——— (1969). "From My Life of Physics." In *From a Life of Physics*. I.A.E.A., Vienna, pp. 59–68. Republished in Abdus Salam, Hans A. Bethe, Paul A.M. Dirac, Werner Heisenberg, Eugene P. Wigner, Oskar Klein, Evgeny M. Lifshitz, eds. (1989). *From a Life of Physics*. Singapore: World Scientific.

Kondo, K. (1955). "Non-Riemannian Geometry of Imperfect Crystals from a Macroscopic Viewpoint." In *RAAG Memoirs of the Unifying Study of Basic Problems in Engineering and Physical Sciences by Means of Geometry*. Volume 1. K. Kondo, ed. Tokyo: Gakujutsu Bunken Fukyukai, p. 459.

Kragh, Helge (1984). "Equation with the Many Fathers. The Klein–Gordon Equation in 1926." *American Journal of Physics* 52: 1024–1033.

Kröner, Ekkehart (1981). "Continuum Theory of Defects." In *Physique des défauts/Physics of Defects*. Les Houches, Session XXXV, Roger Balian, Maurice Kléman, Jean-Paul Poirier, eds. Amsterdam: North Holland, pp. 282–315.

Lee, H.C., ed. (1984). *An Introduction to Kaluza–Klein Theories*. Singapore: World Scientific.

Levi-Civita, Tullio (1917). "Nozione di parallelismo in una varietà qualunque e conseguente specificazione geometrica della curvatura riemanniana." *Rendiconti del Circolo Matematico di Palermo* 42: 173–204.

Lichnerowicz, André (1955). *Théories Rélativistes de la Gravitation et de l'Électromagnétisme*. Paris: Masson.

London, Fritz (1927). "Quantenmechanische Deutung der Theorie von Weyl." *Zeitschrift für Physik* 42: 373–389.

McCrea, William H. and Mikhail, F.I. (1955). "Vector-Tetrads and the Creation of Matter." *Royal Society of London* Proceedings A235: 11–22.

Middleton, Eric W. (1991). "Theodor Kaluza Revisited." Preprint.

Mikhail, F.I. and Wanas, M.I. (1977). "A Generalized Field Theory. I. Field Equations." *Royal Society of London* Proceedings A356: 471–481.

Misner, Charles W. and Wheeler, John A. (1957). "Classical Physics as Geometry." *Annals of Physics* 2: 525–603.

Moeller, Christian (1955). "Old Problems in the General Theory of Relativity Viewed from a New Angle." *Det Kongelige Danske Videnskaberne Selskab. Matematisk-fysiske Meddelelser* Bind 30 nr. 10: 1–29.

Moore, Walter (1989). *Schrödinger, Life and Thought*. Cambridge: Cambridge University Press.

Nordström, Gunnar (1914). "Über die Möglichkeit das elektromagnetische Feld und das Gravitationsfeld zu vereingen." *Physikalische Zeitschrift* XV: 504–506.

Pais, Abraham (1982). *"Subtle Is the Lord... ": The Science and the Life of Albert Einstein*. Oxford: Clarendon Press.

Pauli, Wolfgang (1958). *Theory of Relativity*. With supplementary notes by the author. D. Field, trans. London: Pergamon. Translation of "Relativitätstheorie," in *Enzyclopädie der mathematischen Wissenschaften mit Einschluss ihrer Anwendungen*. Vol. 5, *Physik*, Part 2. Arnold Sommerfeld, ed. Leipzig: B.G. Teubner, 1904–1922, pp. 539–775. [Issued November 15, 1921.]

Pigeaud, Pierre (1963). "Généralization des schémas matière pure et fluide parfait en théorie pentadimensionelle de Jordan–Thiry." *Annales de l'Institut Fourier* 13: 181–217.

Rainich, G.Y. (1925). *Transactions of the American Mathematical Society* 26: 106.

—— (1950). *The Mathematics of Relativity*. New York: Wiley.

Ross, D.K. (1987). "Charge Quantization Without Superheavy Masses in a Kaluza–Klein Description of Electromagnetism." *Journal of Mathematical Physics* 28: 2167–2170.

—— (1989). "Planck's Constant. Torsion and Space-Time Defects." *International Journal of Theoretical Physics* 28: 1333–1340.

Runcorn, S.K., Benson, A.C., Moore, A.F., and Griffiths, D.H. (1950). "The Experimental Determination of the Geomagnetic Radial Variation." *Philosophical Magazine* 41: 783–791.

Salam, Abdus (1980). "Gauge Unification of Fundamental Forces." Nobel Lecture, preprinted by permission of the Nobel Foundation; I.A.E.A., U.N.E.S.C.O., I.C.T.P. IC/80/28.

Schouten, Jan Arnoldus and van Dantzig, D. (1932). "Generelle Feldtheorie." *Zeitschrift für Physik* 78: 639–667.

Schrödinger, Erwin (1943). "The General Unitary Theory of the Physical Fields." *Proceedings of the Royal Irish Academy* 49A: 43–58.

—— (1947). "The Final Affine Field Laws. I." *Proceedings of the Royal Irish Academy* 51A: 163–171.

—— (1948a) "The Final Affine Field Laws. II." *Proceedings of the Royal Irish Academy* 51A: 205–218.

—— (1948b). "The Final Affine Field Laws. III." *Proceedings of the Royal Irish Academy* 52A: 1–9.

—— (1950). *Space-Time Structure*. Cambridge: Cambridge University Press.

—— (1951). "Studies in the Non-Symmetric Generalization of the Theory of Gravitation. I." *Communications of the Dublin Institute for Advanced Studies* Series A, N.6: 1–28.

Sen, D.K. (1968). *Fields and/or Particles*. Toronto: Rierson Press; London and New York: Academic Press.

Sigurdsson, Skuli (1991). *Hermann Weyl, Mathematics and Physics, 1900–1927*. Ph.D. Thesis, Department of the History of Science, Harvard University, Cambridge, Massachusetts.

Thiry, Yves (1948). "Les équations de la théorie unitaire de Kaluza." *Comptes Rendues* 226: 216–218.

Tonietti, Tito (1988). "Four Letters of E. Husserl to H. Weyl and Their Content." In *Exact Sciences and Their Philosophical Foundations*. Wolfgang Deppert, Kurk Hübner, Arnold Oberschelp, Volker Weidemann, eds. Frankfurt am Main: Peter Lang, pp. 343–384.

Tonnelat, Marie-Antoinette (1955). *La théorie du champ unifié d'Einstein et quelques-uns de ses développements* Paris: Gauthier-Villars.

Veblen, Oswald and Hoffmann, Banesh (1930). "Projective Relativity." *Physical Review* 36: 810–822.

Vizgin, Vladimir P. (1989). "The Geometrical Unified Field Theory Program." In *Einstein and the History of General Relativity*. Don Howard and John Stachel, eds. Boston: Birkhäuser, pp. 300–314.

Weyl, Hermann (1918). "Gravitation und Elektrizität." *Königlich Preussische Akademie der Wissenschaften* (Berlin). *Sitzungsberichte*: 465–478. Reprinted in Weyl 1968, Band IV; pages cited from there.

—— (1920). "Elekrizität und Gravitation." *Physikalische Zeitschrift* 21: 649–650. Reprinted in Weyl 1968, Band IV; pages cited from there.

—— (1921). *Raum-Zeit-Materie*, 4th ed. Berlin: Julius Springer; English translation *Space, Time, Matter*. H.L. Brose, trans. London: Methuen, 1922; reprint New York: Dover, 1950. Page numbers cited from the Dover edition.

—— (1950). "50 Jahre Relativitätstheorie." *Die Naturwissenschaften* 38: 73–83. Reprinted in Weyl 1968, Band IV; pages cited from there.

—— (1968). *Gesammelte Abhandlungen*, 4 vols. K. Chandrasekharan, ed. Berlin: Springer-Verlag.

Wiener, Norbert and Vallarta, M.S. (1929). "Unified Field Theory of Electricity and Gravitation." *Nature* CXXIII: 317.

Yang, Chen Ning (1975). "Gauge Fields." *Proceedings of the 6th Hawaii Conference on Particle Physics.*

Yang, Chen Ning and Mills, R.L. (1954). "Conservation of Isotopic Spin and Isotopic Gauge Invariance." *Physical Review* 96: 191–195.

Zaycoff, Raschko (1929a). "Zur Begründung einer neuen Feldtheorie von A. Einstein." *Zeitschrift für Physik* 53: 719–728.

——— (1929b). "Zur Begründung einer neuen Feldtheorie von A. Einstein. (Zweite Mitteilung)." *Zeitschrift für Physik* 54: 590–593.

——— (1929c). "Zur Begründung einer neuen Feldtheorie von A. Einstein. (Dritte Mitteilung)." *Zeitschrift für Physik* 54: 738–740.

——— (1929d). "Zu der neuesten Formulierung der Einsteinschen einheitlichen Feldtheorie." *Zeitschrift für Physik* 56: 717–726.

Vladimir Fock: Philosophy of Gravity and Gravity of Philosophy

Gennady Gorelik

Vladimir Aleksandrovich Fock (1898–1974) was one of the main participants in the history of the general theory of relativity (GTR) in Russia. His teachers, A.A. Friedmann and V.K. Frederiks, were the pioneers of GTR in Russia (Vizgin and Gorelik 1987). He studied the famous Friedmann paper on nonstatic cosmology in manuscript form and translated it, at the author's request, into German. He elaborated a description of the spinor field in GTR (Fock 1929). In 1935 he was an "opponent" (judge) of Matvey Bronstein's thesis—the first deep investigation of quantum gravity. He independently of Einstein, Infeld, and Hoffman (1938) solved the problem of motion in GTR. He is the author of the first Soviet monograph on GTR. And, finally, he was an energetic and tireless participant in the discussions during the 1950s and 1960s on understanding GTR.

However, despite Fock's authority in physics due to his scientific abilities, his attitude toward GTR was not shared by many even in the USSR. This concerned his attitude toward cosmology, the role of coordinate conditions and especially privileged systems of reference, the principles of relativity and equivalence and the philosophical status of GTR.

His stand on the last issue seems to be the most enigmatic because he unequivocally declared his adherence to dialectical materialism and connected it with his understanding of GTR. Such an attitude did not attract sympathy among physicists, although Fock's human dignity and honesty were beyond doubt. Besides, in Stalin's time, Fock was the main defender of quantum and relativistic physics in the USSR.

This situation has already attracted the attention of historians (Graham 1982, 1987). Here I shall try to reveal the roots of Fock's position in his scientific activity proper and to analyze the nature of the communication gap between Fock and his physicist colleagues.

To do this, it is necessary to take into account the following factors: Fock's specific methodological stand, which was intermediate between theoretical physics and mathematics; his predisposition to a philosophical world view; his inclination to schematism or mathematization in life outside natural science; and finally the relatively narrow empirical basis of GTR in the 1930s to 1950s. It is not easy to discuss these factors in academic terms since they manifested themselves in horrible social circumstances. They were embodied concretely in Fock's personality—he was an honest, dignified, fearless and, strange as it may seem, law-abiding person. To reveal these elements of the explanation and to connect them in a united scientific-psychological complex, one should consider the evolution of Fock's views.

1. First Steps in the General Theory of Relativity

Fock (1963) recalled the following of his early acquaintance with GTR:

> A.A. Friedmann and V.K. Frederiks, professors in Petrograd University, were the first who familiarized Russian physicists, who worked in Petrograd, with the theory of gravitation recently created by Einstein. This was at the very beginning of the 1920s when the blockade of Soviet Russia had just been broken and scientific literature from abroad began to arrive. In the Physical Institute of the University a seminar gathered where, among other things, lectures on Einstein's theory were delivered. The participants of the seminar were teachers and students in their last year (and at that time there were very few). The basic speakers on the theory of relativity were V.K. Frederiks and A.A. Friedmann, but sometimes also Yu.A. Krutkov, V.R. Bursian and others spoke. I remember the talks of Frederiks and Friedmann clearly. The style of these talks was different: Frederiks deeply understood the physical side of the theory, but did not like the mathematical computations; Friedmann stressed not physics, but mathematics. He strived for mathematical rigor and attributed great importance to the full and exact formulation of the initial preconditions. The discussions that arose between Frederiks and Friedmann were very interesting.

This recollection, with its "duet of acting characters"—mathematics and physics—as we shall see, can tell us much about Fock's position itself. In spite of the important role of Frederiks in the assimilation of GTR in Russia, one may doubt the depth of his physical comprehension of GTR. On the other hand, "the full and exact formulation of the initial preconditions" was very characteristic of Fock himself.

The earliest documented testimony of Fock's interests in GTR is his handwritten summary of his lecture in a philosophical circle dated 1922.

What should be noted especially is that this summary is half philosophical. Fock begins with:

> I am going to give an account of the physical foundations of the theory of relativity and to point out some contacts with philosophical problems. But I consider myself to be an ignoramus in philosophy, and therefore I do not claim to solve, or even to put philosophical questions in correct form. In this respect I wait for the assistance of my listeners.

After a few words about the historical origins of GTR, Fock turns to

> the leading thread: the search for the really existing in the nature. The really existing is defined as being perceived by all identically. Examples:
> (a) an object being seen from different viewpoints;
> (b) an object being seen by moving observers; its mass and dimensions.
>
> If two observers see differently, it is clear that they see not the whole object but facets of it.
>
> One had to admit as really existing not space and time separately but their combination; instead of r^2 and t^2 the *interval* $r^2 - c^2t^2$ really exists.

Fock summarizes the content of general relativity in the following way:

> Geometry has absorbed physics. From the formal point of view it is all the same, but it is more satisfactory for our intuitions. As well now the physicist believes in the existence of atoms and electrons neither more nor less than in the existence of common "large" objects. If he was a "naive realist," he has remained the same. But he replaced entities that he considered before as really existing. It is possible to go further.
>
> Physics strives to break up phenomena into the simplest elements. But the simplest elements are not commonplace to us; besides that, they (as the simplest) are undefinable. The familiar (i.e., having properties of familiar objects) are only rather complex combinations of these elements. But so far as we have not given definitions for elements, there are no definitions for these combinations either.
>
> And we give a definition for the latter. We take the quantity G_{mn}. We do not say that it is equal to zero when matter is absent, but in another way: being equal to zero means that matter is absent, i.e., absence of matter is a symptom of G_{mn} being equal to zero.

From this short text one can draw some important conclusions.

In 1922 in Soviet Russia the philosophical approach to natural science was still a purely private affair. To lecture at a philosophical circle meant a certain predisposition to a philosophical outlook. It is difficult to attribute Fock's view in 1922 to some *-ism* (e.g., to intersubjective idealism or mathematical realism), but one can definitely say that in Fock's position there is no dialectical materialism (which he would master and appropriate in the mid-1930s). His striving to comprehend the epistemological bridge between physical reality and theory is beyond doubt.

In the philosophical viewpoint of the 24-year-old theorist, it is possible to see some ideas of the classics of relativity. However, Fock's approach to GTR seems highly independent (unlike the first Russian review on GTR by Frederiks (1921)). Instead of the ideas of general relativity, covariance and equivalence, which were usual for the majority of accounts of GTR, in Fock's account what prevails is the geometrical approach, based on the concept of absolute space-time. The other interesting feature is this: Fock mentions "the possibility of finite but boundless space" and does not mention the possibility of a nonstatic universe, while Friedmann's popular (1923) book was finished the day (!) before Fock's lecture, and it talked about this new possibility with enthusiasm. Friedmann's famous paper (1922) was dated May 29.

2. One and a Half Decades under the Symbol *h* and the Year 1937 under the Banner of Marxism

In the following one and a half decades Fock was busy with quantum theory on the whole. His important work on including the Dirac equation into GTR (1929) did not concern questions of principle in GTR.

Some interesting traits for the relativistic portrait of Fock may be revealed in his participation in the defense of M.P. Bronstein's thesis in 1935 (Gorelik and Frenkel 1985). Fock assessed highly this investigation, which was concerned mainly with the quantization of weak gravitational fields. He did not, however, attach importance to one of Bronstein's conclusions, which may have been the most interesting from the general physical and philosophical points of view, but the least definite mathematically. Based on a quantum-relativistic analysis of the measurability of the gravitational field (beyond its weakness and nongeometrical character), Bronstein deduced that, in a complete theory of quantum gravity, the concepts of space and time would have to be generalized radically. In Fock's words of 1935, one can see some distrust of GTR: he admits that the theory (of strong fields) may be changed and doubts the special role of the gravitational radius. (See Gorelik and Frenkel 1985.)

Fock took up the Einstein theory of gravitation in full measure at the end of the 1930s, preceded by some important events in his philosophical and social biography.

At the beginning of the 1930s, Fock discovered dialectical materialism for himself (hereafter I shall use the common Soviet abbreviation *Diamat*). We know from A.D. Aleksandrov's testimony (Aleksandrov 1988, 1989) that Fock read Lenin's book *Materialism and Empirio-Criticism* in 1932 and found in it something interesting and important for him (and he regretted that this book was inculcated in a "police" way). Two decades later Fock, in the introduction to his book on GTR, remarked:

> The philosophical side of my views on the theory of space, time and gravitation was formed under the influence of Lenin's *Materialism and Empirio-Criticism*. The doctrine of dialectical materialism helped me to approach critically Einstein's point of view on the theory created by him and helped to comprehend it anew. This doctrine also helped me to understand correctly and to interpret new results obtained by me. I would like to state this here, though properly philosophical questions are not touched upon in this book. (Fock 1955, p. 16)

Before considering the interaction between Fock, *Diamat*, and the theory of gravity, let us get acquainted with some facts from Fock's biography which may seem to be irrelevant at first glance.

In answering an official questionnaire in 1938, Fock wrote that he was descended from nobles, although he could have given a much less dangerous reply, because his father was a scientist–forester. Fock also informs us: "Since birth I have lived in Leningrad, did not take part in the revolutionary movement, ..., was not repressed by Soviet power."[1]

There are some inexactitudes here. In the first place, having entered Petrograd University in 1916, Fock in 1917 voluntarily joined the artillery school and then went to the war front (in 1918 he was immobilized because of advancing deafness). In the second place, Fock was arrested twice: in 1935 for one day and in 1937 also for only a few days (in the latter case he was released as a result of P. Kapitsa's courageous letter "upstairs").

For Fock, 1937 was also filled with many other events that were not very scientific. He was active in preventing a special session of the Academy of Sciences concerned with the philosophical basis of modern physics (Gorelik 1990). The initiator of this session was the 65-year-old electrical engineer academician, V.F. Mitkevich. Having old-fashioned (meta-)physical views and having obtained new-fashioned political skills, he officially proposed organizing a special meeting "for the struggle for the materialistic foundations of physics and against physical idealism." He had named V.A. Fock (together with I.E. Tamm and Ya.I. Frenkel) as a physical idealist and an opponent of *Diamat*.

Before 1937 Fock did not express his philosophical views publicly but had expressed unequivocally his opinion about the poor scientific level of books by Mitkevich and his fellow campaigner A.K. Timiryazev in a review (Fock 1934) published in the leading popular-scientific journal.

It was this review that was attacked (three years later!) by an aggressive and prolific journalist, V.E. Lvov. He seemed to take into account Fock's arrest and charged him not only with idealism but also with adherence to fascist methods!

Fock had to defend himself and his science. He sent three letters. In the first one, addressed to the Leningrad public prosecutor, Fock demanded prosecution of Lvov for libel and defamation. The second letter was addressed to the Central Committee of Communist Party, and there Fock wrote about the harm done by Lvov to Soviet science. In the third, a seven-page letter of February 13, 1938 to the Presidium of the Academy of Sciences, Fock, without any delicacy, expressed his misgivings about Mitkevich's efforts and insisted on abolishing a (quasi-) philosophical academy session devoted to physics. Judging by all that is known, it was Fock who was the main force in preventing this harmful session (Gorelik 1990).

In the same letter, Fock wrote about the desirability of good philosophical discussion of the new physics based on *Diamat*. He was sincere in writing these sentiments, for we have Fock's manuscript "Does Quantum Mechanics Contradict Materialism?" (23 pp.), dated November 1937. The discussion of 1937 became the subject of the article (Fock 1938a) published in the journal *Under the Banner of Marxism.*

In 1937 Fock was parted from many of his colleagues, who became the victims of the Great Terror. In August, M.P. Bronstein was arrested. When Fock heard about this he went personally to Bronstein's home to learn exactly what had happened (Gorelik and Frenkel 1990). In 1937, this was a very courageous and unusual act. He also signed letters in defense of repressed scientists.

Ten years later Fock wrote the official review of works by D. Ivanenko and A. Sokolov on quantum gravity, which were being presented for the Stalin prize. In this review Fock mentioned the name of M.P. Bronstein, then "an enemy of the people," many times. In particular,

> Whatever causes compelled the authors to avoid mentioning Bronstein's achievements, their work may not be considered as the construction of the quantum theory of gravitation, for this theory was created by Bronstein 11 years earlier.[2]

The facts pointed out here provide a social portrait of Fock and will help us to understand his attitude to GTR better.

3. Fock's Work on Motion in GTR

The year 1939 was a landmark in Fock's biography. He became a full member of the Academy of Sciences and his interests in fundamental physics had moved from quantum theory to gravitation. In 1939, Fock (1939a) published a long paper on motion in GTR. Without discussing the entirety of this important work (Havas 1989), I shall single out only a few essential points for our theme.

Fock begins with the difference between his and Einstein's points of view on GTR as a whole. For Fock, GTR is the foremost theory of gravitation and therefore must be applied to phenomena in which gravitation dominates, "i.e., to phenomena of an astronomical scale," but not to problems on an atomic scale. But at the same time he assesses very skeptically (or worse) "the so-called cosmological problem": "In the modern state of knowledge, any attempt to consider the Universe as a whole has to be of a speculative character."

Fock based his approach on the following: "in the atomic world it is observed that electrical forces greatly dominate the forces of gravity," "the great successes of quantum mechanics during the last 10–15 years and the complete fruitlessness of Einstein's attempt to explain elementary particles by means of a unified field theory." Nevertheless, Fock concludes:

> One should admire the creation of Einstein's genius, which is so rich in physical content in spite of its seeming abstractness. I hope this paper will help to reveal the physical content of this remarkable theory.

In a popular article, dedicated to Einstein's jubilee, Fock (1939b) expressed his position in stronger words, both in criticism in relation to cosmology and in appreciation of Einstein: "one of the greatest modern scientists, whose name is known and dear to every educated man and is shining equally as the name of Newton." In 1939 such enthusiastic appreciation of a "bourgeois" scientist known for his "idealistic" philosophical and "bourgeois-liberal" political views, was already very exotic.

Now let us return to Fock's paper itself, to one of its elements which was of a purely mathematical nature but later acquired considerable physical and even philosophical meaning. This element is the coordinate condition

$$\Box x_\nu = 0,$$

where $\Box$ is the covariant D'Alembertian (there are different forms of this condition). The corresponding coordinate systems are called *harmonic*.

One of the main peculiarities distinguishing Fock's work from Einstein–Infeld–Hoffmann's corresponding work is its view on the choice of coordinate systems. The physical statement of the problem, with the aim of correlation with Newton's equations of motion and post-Newtonian terms, leads to such conditions as the weakness of the gravitational field, its insular character (planetary system), and Euclidean character at infinity. But to solve the field equations of GTR it is necessary to add coordinate conditions, having chosen sufficiently definite coordinate systems.

For Einstein, with his understanding of GTR, the choice of coordinates was a question of technique or mathematics. And Einstein's choice leads to such a laborious pathway to the solution that the computations could not go into the publication and were cited from a complete manuscript in the Princeton Institute.

Fock, having chosen harmonic coordinates, found, as he wrote, "a much simpler" pathway to the solution. Already in the paper of 1939 he attempted to base his choice on more than mathematical grounds.

Having mentioned de Donder and Lanczos, "who first (in 1921, 1923) had pointed out the simplification reached by means of harmonic coordinates," Fock does not limit himself to pure mathematics. In his words,

> Harmonic coordinates are the nearest in their properties to ordinary rectangular coordinates and ordinary time in the Minkowski "world." That is why, in these coordinates, the GTR formulas are the clearest.

Fock supposes that the harmonic coordinate system "deserves the name of inertial," because he considers it

> most probable that from Euclideanness at infinity and from its harmonic character (in connection possibly with some additional conditions like the radiation condition), our coordinate system is defined uniquely, with the indeterminacy of an ordinary Lorentz transformation.... As it seems to me, the possibility of introducing, in the general theory of relativity, uniquely defined inertial coordinate system deserves to be noted.

Judging from the content of Fock's paper (1939a), it seems that the paper is not merely the solution of a certain problem, but the beginning of a large program of work. The war interrupted this work, however, and Fock came to be busy with more applied physical problems.

4. The Contribution of Nikolaus Copernicus to the General Theory of Relativity

The first postwar testimony of Fock's reflections on gravity is his rather short paper (1947) dedicated to Copernicus's jubilee. Why did he write on Copernicus? There are no other traces of Fock's interest in the history of science outside the twentieth century.

It is possible to point out several very different causes. In the ideological life of the postwar USSR, the most "militant materialism" reigned. Idealism (crossed with antipatriotism and cosmopolitanism) was attacked in different fields of science. Debates over the state ideology of Marxism–Stalinism became very leaden, strictly black and white, or, more exactly, red and white. The list of saints and enemies of progress had been formed, and Copernicus had one of the most respectable places among the heroes of science. The state attention to the great Polish founder of the new astronomy was strengthened by state political interests in Eastern Europe.

In Soviet ideology, the positive Copernicus was indissolubly connected with the negative Ptolemy. In a paper, the first in the jubilee volume, Idelson (1947), side by side with a profound analysis of Copernicus's works, had to mention also "the wise words of comrade I.V. Stalin," which consisted only of the phrase "decayed system of Ptolemy."

And it is the pair "Ptolemy–Copernicus" that cast a shadow on general relativity. Of course, the shadow was due not to these classics themselves but to the soldiers of the cause of the one true philosophy. Being ignoramuses in physics, they looked for philosophical mistakes only in popular texts. In such texts dealing with GTR, in order to explain the basic ideas (or for effect), the equal correctness of Ptolemy's and Copernicus's points of view was asserted, e.g., Friedmann 1923; Einstein and Infeld 1938.

In contrast with other articles of the jubilee volume, Fock did not base his article on an appropriate historical interest in 400-year-old events. In spite of the title of his article, it was not his aim to illuminate the relationship between Copernicus and Ptolemy by means of GTR. To the contrary, he preferred to use a controversy, solved long ago, in order to illuminate his understanding of GTR as a geometric theory of gravity. A second and no less important aim was to defend Einstein's theory against ignorant and malicious critics.

In short, in Fock's article of 1947 were present all the main elements of his treatment of GTR (which henceforth he named "Einstein's theory of gravity"):

(1) the radical devaluation of the principles of general relativity, covariance and equivalence;

(2) the possibility of introducing

> as space and time coordinates those variables which are quite analogous to the rectangular Cartesian coordinates and the time coordinates of the special theory of relativity (harmonic coordinates).... The essential condition for this is the requirement of pseudo-euclidean geometry of space-time at infinity...; this requirement is satisfied for systems of masses like a solar system. (Fock 1947, p. 185)

Based on this, Fock firmly rehabilitated "the immortal creation of Copernicus—the heliocentric theory of the solar system."

Fock's argumentation, which has been recounted many times since, is well known due to his monograph of 1955 (2nd edition in 1961 and English editions in 1959, 1964). That is why we may consider only the principal relevant circumstances from 1947 to the mid 1950s.

5. Fock's "Theory of Space, Time and Gravity" against the Background of His Gravity, His Time, and His Space

The next publication on the theory of gravitation that Fock prepared was in 1948. It was based on his prewar work. In addition to concrete results, he also developed his understanding of the fundamental ideas of Einstein's theory. Admitting the historical, heuristic role of the principle of equivalence, Fock denies its validity in the complete theory. He also denies any particular physical role of covariance and general relativity as a more general relativity than in the special theory of relativity. For Fock, Einstein's theory is solely a geometrical theory of gravity. In his words: "I gave a detailed account of my point of view on Einstein's theory of gravity because Einstein's point of view, which I consider as wrong, is dominant up to today" (Fock 1950, p. 70).

The end of the 1940s and the beginning of the 1950s in the USSR were not very suitable years for pure, subtle theory. Soviet physics found itself under the strongest social pressure. After a notorious session of VASHNIL in 1948 had devastated Soviet biology, a similar session was prepared for physics. The unhealthy ambitions of some physicists in unhealthy social conditions were embodied in the struggle against "physical idealism, antipatriotism and cosmopolitanism." In this struggle, V. Fock was of course on the side of genuine science, defending relativistic and quantum physics and scientific ethics (Gorelik 1991). He based this activity also on *Diamat* (Fock 1949).

It was at this same difficult time that discussion of the foundations of relativity was revived by publication in 1950 of lectures delivered by L.I. Mandelstam in the 1930s (Mandelstam died in 1944). In these lectures, in particular, an operational approach to physical concepts was used and attention was paid to conventional elements in the definitions of the special theory of relativity. Fock, in his review (1951), gave a high estimate of the scientific and pedagogic significance of Mandelstam's lectures but criticized the operational and conventional elements.

The question, however, which had been the subject of a methodological analysis of a physical theory for Fock, became a crime for ignorant philosophical overseers. They attacked "bourgeois idealism" on the whole, "reactionary Einsteinism" (Maksimov 1952) in particular and Mandelstam with his school as "the seed-bed of idealism in USSR" especially; to "physical idealism" it was easy to add also anti-"cosmopolitan" arguments because there were a great many Jews among Soviet physicist–theorists. One of the highest ranking among these philosophical overseers was A.A. Maksimov (1891–1976).

The main defender of genuine science was the academician Fock, whose arguments were both scientific and *Diamatic*. He demolished Maksimov in the leading philosophical journal (Fock 1953a)—and Maksimov was on the journal's editorial board.

It was in just such a social atmosphere that Fock went on to elaborate his treatment of Einstein's theory of gravity.

In a sense, 1955 became the year of summation. In that year Fock's monograph *The Theory of Space, Time and Gravity* was published, but for our theme a much less scientific article written by Fock for the principal Soviet newspaper *Pravda* is more interesting. This article was entitled "Half a Century of Great Discovery. About the Theory of Relativity by Albert Einstein." We have the good fortune of looking at this article through the eyes of two remarkable contemporaries—academicians Igor Tamm and Vladimir Fock—because of letters they exchanged on November 13 and 17, 1955.[3]

Judging from the correspondence they kept (the earliest letter is dated 1929) and the testimony of their colleagues, these outstanding Soviet theorists were connected by a mutual respect in both scientific and moral spheres.

The manuscript of Fock's article was sent to Tamm from *Pravda* for his information. It was his reaction to this article that led to Tamm's letter. Having recalled Fock's anti-Maksimov article of 1953 approvingly, Tamm expressed his doubts about the appropriateness of a polemic with Einstein and the discussion of his philosophical errors in a jubilee newspaper article. In the same letter Tamm invited Fock (with great respect for his "fundamental works in quantum electrodynamics and theory of space and time") to take part in investigations on quantizing space-time. This idea attracted Tamm's attention very much then.

Fock's letter, clear and detailed, answers all Tamm's remarks in the following way. He had not intended to write a praising, jubilee article, he explained, but a critical review of a published book (the Russian translation of Einstein's *The Meaning of Relativity*). The article turned out to be rather difficult, but its subject was fairly difficult too. "Einstein is a great physicist, but he is not a very good mathematician," Fock wrote.

> His mathematical errors have to be pointed out. For Einstein's reputation it is useful to cleanse his theory of erroneous statements. It would be a tactical mistake to keep silent about Einstein's erroneous philosophical statements. The only way for physics to get immunity from philosophical attacks is to admit philosophical errors by physicists themselves and to separate these errors from the entity of theory.

Fock concluded:

> Publication in a leading newspaper of an article about Einstein signed by me—independently of its intelligibility—I consider as very useful because:
> (a) it amounts to official recognition in our country of the theory of relativity as a great discovery and great achievement of human genius,
> (b) this recognition is made without grovelling and with reasonable criticism,
> (c) the philosophical sins of Einstein are mentioned but have been forgiven.

In these letters the discussion concerned only the social status of GTR. But, in just a few weeks, the possibility had arisen of giving an account of their true scientific views.

On November 30, 1955 there was an open session of the Academy of Sciences of USSR, dedicated to the 50-year jubilee of the theory of relativity. An introductory speech was made by Tamm. His closest colleague, V.L. Ginzburg, delivered the paper "Experimental Testing of the General Theory of Relativity," and Fock delivered a paper on the equations of motion. These papers, together with some others, comprised the memorial volume *Einstein and Modern Physics*. Fock's viewpoint is represented in the volume very lucidly and in a way that is especially convenient for us. It is in two components: critical and constructive.

The volume includes Einstein's "Autobiographical Notes," which were translated and commented on by Fock. He begins with Einstein's philosophical views; however, his reduction of the many-colored ("extremely inconsequent") philosophical palette of the great physicist into the sharp dichotomy of the terms "materialism–idealism" seems to be a ritual duty. In considering Einstein's pathway to the theory of gravity, Fock does without philosophy at all. He criticizes Einstein's reasonings, "which finally led him to his gravitation theory of genius," and criticizes his "logical inconsistencies," "incorrect use of terms," etc. The questions are, as before, in relativity, covariance, and equivalence (Fock 1956b).

Fock's paper on the equations of motion contains the following constructive statements. For isolated (insular) systems, it is possible to state "conditions determining the coordinate system uniquely, with an indeterminacy up to a Lorentz transformation (harmonic coordinates)." For Einstein's theory, the harmonic coordinate system has a significance in principle, because "the existence of such a system reflects the objective properties of the space-time continuum." The introduction of harmonic coordinates allows the recovery of the equations of motion of masses taking into account their inner structure, all ten classical integrals of motion including relativistic corrections, and the gravitational potentials at large distances.

What reception did Fock's position find with his colleagues?

In Tamm's article there is only one phrase indicating that " 'the special and general theories of relativities' may be not very good terms." In opposition to Fock, one can see the great importance attached by Tamm to cosmology.

A more definite opinion was expressed in Ginzburg's paper, although he also avoided "fundamental questions on space and time, geometry and the theory of field in their connection with the general theory of relativity." According to Ginzburg (1956), GTR is "first of all a relativistic theory of gravity" for which the principle of equivalence is "the basic physical statement," and the principle of general relativity in itself is not physical. Nevertheless, referring to Einstein, who had to admit this last point in 1918, Ginzburg stated that he does not agree with "the opinion of Fock, who says that the 'theory of gravity was incorrectly understood by its author' " (Ginzburg 1956, p. 136).

A straightforward opponent of Fock (and possibly more Einsteinian than Einstein himself) was L. Infeld, who, in the same memorial volume, wrote:

> I am not in agreement with Professor Fock that one should add certain conditions to the theory of relativity, the conditions picking out harmonic systems. During my visit in the Soviet Union (1955), I, to my great surprise, have been convinced that Professor Fock stands apart in this question, and that physicists of such caliber as Landau, Tamm, and Ginzburg are in disagreement with his attitude. (Infeld 1956, p. 238)

This discrepancy, this communication gap between Fock and his physicist colleagues, was maintained for many years, up to his death, in spite of his persistent efforts to elucidate his viewpoint (Fock 1967; Ginzburg 1973). One may find clear testimony of the failure of his efforts in the volume published in the USSR on Einstein's centenary that collected major works in the theory of gravity (*Albert Einstein i teoria gravitatsii*, 1979). This volume contains two of Fock's papers (1929, 1939a), but on the volume's jacket there is the phrase: "... general theory of relativity, i.e., mechanics of arbitrary accelerated system...."

6. Principles of Relativity and Complementarity for the History of Physics

Thinking over the communication gap between Fock and Tamm (as expressed in their letters), or, more generally, between Fock and his physicist colleagues, both in scientific–methodological and in social–philosophical spheres, one should avoid a quick and simple judgement over who was right and who was wrong. Both positions were pure and honest, but the difference that could not be removed stemmed from the difference in their personalities.

To comprehend this situation, a historian of physics might take a lesson from the experience of 20th-century physics. In our case one might take a lesson from a short article written by Fock himself. While summing up the theory of relativity and quantum mechanics epistemologically, and their experience of dealing with nonabsolute truths, Fock concluded:

> As the history of the development of science shows, general principles established for one field of knowledge may be applicable also in other fields. I believe that such general character is possessed by *the principle of relativity to the means of observation.* In this is its philosophical meaning. (Fock 1971)

Of course, in the history of science, to determine a "frame of reference" or "means of observation" for an outstanding scientist is much more difficult than in special relativity or in quantum mechanics. But only after having determined the "direction vectors" of the scientist's world view and having tied these vectors to his unique personality can a historian hope to comprehend his life path.

Now it is time to connect the fairly heterogeneous events from Fock's biography as described above. To connect them with one life line, first of all, it is necessary to describe his frame of life references. One should begin with the vectors characterizing his scientific standpoint, because, for a genuine scientist, and for Fock especially, these vectors are the most important.

According to the prominent experimental physicists P. Kapitsa and D. Rozhdestvensky, who knew Fock very well, "This is a man detached from common life due to his almost absolute deafness. The whole of his life is persistent work with scientific problems"; "Fock thinks by mathematical images and it is very difficult for him to go deeply into the mentality of an experimentalist or average man, in spite of his permanent readiness to help everybody who asks him" (Kapitsa 1989, p. 124; Frenkel 1990, p. 150; [Personal File of V.A. Fock] Archives Russian Acad. Sci. 411-14-127, p. 6

[note 1]). This is why one should turn to Fock's philosophy and social conduct only after having considered his scientific psychology.

If one had to characterize the foundation of Fock's frame of mental references in two words, they seem to be "mathematicity and sobriety." One may consider as key his phrases such as "The correct mathematical framing of a physical problem always must ensure the uniqueness of a solution" (Fock 1956c, p. 160). Mathematicity itself does not exclude a romantic attitude to physics (e.g., H. Weyl), but Fock was an antiromantic.

Without taking this into account, it would be rather comical to see Fock criticizing Einstein's intermediate inferences that led to his theory of genius. Here Fock recalled Einstein's confession that his mathematical intuition was not sufficiently strong (Fock 1956b, p. 79). But a physicist would prefer to recall other words of Einstein: "Unless one sins against logic one generally gets nowhere; or, one cannot build a house or construct a bridge without using a scaffold which is really not one of its basic parts" (Einstein 1953, p. 147).

Fock had "the means of observation" to appreciate Einstein's achievements of genius and he did admit that Einstein had achieved his results by means of these "incorrect" concepts and inferences, but to go mentally into this incorrect practice was, for Fock, precluded by his mathematical powers. It was quite clear to Fock (as well as to his colleague-geometricians) that in Riemann geometry, the zero-curvature case has the most symmetry, that the principle of equivalence cannot be formulated inside GTR.

If there are perfect, exactly defined mathematical structures, why should one not set aside logically dubious constructs without exact mathematical meaning, regardless of their historical merits? If the building is finished, why should one not take the scaffold away?

This applies to the principle of equivalence and to the idea of general relativity, which were important for creating GTR but then dissolved in its mathematical structure. The same applies to the operational analysis of definitions, by means of which L. Mandelstam introduced the special theory of relativity (STR) in his lectures. For a mathematician, in the latter case, to describe Minkowski space is quite enough. But for a physicist, even aside from pedagogics, it is not enough.

Einstein modeled physics with the following epistemological scheme:

$$E \longrightarrow A \longrightarrow S \longrightarrow E,$$

where E is the variety of immediate experiences of the senses, A is a system of axioms, and S are statements deduced (Einstein 1952, p. 137). Mathematical physics (as represented by Fock) reigns over the section $A \longrightarrow S$, while theoretical physics deals with the sections $E \longrightarrow A$ and $S \longrightarrow E$.

Fock reproached Einstein by saying "his general inferences proceed as if they did not take into account that any physical theory is approximative in essence" (Fock 1956b, p. 74). The physicist-theorist looking for a new system of axioms certainly does have to forget that it is approximate. At the same time, in order to be prepared for the coexistence and succession of different axiom systems, he should pay special attention to the section $E \longrightarrow A$.

Fock's attitude toward cosmology was especially revealing, if one remembers that nonstatic cosmology was born in front of him. Of course, his concern was not with the mathematical side of cosmological solutions, but rather with their physical meaning. In 1939 he disassociated himself from cosmological speculations and even reproached them. Later, and up to the end of his life, Fock mentioned formally or described very briefly the mathematics of cosmological applications of GTR, but certainly, in his heart, there were no kind feelings for relativistic cosmology. Cosmology as "a model of the world on the whole" he considered philosophically unsatisfactory; he wrote about "risky extrapolation" and questioned the applicability of GTR to "cosmologically huge regions of space and time" (Fock 1955, p. 464; 1967, p. 33; 1973, p. 72).

What were the causes of such an attitude to cosmology, besides the well-known discrepancy of the Hubble age with the data of geo- and astrophysics? (It had to be especially important for the sober-minded Fock.) One can see the main causes in his "mathematical sobriety" and in the pressure of his own scientific experience.

It is difficult to read without a smile Fock's explanation for his colleagues:

> In any field theory, formulated by means of partial differential equations, boundary conditions (or conditions which can replace them) are as important as the equations themselves; without such conditions the field cannot be determined. (Fock 1956b, p. 79)

—isn't this a student's question?!

To be important for mathematical uniqueness is not equivalent to being important in the history of physics. But for Fock, who was sure that the mathematically correct formulation of a problem is unique, the absence of boundaries and the non-unique extrapolation of cosmological conditions could not replace the clear boundary conditions in the island problem (isolated system).

The theoretical necessity of the relativistic generalization of celestial mechanics was based on a centuries-old, solid foundation, but behind cosmology stood only irresponsible speculations.

Such a position had to be strengthened by Fock's scientific success in solving the island problem. And his attitude toward harmonic coordinates, by means of which he had solved the island problem, had to be strengthened also by his results concerning the conservation laws for the insular system.

Let us pay attention to what Fock said about the ten conservation laws (commonly, in other treatments of GTR, only four laws of conservation of energy and momentum are discussed). In classical mechanics and STR the existence of ten conservation laws is connected with the 10-dimensionality of the Galileo and Poincaré groups, or with the 10-dimensionality of the set of Cartesian inertial frames of reference, and, finally, with the four-dimensionality of space-time. This connection is produced in a most clear and profound way by Noether's theorem (which Fock, however, did not use).

Generalizing the equations of motion and conservation laws for insular systems placed Fock's results on a solid, historically scientific base (in which Fock included also Copernicus's theory). This stimulated Fock to "ontologize" his successful method of solving the problem—harmonic coordinate systems. It is impossible, however, to build Fock's analysis (of the insular system by means of harmonic coordinates) straightforwardly in a cosmological setting. A Euclidean character at infinity is incompatible with any nontrivial cosmology. "So much the worse for cosmology," Fock thought, perhaps.

Some elements of Fock's understanding of GTR were adopted, especially by geometrically orientated physicists (the meaning of the principle of equivalence and general covariance, necessity of coordinate condition). Fock's belief that harmonic coordinates were privileged in principle and comparable with Einstein's equations in significance remained unadopted.

(Fock's interpretation of GTR allows, however, for adaptation to common modern treatments necessarily including cosmology. Harmonic coordinates may be transformed into the idea of a standard coordinate system generated by the inner metrical structure of the given space-time [such coordinates were first introduced by Riemann himself]. Based on metrical coordinates, it is possible in a general geometry to introduce a 10-dimensional quasi-group, generalizing the ordinary Poincaré group for the variable curvature case. With the help of this construction, one can realize the correspondence between GTR, STR, and Newtonian gravity in terms of the island situation and the ten conservation laws of energy–momentum–moment [Gorelik 1988].)

Having described the scientific part of Fock's frame of reference, we can pass to its social-ideological part. The latter occupied, of course, not much time. From Fock's texts and from testimonies of those who knew him (Aleksandrov 1988, 1989; Feinberg 1990; Fock 1991, 1993) emerges the image of a scholar absorbed in his science and, beyond science, honest and self-respecting, responsible and fearless, sympathetic and rather schematic, or mathematical.

If a man belongs to science with his whole mind and heart, it seems probable that, in his life outside science, he is guided by his professional methodology as far as possible. But what if his professional methodology proves to be insufficient in his scientific field? What if, for example, he fails to find a common language with colleagues in spite of great efforts? There is no other way to explain this failure apart from some external factors, though Fock himself hardly would have attributed philosophy to external factors.

Fock learned *Diamat* at the beginning of the 1930s. His textbook was Lenin's *Materialism and Empirio-Criticism.* It is difficult to reconstruct exactly Fock's understanding of *Diamat* from his texts, which contain few quotations. Undoubtedly Fock found in *Diamat* something important and interesting for himself, in spite of police inculcation, a flood of quasi-philosophy, abusive polemics, and anachronisms.

Fock was not alone in his relation with *Diamat.* There is no room in this chapter for a general discussion of Marxist philosophy, its natural-scientific roots, and the socialist prejudices of physicists. Here one must notice only that, among Soviet physicists, there existed various individual combinations of attitudes to different components of Marxist or Soviet ideology, to dialectical and historical materialism, and to the theory and practice of Soviet socialism. Adherence to one part might be accompanied by indifference to another and hostility to a third.

Fock belonged with those who, being predisposed to a philosophical view, found a good base in *Diamat.* Behind Fock's *Diamat,* however, one could, with some imagination, recognize something close to Platonic (true mathematical) idealism: Fock believed in the existence of one true philosophy as the most general scheme or quintessence that uniquely realized the evolution of scientific knowledge.

Such an attitude radically differs from the one of (the physicist) Einstein, who supposed that the physicist has the right (or even obligation) to philosophical opportunism, taking, depending on circumstances, the positions of realist, idealist, positivist... (Einstein 1949).

In speaking about Fock's social psychology, one should take into account that, to Niels Bohr, he seemed to resemble Pier Bezukhov. Perhaps due to Fock's European roots, to the honest, fearless, and profound hero of

L. Tolstoy, one should also add a somehow not-so-Russian respect for law, regularity, and stubbornness.

Fock seemed to be satisfied with the theoretical postulates of Soviet power. To judge the conformity of beautiful schemes with social practice was more difficult for Fock than for his colleagues (most of whom kept social illusions for a long time). Apart from the previously mentioned "detachment from life" and deafness, his own biography might prevent him from seeing social reality. Was he not twice arrested and did not justice "triumph" twice?!

Fock perceived the Stalinist terror (the true scale of which was unknown) as a natural disaster, saying that "cowardice does not influence the probability of arrest" (Aleksandrov 1988, p. 489), and he fearlessly defended those who found themselves under this probability. When social reality (personified, for example, by A. Maksimov) invaded his science, Fock acted resolutely and, as his letter to Tamm shows, rather deliberately. But beyond his own science his judgements were fairly schematic. He wrote in this schematic way, for example, about the conservation of energy in Fock 1949. Some of his judgments in the social field were even more schematic and conforming to the "logic" of Soviet newspapers.

Having limited ourselves to this description of Fock's frame of reference, let us, based on this frame, look at the last three decades of Fock's life in the theory of gravity.

Fock persistently, without sparing effort, explains his (true) understanding of Einstein's theory of gravity, including also certain mathematical clarifications (Fock 1953b, 1956a, 1956b, 1967). The answer to Fock was silence or evasive words or repetition of old words, mathematically meaningless, although sanctified by the great physicist. When, in scientific discussion, scientific arguments are exhausted, additional reasons are sought beyond science. And the direction of the search is prompted by the socio-cultural atmosphere around the scientist through his own world view. As a result,

> It is possible that the difference between the points of views of the two schools [Einstein's and Fock's] on given, concrete questions is not incidental but connected with the difference in their general philosophical directions. (Fock 1955, p. 472)

With regard to another frame of reference (call it "Tamm's"), one must say that, among its inhabitants, there were no specialists in GTR comparable with Fock. These inhabitants were physicist-theorists who could not ignore the laboratory-Newtonian experience behind the abstract Riemannian constructions of GTR and could not look at the physical world from inside Riemannian space-time. Apart from this, these physicists could not see through the mud just recently thrown at "reactionary Einsteinism."

A historian who has attempted to account for a communication gap between outstanding scientists and has found an explanation in the difference of their frames of mental references runs the danger of being accused of superciliousness. After all, he claims to see what the scientists in question failed to see.

To ward off such accusations one can recall once more Fock's article of 1971 and designate as complementary scientific creativity and the ability to shift easily from one scientific frame of reference to another. The former demands of a scientist to stand firmly within his own frame of reference. There is no doubt that Fock's frame of reference led him to outstanding scientific achievements, and the cooperation of different frames of mental reference is necessary for the successful development of science.

Notes

This chapter is an abridged version of Gorelik 1993.

[1] [Personal file of V.A. Fock.] Archives Acad. Sci. USSR 411-14-127.

[2] Ibid. 1034-1-549.

[3] Ibid. 1034-3-691: 31–32; 1034-3-160: 8–10.

References

Anon., ed. (1979). *Albert Einstein i teoriya gravitatsii*. Moscow: Mir.

Aleksandrov, Aleksandr D. (1988). "Vladimir Aleksandrovich Fock" [Russian]. In Aleksandrov, A.D., *Problemy nauki i pozitsiya uchenogo*. Leningrad: Nauka, pp. 489–496.

—— (1989). Interview given to G.E. Gorelik, October 17, 1989.

Einstein, Albert (1949). "Remarks Concerning the Essays Brought Together in This Co-operative Volume." In *Albert Einstein: Philosopher-Scientist*. Paul A. Schilpp, ed. Evanston: The Library of Living Philosophers, pp. 683–684.

—— (1952). Letter to Solovine, May 7, 1952. *Letters to Solovine*. New York, 1987. pp. 135–139.

—— (1953). Letter to Solovine, May 28, 1953. *Letters to Solovine*. New York, 1987. p. 147.

Einstein, Albert and Infeld, Leopold (1938). *The Evolution of Physics*. New York: Simon and Schuster.

Einstein, Albert, Infeld, Leopold, and Hoffmann, Banesh (1938). "Gravitational Equation and the Problem of Motion." *Annals of Mathematics* 39: 65–100.

Feinberg, Evgeniy L. (1990). Interview given to G.E. Gorelik, February 28, 1990.

Fock, Mikhail V. (1991). Interview given to G.E. Gorelik, April 15, 1991.

V.A. Fock
Chronology (Gravitational, Philosophical, Social)

1898	Born in St. Petersburg into family of a noble-scientist
1916	Enters Petrograd University
1917	Volunteers for frontline of World War
1918	Returns to the University
1922	Translates Friedmann's paper on cosmology into German
	Gives lecture at philosophical circle on GTR
1927	Rockefeller grant-holder in Göttingen and Paris
1929	Riemannization of Dirac's equation
1932	Corresponding member of Academy of Sciences USSR
	Professor at Leningrad University
	Publishes the first Russian textbook on QM
	Reads (with interest) Lenin's *Materialism and Empirio-Criticism*
1934	Publishes a very critical review of physics books by the most prominent "materialists" and anti-relativists
1935	First (one-day) arrest
	Examines M. Bronstein's thesis on quantum gravity
1937	Second (five-day) arrest
1938	Successful struggle to prevent quasi-philosophical and anti-relativistic session in Academy of Sciences USSR
1939	Full member of Academy of Sciences USSR
	Paper on problem of motion in GTR
	Jubilee article about A. Einstein
1946	Stalin prize for the work on propagation of radiowaves
1947	Formulates in the main his own attitude to GTR
	Defends Copernicus from super-relativism and relativity from super-materialism
1940s–1950s	Defends quantum and relativistic physics (on behalf of *Diamat*) and defends scientific ethics
1955	Takes part in the Bern jubilee conference on GTR
	Monograph *Theory of Space, Time and Gravitation*
	Article "Halfcentury of the Great Discovery" for newspaper *Pravda*
1960s	Persistently explains his attitude to GTR and expresses his adherence to *Diamat*
1974	Dies

——— (1993). "Recollections about Father" [Russian]. *Voprosy Istorii Estestvoznania i Tekniki.* 2: 80–87; 3: 90–98.

Fock, Vladimir A. (1922). [Summary of the lecture for the philosophical circle, September 6, 1922]. Archives Academy of Sciences USSR1034-1-191: 1–3.

——— (1929). "Geometrisierung der Diracschen Theorie des Elektrons." *Zeitschrift fur Physik* 57: 261–277.

——— (1934). "For Truly Scientific Soviet Book" [Russian]. *Sorena* 3: 132–136.

——— (1937). "Does Quantum Mechanics Contradict Materialism?" (November 1937) [Russian]. Archives Academy of Sciences USSR 1034-1-361.

——— (1938a). "On the Discussion on Problems of Physics" [Russian]. *Pod znamenem marksizma* 1: 140–159.

——— (1938b). Letter to Presidium of Academy of Sciences USSR, February 13, 1938. In Gorelik 1990, pp. 27–29.

——— (1939a). "On Movement of Finite Masses in the General Theory of Relativity" [Russian]. *Zhurnal eksperimental'noi i teoreticheskoi fiziki* 9: 375–410; [in French] *Journal of Physics of USSR* 1: 81–116.

——— (1939b). "Albert Einstein (On His 60-Year-Jubilee)" [Russian]. *Priroda* 7: 95–97.

——— (1940). Letter to the journal "Pod znamenem marksizma" on the review by Ernest Kolman (No. 2, 1940) of Landau 1939. Archives Academy of Sciences USSR 1515-2-98.

——— (1947). "The System of Copernicus and the System of Ptolemy in the Light of the General Theory of Relativity" [Russian]. In *Nikolay Kopernik.* Naum I. Idelson, ed. Moscow: Nauka, pp. 180–186.

——— (1949). "Basic Laws of Physics in the Light of Dialectical Materialism" [Russian]. *Vestnik Leningradskogo universiteta* 4: 34–47.

——— (1950). "Some Applications of the Ideas of Lobachevsky's Non-Euclidean Geometry to Physics" [Russian]. In Aleksandr P. Kotelnikov and Vladimir A. Fock, *Nekotorye primeneniya idey Lobachevskogo v mechanike i fizike.* Moscow: Nauka, pp. 48–86.

——— (1951). "Review of the Book: Mandelstam, L.I., *Polnoe sobranie trudov.* Vol. 5. Moscow, 1950" [Russian]. *Uspekhi fizicheskikh nauk* 45: 160–163.

——— (1953a). "Against Ignorant Criticism of Modern Physical Theories" [Russian]. *Voprosy filosofii* 1: 168–174.

——— (1953b). "Modern Theory of Space and Time" [Russian]. *Priroda* 12: 13–26.

——— (1955). *Theory of Space, Time and Gravity.* [Russian]. Moscow: Nauka.

——— (1956a). "Halfcentury of the Great Discovery" [Russian]. *Pravda* April 15, 1956.

——— (1956b). "Remarks on Einstein's Creative Autobiography" [Russian]. In *Einstein i sovremenaya fizika.* Igor E. Tamm, ed. Moscow: Nauka, pp. 72–85.

——— (1956c). "Equations of Motion of System of Heavy Masses Taking into Account Their Inner Structure and Rotation" [Russian]. *Ibid.* p. 160–162.

——— (1963). "A.A. Friedmann's Works in the Theory of Gravitation by Einstein" [Russian]. *Uspekhi fizicheskikh nauk* 80: 353–356.

——— (1966a). "The Grundprinzipien der Einsteinschen Gravitationstheorie." In *Deutsche Akademie der Wissenschaften zu Berlin. Einstein Simposium* November 2–5, Berlin. pp. 27–37.

——— (1966b). "Comments" [on Graham 1966]. *Slavic Review* 25: 411–413.

——— (1967). *Einstein's Theory and Physical Relativity* [Russian]. Moscow: Znanie.

——— (1971). "Principle of Relativity to Means of Observation in Modern Physics" [Russian]. *Vestnik Akademii Nauk SSSR* 4: 8–12.

——— (1973). "Quantum Physics and Philosophical Problems" [Russian]. In *Fizicheskaya nauka i filosofiya*. Mikhail E. Omel'anovsky, ed. Moscow: Nauka, pp. 55–77.

Frenkel, Viktor Ya., ed. (1990). *Fiziki o sebe*. Leningrad: Nauka.

Frederiks, Vsevolod K. (1921). "General Principle of Relativity by Einstein" [Russian]. *Uspekhi fizicheskikh nauk* 2: 162–188.

Friedmann, Aleksandr A. (1922). "Über die Krummung des Raumes." *Zeitschrift für Physik* 21: 326–332.

——— (1923). *World as Space and Time* [Russian]. Petrograd: Iskra.

Ginzburg, Vitaliy L. (1956). "Experimental Testing of General Theory of Relativity" [Russian]. In *Einstein i sovremennaya fizika*. Igor E. Tamm, ed. Moscow: Nauka, pp. 123, 136.

——— (1973). "General Relativity and Copernicus's Heliocentric System" [Russian]. *Voprosy filosofii* 9, 10.

Gorelik, Gennady E. (1988). "Dimensionality of Space-Time and Poincaré Quasigroup" [Russian]. In *Einsteinovsky sbornik 1984–1985*. Igor Yu. Kobzarev, ed. Moscow: Nauka, pp. 271–300.

——— (1990). "Philosophical Problems of Soviet Physics in 1937" [Russian]. *Voprosy Istorii Estestvoznania i Tekniki* 4: 17–31.

——— (1991). "University Physics and Academy Physics" [Russian]. *Voprosy Istorii Estestvoznania i Tekniki* 1: 31–46.

——— (1993). "V.A. Fock; Theory of Gravitation and Philosophy" [Russian]. *Priroda* 10. (In press.)

Gorelik, Gennady E., Frenkel, Viktor Ya. (1985). "M.P. Bronstein and Quantum Gravity" [Russian]. In *Einsteinovsky sbornik 1980–1981*. Igor Yu. Kobzarev, ed. Moscow: Nauka, pp. 291–327.

——— (1990). *Matvey Petrovich Bronstein (1906–1938)*. Moscow: Nauka.

Graham, Loren R. (1966). "Quantum Mechanics and Dialectical Materialism." *Slavic Review* 25: 381–410.

——— (1982). "The Reception of Einstein's Ideas: Two Examples from Contrasting Political Cultures." In *Albert Einstein. Historical and Cultural Perspectives. The Centennial Symposium in Jerusalem*. Princeton: Princeton University Press, pp. 107–138.

—— (1987). *Science, Philosophy and Human Behavior in the Soviet Union*. New York.

Havas, Peter (1989). "The Early History of the 'Problem of Motion' in General Relativity." In *Einstein Studies*. Vol. 1, Don Howard and John Stachel, eds. Boston: Birkhäuser, pp. 234–277.

Idelson, Naum I. (1947). "Life and Creative Work of Copernicus" [Russian]. In *Nikolay Kopernik*. Naum I. Idelson, ed. Moscow: Nauka, pp. 5–42.

Infeld, Leopold (1956). "My Recollections about Einstein" [Russian]. In *Einstein i sovremennaya fizika*. I. Tamm, ed. Moscow: Nauka, pp. 197–260.

Kapitsa, Pyetr L. (1989). *Pis'ma o nauke*. Moscow: Nauka.

Kozhevnikov, Aleksey B. (1988). "V.A. Fock and the Method of Secondary Quantization" [Russian]. In *Issledovaniya po istorii fiziki i mekhaniki, 1988*. Grigorian, Ashot T., ed. Moscow: Nauka, pp. 113–138.

Maksimov, Aleksandr A. (1952). "Against the Reactionary Einsteinism in Physics" [Russian]. *Krasny Flot* June 12.

—— (1973). "About Some Negative Phenomena at the Frontline of Philosophy " [Russian]. Archives Academy of Sciences USSR 1515-4a-327: 72.

Mandelstam, Leonid I. (1950). *Polnoye sobranie trudov*. T.5. Moscow: Nauka.

Markov, Moisey A., ed. (1979). *Einstein i filosofskie problemy fiziki XX veka*. Moscow: Nauka.

Medvedev, Boris V. (1977). *Nachala teoreticheskoy fiziki*. Moscow: Nauka.

Miroshnikov, Mikhail I. ed. (1978). *V.A. Fock. K 80-letiyu so dnya rozhdeniya*. In *Trudy Gosudarstvennogo Opticheskogo Instituta*. Vol. 43. Leningrad.

Tamm, Igor E., ed. (1956). *Einstein i sovremennaya fizika. Sbornik pamyati F. Einsteina*. Moscow: Nauka.

Vizgin, Vladimir P. and Gorelik, Gennady E. (1987). "The Reception of the Theory of Relativity in Russia." In *The Comparative Reception of Relativity*. Thomas F. Glick, ed. Boston: Reidel, pp. 265–326.

S. Chandrasekhar's Contributions to General Relativity

Kameshwar C. Wali

> Chandrasekhar's approach to relativity is his own: He entered it as a mature scientist, unlike most of the rest of us who were educated in one of the major "schools" of the subject.
>
> Kip S. Thorne

S. Chandrasekhar (known simply as Chandra to most of the scientific world) was introduced to general relativity in his first year as a graduate student at Trinity College, Cambridge, England, in 1930–1931, by none other than Sir Arthur Stanley Eddington. Charmed though Chandra was by Eddington's exposition of relativity, full of fun and humor, he shied away from a serious study of relativity for over 30 years. In his student years and afterwards, Chandra distanced himself from general relativity. This was, as he recalls, partly because of "the veiled contempt" that physicists like Bohr and others had for the work of Eddington related to his fundamental theory and of Milne for his kinematical relativity, and partly because, at the time, relativity did not seem to be relevant for problems of stellar structure, the internal constitution of stars, and other down-to-earth problems in astronomy. Recalling an occasion from those years, Chandra remembers Dirac asking him why he was doing astrophysics, remarking that if he (Dirac) ever became interested in astronomy, he would engage himself in cosmology. Chandra's reply was, "I would rather have my feet on the ground" (Chandrasekhar 1990). Subsequently, even until the late 1950s, Chandra continued to shy away from relativity. Once when the physicist Gregor Wentzel, Chandra's colleague and friend at the University of Chicago, asked him why he had not worked in this field, he replied, half jocularly, that relativity had proved to be the graveyard of many theoretical astronomers and that he was not prepared for a burial—not yet (Chandrasekhar 1990). In a more serious vein, he felt that astronomers who went into general relativity were prone to play for high stakes and that his own approach to science was more conservative.

Notwithstanding this early attitude, Chandra could have made an official entry into the circle of relativistic astrophysicists as early as 1935, if not for the unexpected encounter and controversy with Eddington, concerning the role of special theory of relativity in understanding the structure of white dwarfs and the discovery of the celebrated "Chandrasekhar limit." Since an account of this encounter and controversy are documented in sufficient detail elsewhere (Wali 1990), I will not dwell on it here, except for the sake of historical interest to remark that, in all probability, the chronological account of the role of relativity in astronomy and astrophysics might have been different if this controversy had not occurred and if Eddington had recognized the validity of the Chandrasekhar limit, instead of dismissing it as "*Reductio ad absurdum* behavior and 'stellar buffoonery.'" As Chandra says,

> Suppose, just for a moment, Eddington had accepted my result, suppose he had said, "Yes, clearly the limiting mass does occur in the Newtonian theory in which it is a point-mass. However, general relativity does not permit a point-mass. How, then, does general relativity take care of that?" If he had asked this question and worked on it, he would have realized that the first problem to solve in that connection is to study radial oscillations of the star in the framework of general relativity. It's a problem I did in 1964, but Eddington could have done it in the mid-1930s! Not only because he was capable of doing it—he certainly had mastered general relativity—but also because his whole interest in astrophysics originated from studying pulsations of stars. And if he had done it, he would have found that the white dwarf configurations constructed on the Newtonian model became unstable before the limiting mass was reached. He would have found that there was no *reductio ad absurdum*, no stellar buffoonery! He would simply have found that stars become unstable before they reached the limit and that a black hole would ensue. Eddington could have done it. (Wali 1990, p. 143)

Chandra could have done it, too, but he made a personal decision at the time. He felt that astronomers without exception thought he was wrong and that they considered him to be a sort of Don Quixote trying to kill Eddington. Faced with the very discouraging experience of finding himself in a controversy with the leading figure in astronomy and having his work completely and totally discredited, he decided to discontinue research connected with white dwarfs altogether and went on to do something else. He had begun some work with John von Neumann using the fully relativistic equations of state, which would certainly have led to the study of equilibrium conditions of stellar configurations within the framework of general relativity. This study, as we know, was undertaken a few years later by J. Robert Oppenheimer and George M. Volkoff who wrote their classic paper on neutron stars (Oppenheimer and Volkoff 1939). Chandra's own entry into general relativity was postponed for over 30 years until the early 1960s.

Chandra's distinctive pattern of research, as is widely known, has encompassed several areas, each of which occupies a particular period in time.[1] In each epoch, as Goldberger says, "Chandra has produced an infinite series of papers followed by an infinitely thick book on the subject" (Wali 1990, p. 23). After the completion of his work on hydrodynamic and hydromagnetic stability (1952–1961), Chandra decided to turn to general relativity and, in his typical fashion, began the summer of 1960 with an intense study of the subject followed by teaching an advanced course on it the following fall semester. In 1962, he attended the Warsaw Conference on general relativity, as an observer, mainly to get a feeling of "what the experts were thinking." In 1964, he produced, what was to prove to be an extremely important paper titled, "The Dynamical Instability of Gaseous Masses Approaching the Schwarzschild Limit in General Relativity" (Chandrasekhar 1964). After that a steady stream of papers followed, and, as Kip S. Thorne has said, "Nobody has done more than S. Chandrasekhar to bring general relativity to its 'natural home,' astronomy." Volumes 5 and 6 of *Selected Papers, S. Chandrasekhar* are devoted to this vast body of work. In what follows, I can only provide an overview.

A convenient source for Chandra's research publication is the set of six volumes of *Selected Papers, S. Chandrasekhar* (Chicago: University of Chicago Press, 1990). I have used this source, referred to hereafter as SP *x*: p. *y*, unless specifically stated otherwise. For some accounts, I have also used material from S. Chandrasekhar, "A Scientific Autobiography, 1943–1990," unpublished manuscript, S. Chandrasekhar Papers, Box 1, Folder 1, The University of Chicago Archives.

1. The 1960s: Relativistic Instabilities and Post-Newtonian Approximations

1.1 Relativistic Instabilities

Chandra's entry into general relativity could not have been more opportune than when it occurred in 1964. First of all, rapid discoveries were taking place in astronomy; quasars, pulsars, radio galaxies, cosmic x-ray sources, cosmic microwave background, and proclaimed detection of gravity waves created a new arena of research for practical-minded relativists. On the theoretical side, a new discipline, relativistic astrophysics, was shaping up; it was dominated by youthful personalities, including Kip Thorne, Roger Penrose, James Hartle, James Bardeen, Stephen Hawking, Brandon Carter, and others. "Chandra was our young-at-heart co-worker," says Thorne, "as new to relativity as we. We had the flexibility of youth, freedom from preconceived notions that is a modest compensation for lack of experience. Chandra had the wisdom of decades of research in fundamental, Newtonian physics and astrophysics—a wisdom that gave him guidance on what problems were worth studying and how to approach them" (SP 5: xii). Secondly, the problem mentioned earlier, the problem of radial oscillations of a star within the framework of general relativity, which Eddington should have and could have studied in the 1930s, was still waiting! Now, with new discoveries, it had assumed immensely increased significance.

In Newtonian theory, a nonrotating spherical gaseous mass of perfect fluid in equilibrium under its own gravitational forces and internal pressure and energy would be stable against radial perturbations provided γ, the adiabatic exponent (the average ratio of specific heats), was $>4/3$. Since this was likely to be the case in stellar configurations, especially in massive stars, Newtonian theory predicted that, no matter what the mass of the star, it could be in a stable configuration with finite radius that decreased with increasing mass, reaching zero only when the mass was infinite. *Chandra showed that this was no longer the case within the framework of general relativity.* In addition to γ, the stability depended on the radius of the star, as well. Therefore, stars that could be considered stable in Newtonian theory would become unstable in general relativity!

To be more specific, it was known that a spherical mass in hydrostatic equilibrium giving rise to the well-known Schwarzschild metric external to itself would become unstable, no matter how high the value of γ, if the radius

$$R \leq 9/8 R_S,$$

where R_S (Schwarzschild radius) $= 2GM/c^2$. As $\gamma \to \infty$, the radius $9/8\, R_S$ defined, in fact, the minimum radius that any gravitating mass in hydrostatic equilibrium could have in the framework of general relativity. Chandra was able to sharpen this result further and show that if γ differed from and was greater than $4/3$ only by a small positive constant, then the instability would set in for a radius R much larger than $9/8\, R_S$. He showed that, as $\gamma \to 4/3$, instability for radial perturbations will set in for all

$$R < \left(\frac{K}{\gamma - 4/3}\right) R_S,$$

where K is a constant that depends on the structure of the star.[2] Consequently, for values of γ slightly in excess of $4/3$, dynamical instability would set in well before the mass reached the Schwarzschild limit, and also well before degenerate configurations such as white dwarfs reached their limiting mass. For example, for a super massive star of $M = 10$ solar mass and an estimated value of $(\gamma - 4/3) \sim 7.2 \times 10^{-5}$, it followed that $R_C = 1.6 \times 10^4 R_S$, which was approximately 0.5 light years, a radius of correct order for a quasar of 10 solar masses. This estimate of radius was in agreement with estimates from other considerations. Further, Chandra's analysis of normal modes of the radial oscillations of electron-degenerate configurations (white dwarfs) showed that, due to the relativistic instability, the period of oscillations reached a minimum and then tended to infinity as the degenerate mass approached its limiting mass. Such a minimum period, calculated to be about seven-tenths of a second, was nonexistent in a Newtonion framework. Since pulsars of a much shorter period were known to exist, one could rule out the possiblity that they were white dwarfs, a result of obviously great significance in understanding the nature of quasars.

Although explicit results in these pioneering papers were derived starting from the idealized model of a star consisting of homogeneous compressible fluid, the principle conclusion—namely, according to general relativity, a massive star would become unstable long before its mass contracted anywhere near the Schwarzschild limit—became incontrovertible because most of the estimates of the instabilities were underestimates and not likely to be altered in more realistic models. Such a picture had tremendous implications on the fate of massive stars. A massive star must collapse once it has exhausted its nuclear source of energy. If it had to collapse into the dimensions of some 10 to 20 kilometers to form the stable configuration of a neutron star, it had to eject a substantial fraction of its mass (which is processed matter through nuclear reactions) into interstellar space. Such ejection could be a cataclysmic event, such as a supernova explosion. If the remnant mass was within a narrow permissible range, it would then settle into a stable state of a neutron star and become a pulsar. But why would every massive star of 10 or more solarmasses eject the right amount of its material so that it is left behind with a mass of the right range? As Chandra says, "It is more likely that the star ejects an amount of mass that is either too large or too small. In such [latter] cases the residue will not be able to settle into a finite state; *and the process of collapse must continue indefinitely until the gravitational force becomes so strong that what Eddington concluded as a* reductio ad absurdum *must in fact happen: the gravity becomes strong enough to hold the radiation. In other words, a black hole must form.*"[3] [Italics are mine.] Thus, if general relativity had a say in the matter, the existence of black holes in the astronomical universe had to be accepted as a reality.

1.2 Post-Newtonian Approximation Scheme

The dynamical instability due to relativity that Chandra discovered was one in which relativistic corrections were in effect small. That is, there were no constituent motions that involved relativistic energies or other relativistic kinematical or dynamical factors. Further, the problem of radial oscillations of a spherically symmetric perfect fluid body that he had solved exactly was a particularly simple one and also far from a realistic working model for stars that occur in nature. For further progress, it was imperative that one consider nonradial oscillations and rotating perfect fluid bodies to serve as models for rotating stars, as well as the all too important problem of gravitational radiation and its reaction on the emitting bodies.

In the Newtonian theory, given the interactions and external forces, one can in principle write a complete set of equations of motion for every constituent particle if one is dealing with a system of discrete particles or for an element of fluid if one is dealing with a hydrodynamical system. Chandra realized that such an exact set of equations cannot be written—at least explicitly—in the framework of general relativity. Even if such equations were written, finding their solutions presented formidable difficulties if one had no presupposed symmetry as in the case of nonradial oscillations. Under the circumstances, a more modest inquiry would be to ask specific questions:

(1) Since relativistic corrections were small, could one develop a well-defined scheme of successive post-Newtonian approximations, in which a set of explicit equations would govern the departures from the Newtonian motions resulting from the effects of general relativity?

(2) Since it is generally believed that gravitating systems emit gravitational radiation, could one write these approximate equations to a high-enough order that terms representing the radiation reaction of the system occur explicitly in them and could be unmistakably recognized as such?

A scheme of this sort existed in the pioneering work of Einstein, Infeld, and Hoffman (1938) in the case of a gravitating system of N point-particles

(N-body problem).[4] They had, for example, derived a post-Newtonian Lagrangian that differed from a Newtonian Lagrangian by quantities of the first order in v^2/c^2 and U/c^2 (U = gravitational potential); this was sufficient to derive precession of the Keplerian orbit of two finite mass points about one another. However, there was no satisfactory treatment of higher-order terms. They had attempted to calculate radiation reaction but had either failed or obtained ambiguous results. Their treatment suffered from mathematical difficulties because of a point-particle assumption alien to general relativity. Chandra's idea was to develop a relativistic hydrodynamics of a perfect fluid and devise a post-Newtonian approximation scheme to answer the question raised above, which he did (Chandrasekhar 1965).

The physical basis for the approximation scheme was the fact that under conditions of common occurrence in the universe, the rest energy of the systems by far dominated other forms of energy such as the kinetic energy of mass motion, the gravitational potential energy, or the internal energy. Therefore, one could use "smallness" parameters and distinguish orders of successive approximations by the powers of $1/c$ terms retained in the ensuing equations of motion. Secondly, the equivalence principle provided the starting point for the metric in the space-time geometry to be associated with the Newtonian theory of gravitation, namely, the departure of the metric from the Minkowskian one. The dominance of rest energy over all other forms of energy and the minimum departure from the Minkowskian metric dictated by the equivalence principle provide a starting point; then the corrections of successive higher powers in $1/c$ in the metric coefficients and the energy momentum tensors are played against each other to reveal departures from the Newtonian theory due to general relativity. Having established the formalism, Chandra applied it to uniformly rotating perfectly fluid bodies as models for rotating stars. He had concurrently just completed the classical work with Norman Lebovitz on the equilibrium of rotating Newtonian spheroidal and ellipsoidal bodies within the framework of Newtonian theory.[5] In a series of papers, he derived the consequences of post-Newtonian effects due to general relativity on uniformly rotating Maclaurin spheroids, Jacobian and Dedekind ellipsoids, and on the model of a rotating star due to Roche, consisting of a tenuous, centrifugally deformed envelope in the gravitational field of a massive, undeformed core.[6] In all cases, he found huge departures from the Newtonian theory and relativistic instabilities where Newtonian theory predicted stable neutral modes.

Two other important problems that Chandra solved successfully in the post-Newtonian framework need to be mentioned. One was the identification of conserved quantities in successive approximations,[7] and the second, more important, was the radiation reaction terms. In the latter case, all previous attempts had failed or led to ambiguous results. Chandra, along with his student, Paul Esposito, was able to carry out the post-Newtonian scheme to the necessary high order (2.5 order, meaning retaining consistently all terms of order c^{-5}), was able to derive these terms correctly, and to discover a new dramatic kind of instability in rotating bodies caused by these terms (Chandrasekhar and Esposito 1970). In the case of the Maclaurin spheroid, for instance, radiation reaction made it unstable beyond the bifurcation point. Likewise, the triaxial Jacobian ellipsoid was driven toward increasing angular velocity exponentially approaching the bifurcation point where it ceased to radiate. Commenting upon the astrophysical significance of these findings and their relevance to the theory of gravitational collapse after a supernova outburst, Chandra, toward the end of this classic paper, says,

> A rapidly rotating highly condensed configuration may, in the first instance, form as a result of the collapse; and it is not unlikely that the rotating configuration may, in fact, be similar to a Jacobian ellipsoid at the limit of its stability. Then by gravitational radiation, its angular velocity will increase and the object will approach a point of bifurcation where the object becomes spheroidal and nonradiating. But once it reaches the point of bifurcation, radiation reaction will make the configuration secularly unstable, and it is possible that further development may proceed in the direction of fragmentation. In any event, the fact that radiation reaction can induce secular instabilities must have an important bearing on what may happen during the late stages of gravitational collapse.[8]

In the relativistic theory of stellar pulsations with which Chandra earnestly began his journey into relativity in 1964, two discoveries are identified as major in review literature of the late 1980s: *the relativistic instability against gravitational collapse in massive star and the radiation reaction induced instability in rotating stars*. It is remarkable that both these discoveries are Chandra's (see Schutz 1986, p. 123). Further, as Kip Thorne says, "The post-Newtonian and post-post-Newtonian formalisms that he (Chandra) developed have become standard working tools of physics and astrophysics. Over the past two decades, they have been used in studies of stars, star clusters, gravitational-wave generation, and the motions of the planets and the moon" (SP 5: xvii). In the form of parameterized post-Newtonian formalism, it has become an invaluable tool to confront not only Einstein's general theory of relativity to experimental tests, but other metric and nonmetric theories of gravity (see Will 1981). In spite of this monumental contribution, Chandra did not edify the work with a monograph as he did with the ending of his other "periods." According to him, he did not have that "aesthetic feeling of completeness and coherence" in this sphere of his research.

2. The 1970s: Rotating Stars and Black Holes

2.1 Relativistic Stars

In the summer of 1970, with the work on post-Newtonian approximation carried out as far as he was interested, Chandra began to think about a future direction for his research. Partially toward this end, he organized a private summer school with Brandon Carter, George Ellis, and Robert Geroch. While two months of intense "schooling," occupied with seminars and writing notes of lectures by Geroch and Ellis, proved strenuous, Chandra says, it did not prove very helpful at the time to chart his next "period." It had already occurred to him that a systematic exploration of homogeneous uniformly rotating masses within the framework of general relativity was the next venture to undertake. The intimate knowledge of the Newtonian situation would prove again to be of immense help in formulating the problem. He could project the extensive work involved and sought the collaboration of John Friedman, who had just completed his first year of graduate studies. Together they wrote a series of papers in the first half of the 1970s, setting up a formalism for studying axisymmetric perturbations of rotating stars in the framework of general relativity, a formalism that closely paralleled the Newtonian theory and revealed departures from it.

In the Newtonian theory, there existed a rigorous formula for the fundamental frequency σ of the axisymmetric pulsation of a slowly rotating star in the form

$$\sigma^2 = \sigma_0^2 + \Omega^2\sigma_1^2 + O(\Omega^4),$$

where Ω was the frequency of uniform rotation of the body and σ_0 was the frequency of radial pulsation of the nonrotating star; σ_1 depended only on the amplitude of the radial pulsation associated with σ_0 and the spherically symmetric distortion caused by the rotation. And the instability condition for the nonrotating star was modified to be

$$\bar{\gamma} - \tfrac{4}{3} + \text{const}\,\frac{\Omega^2}{\pi G\rho} < 0.$$

The rotation had a stabilizing influence on instability. Chandrasekhar and Friedman found that a formula exactly analogous to the above inequality could be derived in the general relativistic case for slow enough uniform rotations and establish quantitatively the stabilizing influence on the relativistic instability. This was possible because under the assumption of slow rotation, neglecting gravitational radiation reaction terms that were normally an integral part of the theory was justified. The slow rotation case was first studied by Hartle (1967), and Hartle, Chitre, and Thorne (1972) used the theory to model rotating neutron stars. Commenting upon the comparison of their work with that of Chandrasekhar and Friedman, Kip Thorne says, "The slow rotation case was studied independently by James Hartle, Kumar Chitre, and me, using the computer to do the complex algebra. The ability of Chandrasekhar and Friedman to do their calculation by hand and get results which agreed with ours is an amazing tribute to their computation abilities. We had not thought it possible" (SP 5: xviii).

In the process of setting up the formalism for relativistic rotating stars, the Chandrasekhar–Friedman papers also gave rise to the idea of studying deformations of vacuum solutions external to a black hole to which Chandra turned his attention.

2.2 The Mathematical Theory of Black Holes

Chandra's study of black holes, which began with an analysis of the equations governing the perturbations of the Schwarzschild black holes (Chandrasekhar 1975) was to develop into a complete body of work of his own to be edified in the form of a treatise (Chandrasekhar 1983). This study was important, since one of the best ways to find some of the physical attributes of a system is to find out how it reacts to external perturbations and, in the first instance, to infinitesimal perturbations. The study of such infinitesimal perturbations by studying how a black hole reacts to incident waves of different sorts throws light on the stability of the black holes. In spite of a great deal of earlier work, there were elements of mystery shrouding the subject. There was more than one way of analyzing the perturbations leading to radial equations. One was known as the Zerilli equation that was of Schrödinger type with a real potential. Another equation known as the Bardeen and Press equation was characterized by a complex potential. And a determination of the reflection and transmission coefficients for incident plane waves with varying wave numbers enabled one to determine the evolution of any initial perturbation of the black hole. This prompted Chandra to seek a "coherent, self-contained theory of the perturbations of the Schwarzschild black hole," which he did, and clarified the relation between the Zerilli and Bardeen–Press equations and also the relation between the Regge–Wheeler and Bardeen–Press equations. He followed that up, along with S. Detweiler, by studying the quasi-normal modes of the

Schwarzschild black hole (Chandrasekhar and Detweiler 1975). With the study of Schwarzschild black hole perturbations completed, Chandra then set forth to analyze the Kerr and the Reissner–Nordstrom black holes. Two major papers on the Newman–Penrose gravitational perturbations of the Kerr space-time followed (Chandrasekhar 1978). As Basilis Xanthopoulos says, "He [Chandra] considered both the metric and Newman–Penrose perturbations, established the equivalence of the axial and the polar scatterings, and investigated issues such as the decoupling of gravitational and electromagnetic waves and the transformation of one kind of wave to the other in the scattering process" (SP 6: xii).

A problem of side interest, but one that gave Chandra immense satisfaction, was his successful separation of the Dirac equation. This would later lead to the separation of the Dirac equation in Minkowskian space-time in prolate spheroidal coordinates and the development of a new theory on the separability of partial differential equations. It also led to his own work on the two-component neutrino equation and the study of reflection and transmission of neutrino waves by a Kerr black hole (Chandrasekhar and Detweiler 1977).

To complete the story of this period, one must mention Chandra's one failure, namely, the failure to separate the Kerr–Newman perturbations. Remarking on this failure, Basilis Xanthopoulos says, "Since Chandra failed, no one seems to be willing to give this problem a serious try, and the perturbations of the Kerr–Newman solution have remained an unsolved problem for the dozen years since Chandra gave up. Perhaps for the sake of science, he should have kept his failure a secret. On the other hand, it is very likely that the KN perturbations cannot be separated and his documented failure has saved many scientific years of fruitless effort" (SP 6: xii).

3. The 1980s: Colliding Waves and the Two-Center Problem

After the completion of *Mathematical Theory of Black Holes*, Chandra briefly entertained the notion of separating himself from serious research, but that expectation and hope did not last very long. Soon thereafter he got himself involved in as serious a scientific effort as he ever had. One may at least partially credit or blame this renewal of effort on the two young collaborators he found in Basilis Xanthopoulos and Valeria Ferrari.

The best account of Chandra's renewed effort during the 1980s is his own, to be found in the proceedings of the Yale symposium in honor of the 150th anniversary of the birth of J. Willard Gibbs (Chandrasekhar 1989). Entitled "How One May Explore the Physical Content of the General Theory of Relativity," Chandra discusses the inner coherence and richness of the exact general theory of relativity and how its physical content can be explored by allowing "one's sensibility to its aesthetic base guide in the formulation of problems with conviction in the harmonious coherence of its mathematical structure" (Chandrasekhar 1989, p. 250).

3.1 Colliding Plane Waves

Khan and Penrose (1971) had discovered, in the collision of two gravitational waves with plane wave fronts and parallel polarizations, a space-like singularity as a result of collision. The nature of this singularity was very much like the one in the interior of a black hole. This was the consequence of the exact theory and not to be found in the linearized version. Penrose had emphasized this fact and had suggested that possibly new physics of general relativity had yet to be explored. Matters stood there until Chandra became seriously interested in the problem in 1984 after he completed his book. In the meantime, the work of Khan and Penrose was extended to include nonparallel polarizations of the impulsive waves by Chandra's former student Yavuz Nutku along with Halil (Nutku and Halil 1977). Furthermore, Penrose, in a letter in 1984, had raised the problem of how to describe coupled gravitational and electromagnetic impulsive waves. Unlike the case of gravitational waves, one could not construct in a straightforward manner impulsive electromagnetic wave fronts since it required the square root of a δ-function in the field variables in order to have a δ-function in the energy profile of the impulsive wave front. The square root of a δ-function, however, was not a mathematically permissible or physically sensible concept.

Since this did not make sense, some new idea was necessary. What was needed, Chandra thought, was a rigorous mathematical theory of colliding waves patterned exactly after the mathematical theory of black holes. With Ferrari and Xanthopoulos, he reformulated the theory for colliding waves showing the underlying structural similarity of the mathematical theories describing the colliding waves and the black holes. The same set of two equations known as Ernst equations governed the two cases. One could write identical solutions in both the cases using different combinations of metric functions. Thus the Khan–Penrose solution for colliding waves corresponded exactly to the Schwarzschild black hole solution and the Nutku–Halil solution to that of the Kerr–Newman black hole (Chandrasekhar and Ferrari 1984). The $\sqrt{\delta}$-type singularity problem in the case of coupled gravitational and electromagnetic colliding waves was also sidestepped by seeking solutions to the Einstein–Maxwell equations that reduced to an appropriate black hole solution when the electromagnetic field was switched off. "The problem is not a straightforward one," says Chandra, "since in the framework of the Einstein–Maxwell equations, we do not have an Ernst equation which reduces to the Ernst equation for the particular combination of the metric functions that is appropriate for the Nutku–Halil vacuum solution" (Chandrasekhar 1989, p. 235). The technical problems, however, were successfully overcome, and a physically consistent, satisfactory solution was obtained (Chandrasekhar and Xanthopoulos 1985).

The discovery of an underlying unity in the mathematical description of black holes and colliding waves has led Chandra and his young collaborators to the discovery of new solutions to the Einstein-vacuum and Einstein–Maxwell equations describing space-times with totally unexpected features. They have found colliding wave solutions that have no curvature singularities during the process of collision, but solutions that are characterized by the formation of event horizons and a space-time domain which is the mirror image of the space-time that was left behind. "It is remarkable," Chandra says, "that a space-time resulting from the collision of waves should bear such a close resemblance to Alice's anticipations with respect to the world through the looking glass. 'It [the passage in the Looking-Glass House] is very much like our passage as far as we can see, only you know it may be quite different on beyond'" (Chandrasekhar 1989, pp. 236–237). Indeed, one may expect a great deal of new physics of general relativity implied by these solutions summarized in a three-page table in Chandrasekhar 1989, pp. 239–242.

3.2 Binary Black-Hole Solutions

If gravitational forces were the only forces, it is clear that one cannot have a completely static configuration of matter. Problems with fixed centers of gravitation, therefore, are somewhat artificial, although their solubility in certain cases have attained a certain amount of celebrity (see Whittaker 1937). The problem, however, can be made conceptually plausible by introducing Coulomb electric forces of repulsion. Thus, we can envisage static configurations of any number of mass points $M_1, M_2, M_3, \ldots, M_n$, at arbitrary locations, with charges $Q_1, Q_2, \ldots, Q_n$, all of the same sign such that $M_i\sqrt{G} = Q_i$ ($i = 1, 2, \ldots, N$, and where G denotes the constant of gravitation). The Newtonian attraction is then balanced by the Coulomb repulsion leading to a static equilibrium configuration. The same static configuration is allowed in the framework of general relativity as a solution of the Einstein–Maxwell equations as shown by Majumdar (1947) and Papapetrou (1947) and is the only static multiple black-hole solution compatible with the smoothness of the space-time external to the event horizons and asymptotic flatness. (Hartle and Hawking [1972] have interpreted the solution as representing an assemblage of extreme Reissner–Nordstrom black holes.)

Having discovered the unity in the underlying mathematical description of black holes and impulsive colliding waves, Chandra was attracted to the fixed center problem in the framework of general relativity. The unity was displayed in the solutions of the complex Ernst equation governing the axisymmetric vacuum. Chandra had an alternate formulation of the same problem in terms of two real equations denoted by him as the X- and Y-equations. Chandra believed that the solutions to the X- and Y-equations should provide, in a suitable context, a space-time of some real significance. But the X- and Y-equations had remained "like Cinderella," says Chandra, "the ignored stepsister of the Ernst equation" (Chandrasekhar 1989, p. 243). Like Cinderella, they had to be rescued.

The rescue came about in the context of the Majumdar–Papapetrou solution for stationary or static Reissner-Nordstrom black holes alluded to before. Imagine for simplicity two such black holes placed at a finite distance apart at rest. One can then specify the metric for a static Einstein–Maxwell space-time in terms of a scalar potential function that obeys a three-dimensional Laplace equation. Chandra has shown that this scalar potential is his X- or Y-function that satisfies equations following from a metric that corresponds to a stationary axisymmetric vacuum space-time. *This meant that one could pass freely from the Majumdar–Papapetrou metric appropriate as a solution of the Einstein–Maxwell equations to a metric appropriate as a solution of the stationary Einstein vacuum equations, and conversely.* Describing the one-to-one correspondence as a manifestation of a natural and harmonious blending of Einstein's relativity and Maxwell's electrodynamics in a single unified structure, Chandra makes an analogy with the description of a mythical hall in the Indian epic *The Maha-bharata*, and says, "when wandering through the great hall of general relativity, that what one had believed to be Einstein's hall, is in fact a corridor leading to Maxwell's hall; and when one is certain that one is examining the gems in Maxwell's hall, one has inadvertently slipped into Einstein's hall. 'So much is the beauty' that Einstein has 'imparted' to it" (Chandrasekhar 1989, p. 248).

Extending this work further, Chandra and his collaborators have found two classes of solutions. In one case, solutions describing two stationary Reissner–Nordstrom black holes go over to solutions that describe two static non-Abelian magnetic monopoles. In the second case, the solutions that describe two static Dirac monopoles of opposite charge held in place by a connecting string go over to binary black hole solutions with strings.

4. Concluding Remarks

The broad overview presented here, needless to say, does only partial justice to the enormous body of Chandra's work and its broad implications. As an astrophysicist, he began his research in general relativity by considering perfect fluid solutions. The new type of relativistic instabilities led him to develop the needed post-Newtonian approximation scheme to study them in detail. When he became convinced of the reality of black holes in the astronomical universe, he delved into the vacuum Einstein equations with new mathematical and physical insights. Won over by the geometrical structure and the richness of the general theory, his innate and characteristic drive for completeness drove him from approximate techniques to exact solutions, from charged black holes to colliding waves, and from colliding waves to black holes with strings. The saga is by no means at an end. The newfound techniques in the study of colliding waves find themselves under a new harness in the study of nonradial oscillations of a star.

A unifying mathematical structure governing a diverse set of phenomena emerged from Chandra's exploration of general relativity. Studying this structure, one is reminded of the well-known Monet serial paintings—paintings in which the same scene is depicted over and over again under different natural illuminations and seasonal variations. The valley, the trees, or the fields or the grain stacks are the same. The paintings, however, radiate totally different aesthetic content. In his case, seemingly the same equations and solutions describe different physics.

ACKNOWLEDGMENTS. I would like to thank Professor Abhay Ashtekar for a critical reading of the manuscript and helpful comments. I have also benefited from discussions with Chandra. This work is partially supported by the U.S. Department of Energy under contract number DE-FG02-85ER40231.

Notes

This chapter is based on a talk at the Third International Conference on the History and Philosophy of General Relativity, University of Pittsburgh at Johnstown, June 27–30, 1991.

[1] In the Nobel lecture reprint that includes a brief autobiographical account, Chandra says, "There have been seven periods in my life. They are briefly: (1) stellar structure, including the theory of white dwarfs (1929–1939); (2) stellar dynamics, including the theory of Brownian motion (1938–1943); (3) the theory of radiative transfer, the theory of illumination and the polarization of sunlit sky, the theories of planetary and stellar atmospheres, and the quantum theory of the negative ion of hydrogen (1943–1950); (4) hydrodynamic and hydromagnetic stability (1952–1961); (5) the equilibrium and the stability of ellipsoidal figures of equilibrium

(1961–1968); (6) the general theory of relativity and relativistic astrophysics (1962–1971); and (7) the mathematical theory of black holes (1974–1983)."

[2] For more details, see Chandrasekhar 1984.

[3] Chandrasekhar 1972, p. 524. It should be pointed out, however, that the precise manner in which extreme conditions develop in the interior of stars leading to instabilities of various sorts and supernova phenomena are not completely understood. For massive stars, theory suggests that a relativistically degenerate core with a mass approximately that of the limiting mass (equal to 1.4 $M_\odot$ for μ_e = mean molecular weight per electron = 2) is formed at the center. Then instability of some sort is expected to set in followed by gravitational collapse and the phenomenon of the type II supernova. In some instances, the highly degenerate core of approximately 1.4 $M_\odot$ will be left behind as a neutron star. That this does indeed happen sometimes is confirmed by the fact that in those cases where reliable estimates of the masses of pulsars exist, they are remarkably close to 1.4 $M_\odot$. In other instances, what is left behind will have masses in excess of that allowed for stable neutron stars and the formation of black holes is an inevitability. For further details, see Chandrasekhar 1984.

[4] Lorentz and Droste were the first to obtain the post-Newtonian equations of motion for a number of bodies interacting gravitationally (Lorentz and Droste 1917). For an outline of the history of the post-Newtonian problem, see, for instance, Pascoe et al. 1976. I am indebted to Professor Stachel for bringing to my attention the early history of the problem.

[5] For original papers on this subject, see SP 4, Part 3. See also Chandrasekhar 1969a.

[6] A series of six papers beginning with paper 18 in SP 5, p. 234.

[7] While Chandrasekhar obtained the conservation laws by direct integration of the equations of motion (see Chandrasekhar 1969b), Pascoe and Stachel (1969) obtained these laws from space-time symmetry properties of a Lagrangian known as the Plebanski–Bazanski Lagrangian.

[8] Chandrasekhar 1970, p. 195

References

Chandrasekhar, Subrahmanyan (1964). "The Dynamical Instability of Gaseous Masses Approaching the Schwarzschild Limit." SP 5: 6–25.

——— (1965). "The Post-Newtonian Equations of Hydrodynamics in General Relativity." SP 5: 51–75.

——— (1969a). *Ellipsoidal Figures of Equilibrium*. New Haven: Yale University Press.

——— (1969b). "Conservation Laws in General Relativity and in the Post-Newtonian Approximation." SP 5: 76–85.

——— (1970). "Solutions to Two Problems in the Theory of Gravitational Radiation." SP 5: 192–196.

——— (1972). "The Increasing Role of General Relativity in Astronomy." SP 5: 517–531

——— (1975). "On the Equations Governing Perturbations of the Schwarzschild Black Hole." SP 6: 3–12.

——— (1978). "The Gravitational Perturbations of the Kerr Black Hole." SP 6: 138–156, 157–181.

——— (1983). *The Mathematical Theory of Black Holes*. Oxford: Clarendon Press.

——— (1984). "On Stars, Their Evolution and Their Stability." *Reviews of Modern Physics* 56: 137–147.

——— (1989). "How One May Explore the Physical Content of the General Theory of Relativity." In *Proceedings of the Gibbs Symposium*, Daniel G. Caldi and C.D. Mostow, eds. American Mathematical Society and American Institute of Physics.

——— (1990). "A Scientific Autobiography, 1943–90." Unpublished manuscript, S. Chandrasekhar Papers, Box 1, The University of Chicago Archives.

Chandrasekhar, Subrahmanyan and Detweiler, Steven (1975). "The Quasi-Normal Modes of the Schwarzschild Black Hole." SP 6: 13–24.

——— (1977). "On the Reflection and Transmission of Neutrino Waves by a Kerr Black Hole." SP 6: 124–137.

Chandrasekhar, Subrahmanyan and Esposito, F. Paul (1970). "The $2^1/_2$ Post-Newtonian Equations of Motion of Hydrodynamics and Radiation Reaction in General Relativity." SP 5: 111–137.

Chandrasekhar, Subrahmanyan and Ferrari, Valeria (1984). "On the Nutku–Halil Solution for Colliding Impulsive Gravitational Waves." SP 6: 287–306.

Chandrasekhar, Subrahmanyan and Xanthopoulos, Basilis C. (1985). "On Colliding Waves in the Einstein–Maxwell Theory." SP 6: 307–343.

Einstein, Albert, Infeld, Leopold, and Hoffman, Banesh (1938). "The Gravitational Equations and the Problem of Motion." *Annals of Mathematics* 39: 65–100.

Hartle, James B. (1967). "Slowly Rotating Relativistic Stars. I. Equations of Structure." *Astrophysical Journal* 150: 1005–1029.

Hartle, James B. and Hawking, Stephen W. (1972). "Solutions of the Einstein–Maxwell Equations with Many Black Holes." *Communications in Mathematical Physics* 26: 87–101.

Hartle, James B., Thorne, Kip S., and Chitre, S.M. (1972). "Slowly Rotating Relativistic Stars. VI. Stability of the Quasi-Radial Modes." *Astrophysical Journal* 176: 177–194.

Khan, K. and Penrose, Roger (1971). "Scattering of Two Impulsive Gravitational Plane Waves." *Nature* 229: 185–186.

Lorentz, Hendrik Antoon and Droste, Johannes (1917). "De beweging van een stelsel lichamen onder den invloed van hunne onderlinge aantrekking, behandeld volgens de theorie van Einstein I, II." *Koninklijke Akademie van Wetenschappen te Amsterdam. Verslagen van de Gewone Vergaderingen der Wis- en Natuurkundige Afdeeling* 26(1917–1918): 392–403, 649–660.

Majumdar, S.D. (1947). "A Class of Exact Solutions of Einstein's Field Equations." *Physical Review* 72: 390–398.

Nutku, Yavuz and Halil, M. (1977). "Colliding Impulsive Gravitational Waves." *Physical Review Letters* 39: 1379–1382.

Oppenheimer, J. Robert and Volkoff, George N. (1939). "On Massive Neutron Cores." *Physical Review* 55: 374–381.

Papapetrou, Achilles (1947). "A Static Solution of the Equations of the Gravitational Field for an Arbitrary Charge-Distribution." *Proceedings of the Royal Irish Academy* 51: 191–204.

Pascoe, Thomas and Stachel, John (1969). "Variational Principle and Conservation Laws in Post-Newtonian Hydrodynamics." *Bulletin of the American Physical Society* 14: 69.

Pascoe, Thomas, Stachel, John, and Havas, Peter (1976). "Center-of-Mass Theorem in Post-Newtonian Hydrodynamics." *Physical Review* D14: 917–921.

Schutz, Bernard F. (1986). "Relativistic Gravitational Instabilities." In *Gravitation in Astrophysics*, B. Carter and J.B. Hartle, eds. New York: Plenum Press, pp. 123–153.

Wali, Kameshwar (1990). *Chandra: A Biography of S. Chandrasekhar*. Chicago: University of Chicago Press.

Whittaker, Edmund T. (1937). *Analytical Dynamics*. Cambridge: Cambridge University Press.

Will, Clifford (1981). *Theory and Experiment in Gravitational Physics*. Cambridge: Cambridge University Press.

Part V

Cosmology and General Relativity

Lemaître and the Schwarzschild Solution

Jean Eisenstaedt

1. Introduction

In several papers (Eisenstaedt 1982, 1987, 1989a), I discussed the early *pragmatic* (as I called it) interpretation of the Schwarzschild solution. In this paper, I want to turn my attention to the emergence of the modern interpretation that came to be accepted in the 1960s. In particular, I want to look at work done by Georges Lemaître in the early 1930s that, I claim, was of fundamental importance for this new interpretation.[1]

The Schwarzschild solution, the reader will recall, is the spherically symmetric exterior solution of the field equations of general relativity.[2] One of the most interesting aspects of the solution is what in the old days was called the "Schwarzschild singularity," the early, pragmatic, interpretation of which was that of an impenetrable sphere of radius $2Gm/c^2$ at the center, a singular sphere on which matter and light aggregate without penetrating it; a "magic circle" as Eddington called it (Eddington 1920, p. 98). It is now generally called the "Schwarzschild horizon."

In the papers mentioned above, I discussed the arguments underpinning this interpretation in considerable detail. Here I just want to make a few comments. First, most of these arguments, it turns out, strongly depend upon the particular choice of coordinates.[3] Not all of them do, though. In particular, there is an argument based on the interior Schwarzschild solution that does not. This argument will be of some importance to our present analysis. Second, since one believed that the density in nature, in stars and atoms alike, was too low—and this still is an open question—the "Schwarzschild singularity" was thought to stay deeply hidden in the material; it was thought to remain virtual.

In this paper, I will discuss the first steps toward the modern interpretation of the Schwarzschild solution. My main point will be that Lemaître's work, especially his 1932 article,[4] constitutes one of very few milestones in the comprehension, in the reinterpretation, of the Schwarzschild solution; not just because he demonstrated, as is well known, that the Schwarzschild singularity is fictitious, but also because he invented a tool—the so-called dust solution[5]—that enabled Oppenheimer and Snyder to give a correct and simple description of star collapse at zero pressure. In fact, I will show that Lemaître stood at the origin of some of the main developments of general relativity in the 1930s. The fact that he was a cosmologist is, I think, and I will develop the point below, no coincidence in this context. But first, let me go over some points concerning the history of cosmology around the 1920s.

2. Singularities and Mass-Horizons

The issue of *singularities, discontinuities, horizons*—as such objects have been called—has been of central importance to cosmology. To provide some background to Lemaître's work, I will briefly discuss how the issue came up and how it developed in the early years of general relativity.[6]

Einstein was puzzled by the fact that solutions of his general relativity—Schwarzschild's and De Sitter's—could contradict his interpretation of Mach's principle. Thus, in a letter to Schwarzschild in January 1917, Einstein expressed his concern about the incompatibility of Schwarzschild's solution of the vacuum equations of general relativity on the one hand and his interpretation of Mach's principle on the other.[7] On the basis of Einstein's Machian prejudices, one would not expect to find solutions of the field equations describing curved space-times in the absence of matter or in the presence of just one single body. In Einstein's view, the existence of such solutions was—as John Stachel put it (Stachel 1979, p. 440)—a "scandal."

The issue of singularities came up explicitly in a cosmological context, just after the publication of the *Kosmologische Betrachtungen* (Einstein 1917), in the Einstein–De Sitter controversy. Einstein had just received De Sitter's letter of March 20, 1917 (EA 20-545) in which De Sitter communicated his solution to Einstein. Einstein quickly responded (Einstein to De Sitter, March 24, 1917, EA 20-547). He argued that De Sitter's model has a closed singular surface at a physically finite distance, and that the solution therefore "does not correspond to any physical possibility."[8] What lies behind this objection are Einstein's early Machian conceptions, as is clear from a paragraph in the same letter quoted by De Sitter at the end of the article in which he published his solution. Einstein wrote:

In my opinion, it would be unsatisfactory if there were a possible world without matter. The $g_{\mu\nu}$-field must rather *be determined by matter without which it can not exist*. This is the core of what I mean by the demand of relativity of inertia.[9]

The following year, Einstein published a paper attacking De Sitter's solution (Einstein 1918). Although far more precise, it is pretty much along the same lines as his 1917 letter to De Sitter. Once again, Einstein raised the question of whether the discontinuity in De Sitter's solution occurs at a "physically finite" distance.

In his paper, Einstein gave an interesting definition of regularity. As he put it, the field equations must be valid at every point at a finite distance and this could only be the case "if the $g_{\mu\nu}$ as well as the corresponding contravariant components $g^{\mu\nu}$ (and their first-order derivatives) are continuous and differentiable; thus, in particular, the determinant $g = |g_{\mu\nu}|$ must never vanish at a finite distance."[10] After having defined what he meant by a "finite distance,"[11] Einstein elaborated on his definition of regularity to make it compatible with general covariance:

Moreover, the condition of continuity for the $g_{\mu\nu}$ and the $g^{\mu\nu}$ should not be taken as saying that there has to be a coordinate system such that continuity holds throughout space[-time]. Clearly, one only has to require that in the neighborhood of every point there *exists* [my emphasis] a coordinate system such that continuity holds in this neighborhood.[12]

Of course, the important point here is the word *exists*: it implies the demand of regularity in at least one coordinate system.[13] Einstein is fully aware of this point: "this restriction on the demand of continuity follows naturally from the general covariance of the [field] equations."[14]

Einstein wrote the line element of the De Sitter solution in the form[15]

$$ds^2 = \cos^2\chi\, c^2\, dt^2 - a^2\big(d\chi^2 + \sin^2\chi\,(d\theta^2 + \sin^2\theta\, d\theta^2)\big). \qquad (1)$$

He then pointed out that its determinant,

$$g = -c^2 a^6 \sin^2\theta \sin^4\chi \cos^2\chi, \qquad (2)$$

vanishes for $\chi = 0$ and $\theta = 0$, at the origin of the coordinate system. Einstein noted that this behavior is only apparently problematic, and that it is simply due to the use of polar coordinates. It is easy to find a different coordinate system in which the discontinuity does not appear. This is a simple but ingenious application of his definition of regularity. Unfortunately, there is another discontinuity. The determinant (2) also vanishes at

$\chi = \pi/2$, and this discontinuity occurs at a finite distance. Hence, in Einstein's words, "it looks as if this discontinuity cannot be eliminated by any choice of coordinates."[16] Of course, there is still room for doubt. After all, Einstein has not shown that no such coordinate system can be found. He had to be careful in conclusion: "Until there is evidence to the contrary, one has to accept that the De Sitter solution has a real singularity."[17] Einstein noted that on the singular surface $\chi = \pi/2$ in the De Sitter solution "the component g_{44} of the gravitational potential vanishes" just as in "the immediate neighborhood of a gravitating mass point."[18] This comparison seems to be an obvious reference to the Schwarzschild solution. Einstein concluded that "De Sitter's system does not correspond to the case of a universe without matter, but rather to a universe in which all matter is concentrated on the surface $\chi = \pi/2$."[19] The essential point for Einstein was not so much to give a clear definition of regularity, but rather to show that some matter existed on the horizon of a De Sitter universe. Rather than allowing for the possibility that the De Sitter singularity might be an artifact of the chosen coordinate system, a possibility opened up by his own careful definition of what constitutes a singularity, he jumped at the seemingly singular surface to save his beloved Machian principle. This principle is reiterated at the end of the paper: "no $g_{\mu\nu}$-field, i.e., no space-time continuum, is possible without the matter that generates it."[20] Basically, it was Einstein's view that gravitationally structured solutions of the exterior field equations can only exist if they have singularities. These singularities could then be interpreted either as a sign that matter was present or as a sufficient ground for discarding the solution as being nonphysical. Einstein's stance on this issue shows his strong commitment to the Machian notion that gravitational or inertial fields should be determined by matter, matter that can somehow be hidden by singularities. This was the reason for Einstein's search for matter—and singularities—in a De Sitter universe.

Questions about singularities touch on the very foundations of general relativity and would become the object of important subsequent research. Einstein himself came back to these issues at various times, as did a number of other experts. It was not until 1939, however, that André Lichnerowicz demonstrated in his dissertation[21] that there can be no nonsingular (spatially) asymptotically flat stationary solutions to the exterior equations of general relativity.[22]

The evidence that Einstein's considerations on the singular character of the surface $\chi = \pi/2$ were not correct was to come soon. In a letter from Göttingen written on June 16, 1918, Felix Klein, referring to Einstein's criticism of De Sitter, which had just been published, showed that the surface was not singular at all. Einstein immediately accepted Klein's results. "You are completely right," he wrote[23] in reply to Klein on June 20, 1918. He even agreed that his own article needed a rectification, although he never actually published one. And he did stress that De Sitter's solution cannot correspond to a physically possible universe: "So there really exists a singularity-free solution of the gravitational [field] equations without matter. Such a world, however, cannot be considered as a physical possibility."[24] In two papers published in late 1918 (Klein 1918a, 1918b), Klein would rewrite the De Sitter solution as a four-dimensional hypersphere of constant curvature

$$x^2 + y^2 + z^2 + w^2 - c^2T^2 = a^2,$$

in a five-dimensional pseudo-Euclidean manifold, with a line-element that can be written as

$$ds^2 = c^2\, dT^2 - (dx^2 + dy^2 + dz^2 + dw^2).$$

Klein drew the following conclusion:

> All of these results are in complete agreement with De Sitter's own exposition. However, they contradict the objections that Einstein raised against De Sitter in his contribution of March 1918 and that Weyl supported with detailed calculations in his book as well as in a remarkable article in *Physikalische Zeitschrift*.[25]

As can be gathered from the passage quoted above, Weyl developed Einstein's interpretation of the surface $\chi = \pi/2$ as a "mass-horizon" at length. The source of the gravitational field of the De Sitter solution is matter that is supposed to be concentrated on the singular horizon.[26] In the article quoted by Klein, Weyl asserted that the velocity of light vanishes on the equator of the sphere and that the "fundamental metrical form of the [De Sitter] universe will thus be singular" (Weyl 1919, p. 31). Then he performed an—erroneous—calculation of the mass present in the immediate vicinity of the horizon. As Klein pointed out, this line of reasoning can still be found in the first edition of his *Raum-Zeit-Materie*, where he concludes his calculation saying that "there must at least be masses on the horizon."[27]

De Sitter reacted cautiously to Einstein's criticism. Essentially, he accepted that the surface $\chi = \pi/2$ is singular, but insisted that it was "physically inaccessible."[28] In a letter to Einstein, he made it clear that he doubted the Einstein–Weyl interpretation in terms of a "mass-horizon" (De Sitter to Einstein, April 10, 1918, EA 20-565).

In 1922, Lanczos wrote two papers (Lanczos 1922a, 1922b) which are essential to the history of singularities. In the first one—which dealt primarily with the well-known coordinate condition now known by his name—he showed how a mere coordinate transformation could turn a regular line element into a singular one (Lanczos 1922a).[29] This cast some doubt on the accepted interpretation of the "Schwarzschild singularity." In a second paper (Lanczos 1922b), written shortly afterward and entitled "Bemerkung zur De Sitterschen Welt," Lanczos showed that the De Sitter solution was regular everywhere and concluded that Weyl's result concerning the "mass-horizon" in the De Sitter universe was mistaken. Referring to Klein 1918b, he performed a coordinate transformation from Klein's form of the De Sitter line-element to

$$ds^2 = c^2\,d\tau^2 - \cosh^2\tau\left(d\psi^2 + \sin^2\psi\,(d\theta^2 + \sin^2\theta\,d\phi^2)\right).$$

In this coordinate system the singularity is eliminated, which means that the singularity, as Lanczos pointed out, "can only come from the system of coordinates used."[30] In Lanczos's global coordinate system, the De Sitter universe is clearly seen to be an FLRW[31] geometry with positive curvature.

3. Eddington on Singularities

Before turning to Lemaître, I briefly want to discuss some of Eddington's contributions just because Eddington greatly influenced the young Lemaître.[32] In 1923, Eddington published *The Mathematical Theory of Relativity*, perhaps the most important textbook on relativity published between the two world wars. Lemaître, who was in Cambridge at the time of its publication, carefully studied Eddington's book. In the book, Eddington brought up two fundamental issues that Lemaître would develop later on: the issue of singularities and what became known as "Eddington's problem."

In his 1923 book, Eddington was very cautious in his discussion of the issue of singularities. What we get are not results from research he himself did on the topic but rather an exposition of the results of others, not always mentioned by name, as seen from Eddington's point of view. In the chapter devoted to the "properties of De Sitter's spherical world" (Eddington 1923, pp. 164–166), Eddington gave Klein's embedding of the De Sitter line element in order "to obtain a clearer geometrical idea of De Sitter's world."[33] Curiously enough, he did not seem to realize that this embedding shows directly that the horizon in the De Sitter solution is regular. Instead, he embarked on a discussion of the alleged mass-horizon in De Sitter's world. In an earlier chapter of his book, Eddington had given the solution representing the gravitational field in a De Sitter universe containing just one single particle. This solution is, in fact, just the Schwarzschild solution for the case where the cosmological constant has a nonvanishing value.[34] Eddington compared the Schwarzschild singularity to the De Sitter singularity. Given the apparent mass-horizon in the latter case, he asked, "must we not suppose that the former singularity also indicates matter—a 'mass horizon' or ring of peripheral matter?" (Eddington 1923, pp. 165). This is an implicit reference to Eddington's own interpretation of the Schwarzschild singularity. This interpretation to which he generally referred as "the magic circle" (Eddington 1920, p. 98),[35] is very similar to the mass-horizon interpretation of the De Sitter singularity. In both cases, matter is supposed to be concentrated on the singularity (Eddington 1923, pp. 100–101).

However, after having introduced the mass-horizon theme, Eddington—following Lanczos?[36]—went on to stress that "a singularity of ds^2 does not necessarily indicate material particles, for we can introduce or remove such singularities by making transformations of coordinates." Eddington continued, "It is impossible to know whether to blame the world-structure or the inappropriateness of the coordinate-system" (Eddington 1923, p. 165). It would seem, however, that Eddington is fundamentally stuck because he only allowed himself to use well-behaved one-to-one transformations.[37] As he put it, "all the transformations (even a change of origin) introduce a singularity somewhere" (Eddington 1923, p. 166), a remark from which he quickly drew the conclusion that "it is impossible to find any coordinate-system which represents the whole of real space-time regularly" (Eddington 1923, p. 166).

Still, Eddington clearly realized that even though the coordinate expression of the De Sitter line element may look singular, the De Sitter world itself is not. "The whole of De Sitter's world can be reached by a process of continuation," he wrote, and he concluded the chapter stating his belief "that the mass-horizon is merely an illusion of the observer at the origin" (Eddington 1923, p. 166). Thus, it looks as if Eddington was keenly aware of the fact that he was dealing with a question on the border between topology and coordinate representation.

Eddington's discussion of this issue is typical of his general style of doing physics. He had a broad vision of the subject, and, in a very creative manner, approached it from various different angles. Curiously enough, Eddington did not try to reach a definite point of view, a strict coherence of the subject; he allowed for some imprecisions and even some contradictions; still, his opinions, though a far cry from a formal solution of the problem, were rich and to the point, his conclusions accurate and fair. Of course, he did not come to some final verdict on the issue; nobody did at the time. One can see how Eddington's discussion would set a student reading his book (such as Lemaître) thinking about the issue.

In 1924, in an article in which he compared the line elements in Einstein's and Whitehead's theories for a particle at rest at the origin, Eddington gave a regular form of the Schwarzschild line element—essentially Finkelstein's 1958 line element—without making any special comment on it.[38] Moreover, in the German edition of his *Mathematical Theory of Relativity*, published in 1925, he worked out the radial light trajectories in these coordinates (Eisenstaedt 1987, p. 324).

4. Lemaître: From De Sitter's Universe to Einstein's

Just after having been ordained a priest in the fall of 1923, Georges Lemaître,[39] went to Cambridge, England, with a traveling grant from the Belgian government, to study general relativity with Eddington. As early as 1924, he published his first paper on the subject in the *Philosophical Magazine* with a foreword by Eddington (Lemaître 1924). Eddington was impressed with Lemaître, and, after Lemaître had left, he wrote a letter to de Donder praising him highly.

In the fall of 1924, Lemaître went to Cambridge, Massachusetts, where he worked with Shapley on the theory of variable stars. During this period, he attended several conferences in the United States and in Canada. Lemaître was affiliated with the Harvard College Observatory but worked on his Ph.D. in astronomy at M.I.T. under Vallarta. He gave a presentation on his thesis on November 19, 1925, although he would only submit it in 1927 during a second trip to the United States (Lemaître 1927a). It consisted of three parts, respectively entitled "The Gravitational Field in a Fluid Sphere of Uniform Invariant Density according to the Theory of Relativity," "Note on De Sitter's Universe," and "Note on the Theory of Pulsating Stars." The last part seems to have gotten lost.

Thus, already in his thesis, Lemaître took up the difficult subject of the interpretation of the De Sitter solution. He published this part of his thesis in 1925 in the *Journal of Mathematics and Physics* (Lemaître 1925a).[40] Lemaître pointed out that the form (1) of the De Sitter line element is somewhat misleading, since it suggests that the solution has some preferred center. Looking for an alternative expression that would reflect the symmetry of the solution, he was led to a homogeneous, nonstatic, Euclidean metric field corresponding to the line element:

$$ds^2 = c^2\,d\tilde{t}^2 - e^{2\tilde{t}/a}\left(d\tilde{r}^2 + \tilde{r}^2(d\theta^2 + \sin^2\theta\,d\phi^2)\right),$$

which is the FLRW Euclidean ($k = 0$) form of the De Sitter solution. The most important point is its nonstatic character,[41] a feature Lemaître thought to speak "perhaps rather in favor of De Sitter's theory" because it "gives a possible interpretation of the mean receding motion of spiral nebulae" (Lemaître 1925a, p. 192).[42] On the other hand, Lemaître thought that it was "completely inadmissible" (Lemaître 1925a, p. 192) that the solution was Euclidean.

It is not entirely clear which sources Lemaître used for his 1925 paper. He probably did not read Lanczos 1922b; he certainly was strongly influenced by Eddington 1923, the only reference he gave.[43] What is more important, though, is that by 1925 Lemaître had already developed his own views on the subject, views that would become more pronounced in the following years. Lemaître had a combination of talents that was fairly rare in general relativity at the time. He had an excellent background in mathematics, especially in differential geometry, and a strong physical intuition. It seems very likely that this first piece of work had a major impact on Lemaître's subsequent thinking about singularities. We will keep that in mind when analyzing Lemaître's 1932 article.

In 1927, independently of Friedman's 1922–1924 articles, Lemaître published a paper—in French and in an "obscure journal"[44]—in which he proposed a dynamical model of the world (Lemaître 1927b).[45] The properties of Lemaître's model—which starts off as a static Einstein universe and asymptotically changes into a De Sitter universe—are remarkably close to what Eddington expected in 1923: "It seems natural to regard De Sitter's and Einstein's forms as two limiting cases, the circumstances of the actual world being intermediate between them" (Eddington 1923, p. 160). Despite this comment, Eddington did not pay any attention to Lemaître's paper at the time. At the meeting of the Royal Astronomical Society on January 10, 1930, however, he "called attention to the need for intermediate solutions."[46] Eddington then decided to work on the question of the stability of Einstein's universe in collaboration with McVittie. It was "at once apparent" from Lemaître's paper that Einstein's universe would be unstable (Eddington 1930, p. 668). In fact, Lemaître, upon reading the issue of the *Observatory* in which Eddington suggested to investigate intermediate solutions, had sent a few copies of his 1927 paper to Eddington. The letter Lemaître enclosed—a draft of which survives in the Lemaître Archives—sheds some light on discussions he had on the topic with Einstein:[47]

> I had occasion to speak of the matter with Einstein two years ago. He told me that the theory was right and is all which [needs] to be done, that it was not new but had be[en] considered by Friedman, he made critic[ism]s against which he was obliged to withdraw but that from the physical point of view it was "tout à fait abominable."[48]

Eddington decided to publish Lemaître's paper in an English translation (Lemaître 1931).[49] From that point onward, the problem of nonstationary universes got more attention, and Lemaître's 1927 solution, which had been neglected for more than two years, finally got the praise it deserved. Eddington suggested that the actual universe may be considered as expanding from an initial state like an Einstein universe toward a state that in the limit becomes a De Sitter universe. If this picture were accurate, any initial disturbance would cause the universe either to expand or to contract. Eddington suggested that local condensation of matter in the universe actually provided such initial disturbances. In the following years, McCrea and McVittie studied nonstationary universes with a single instance of matter condensation at the origin (McCrea and McVittie 1931).[50] A substantial number of papers were published on this topic. It was probably this work that sparked Lemaître's interest in the problem of condensation in an expanding universe, a problem discussed at length in his 1932 paper.

Another theme of Lemaître's 1932 paper—a theme he already addressed in his Ph.D. thesis—is what is known as "Eddington's problem." This has to do with the so-called Schwarzschild limit. In the interior Schwarzschild solution, matter is described by a fluid sphere of constant density ρ and radius $r = a$ (Schwarzschild 1916). Schwarzschild noticed that, in his model, the pressure becomes infinite at the center as soon as the radius of the sphere is equal to the Schwarzschild limit. Since the radius $2Gm/c^2$ of the Schwarzschild singularity is smaller than the Schwarzschild limit, the latter could be used to lay worries about the former to rest. The Schwarzschild singularity, it seems, will be physically inaccessible. Schwarzschild availed himself of this strategy: "Thus, there is a limit concentration above which an incompressible fluid sphere cannot exist."[51] For a long time, Schwarzschild's limit was considered to be very important because it seemed to provide a physical explanation of the inaccessibility of the Schwarzschild singularity. In the early 1920s, Einstein himself defended this interpretation of the Schwarzschild limit.[52] It seems that Lemaître did not believe in this interpretation of the Schwarzschild limit. He felt that the Schwarzschild limit was not physically invariant in the sense that it depended on a particular solution, namely the interior Schwarzschild solution. In the introduction of his thesis, he wrote,

> The gravitational field within a fluid sphere of uniform density has been the subject of many investigations, especially by Schwarzschild (1916), Nordström (1918) and de Donder (1921, p. 169) and was considered as a solved problem until Eddington made some fundamental objections against the solution of these authors. (Lemaître 1927a, pp. 1–2)[53]

In short, the question raised by Eddington was whether the invariant density $T = \rho - 3p$, the trace of the stress-energy tensor, or just ρ, the "Schwarzschild density" as he called it, should be taken as the "true representation of the density" (Lemaître 1927a, p. 2). What really interested Lemaître in his dissertation, I think, was not Eddington's problem, merely formal, but rather the consequences of the answer to the question for the interpretation of the Schwarzschild singularity. Lemaître would have liked to show that no such limit as Schwarzschild's existed at all in the context of a solution of Eddington's problem. This, however, could not be done in 1925. Not only did Eddington's problem not have an exact solution, it turns up another fundamental "difficulty more striking than in Schwarzschild's solution."[54] As Lemaître put it in the introduction of his thesis, "It is unfortunate that the solution breaks down for large spheres, because the existence of a limit to the size of the sphere is one of the most interesting objects of the research" (Lemaître 1927a, p. 3). This clearly shows Lemaître's disappointment. But it shows, too, that what lies behind Lemaître's interest in Eddington's problem is his interest in the Schwarzschild limit. Thus, his dissertation gave Lemaître two reasons to believe that the Schwarzschild singularity is apparent: from his study of the De Sitter solution he knew that a singularity may be only an horizon, and from his study of Eddington's problem he knew that the Schwarzschild limit is not a necessary feature of any interior solution to Schwarzschild's exterior solution.

5. Lemaître's 1932 Paper and the Schwarzschild Solution

By 1932, Lemaître had become very expert in the field of the general theory of relativity. The 1931 translation of his 1927 paper had just been published, and had brought him some attention. His paper "L'univers en expansion" (Lemaître 1932)[55] gives an overview of the development of Lemaître's specific interests in the field. The main focus of the paper, however, is on the question of condensation in an expanding universe.

In the first chapters, Lemaître looked for a convenient formulation of the field equations of general relativity in the case of spherical symmetry—including, of course, the cosmological constant λ of which he was a lifelong supporter.[56] Lemaître was looking for general dynamical solutions for the spherical case and as his energy-momentum tensor he used the energy-momentum tensor for a perfect fluid with energy density $\rho(\chi, t)$ and pressure $p(\chi, t)$.[57] In the co-moving coordinate system he chose, the line element has the form

$$ds^2 = c^2\,d\tau^2 - a^2\,d\chi^2 - r^2(d\theta^2 + \sin^2\theta\,d\phi^2), \tag{3}$$

where c, a, and r are functions of χ and τ, and the only nonvanishing components of the energy-momentum tensor are

$$T_4{}^4 = \rho, \quad T_1{}^1 = T_2{}^2 = T_3{}^3 = p.$$

Lemaître now wrote down the field equations in a very elegant fashion; it was the first time that general nonstatic equations in the spherical symmetrical case were given.[58] What is more, the equations were given in a very simple form.[59]

Lemaître also found some interesting solutions to these equations. These solutions were among the first exact nonstatic solutions of the field equations of general relativity, together with the Friedman–Lemaître cosmological solutions and the Einstein–Rosen cylindrical waves solution (Beck 1925; Einstein and Rosen 1937). In the introduction to his paper, Lemaître, in fact, stressed his special interest in exact solutions:

> The theory [of general relativity] may be developed in two different ways: through the study of exact solutions of the equations of gravitation, using simplified models, or through approximations to the solution for more complicated problems. I think it is important not to mix up these two methods. In this paper, we will concern ourselves only with mathematically exact solutions.[60]

This emphasis on exact solutions is a good example of Lemaître's realistic and precise approach to problems in general relativity. It is also a further reason for Lemaître to be interested in the Schwarzschild solution, which at that time was still one of the few exact solutions known in general relativity.

Lemaître now looked at some special cases. First, he looked at the case in which the energy-density is uniform, while the pressure is a function of χ and t. With the help of his clever formulation of the field equations, he found another new solution (with uniform energy-density and a transverse pressure) that he used to prove that the Schwarzschild limit "vanishes when we do not impose the condition on matter that it is in the fluid state,"[61] a result, I trust, that he had long felt should hold.

In the next sections of his paper, Lemaître gave a short exposition of the main results of his Ph.D. thesis and discussed the instability of Einstein's universe. In section 8, entitled "Condensations in the Expanding Universe," Lemaître gave—for the first time—the general solution of the field equations when we have spherical symmetry and no pressure. It is interesting to see Lemaître's arguments dealing with the pressure-free case. As is often the case with Lemaître, they are ingenious and lucid at the same time:

> In applications involving the actual universe, the pressure is generally negligible compared to the density. In cases where we have equilibrium, we had to take it into account, since the study of the breakdown of equilibrium will naturally involve very small forces. When we study the expansion of the universe, however, and the occurrence of condensation in the course of the expansion, we can neglect it.[62]

Lemaître was able to integrate his general field equations for the pressure-free case in a very simple way, and to express his results in a surprisingly simple form.[63] In fact, what Lemaître found is just what has become known as the Tolman–Bondi solution, because of two papers by these authors dealing with this solution (Tolman 1934b; Bondi 1947).[64] Lemaître then showed that one obtains Friedman's universe (the elliptical case) as a special case of his general solution. Then he finally turned to the study of condensation in an expanding universe.[65]

For our purposes, the most interesting part of Lemaître's 1932 article is his demonstration that the Schwarzschild singularity is only an apparent singularity. Lemaître was the first to prove this explicitly and consciously.[66] As he stated the result himself, "The singularity of the Schwarzschild field then is a fictitious singularity, analogous to the one appearing on the horizon of the center in the original form of the De Sitter universe."[67] Before giving his demonstration, Lemaître explained what motivated him to look at this problem: "The equations of the Friedman universe admit... solutions in which the radius of the universe goes to zero. This contradicts the generally accepted result that a given mass cannot have a radius smaller than $2Gm/c^2$."[68] Thus, on the basis of his results, Lemaître understood that Friedman's solution was a possible interior solution to Schwarzschild's exterior solution. It follows from this simple but important observation that there cannot be a physical or mathematical limit that prevents such an interior solution from collapsing. On the other hand, the Schwarzschild singularity seems to impose a lower boundary on the radius of a star that does not show up in the (Friedman) interior solution. Lemaître's way out of this dilemma was to deny the latter claim. The question then becomes: what is the status of the space between a collapsing interior mass and the Schwarzschild singularity, i.e., the space between $r = 0$ and $r = 2Gm/c^2$? Lemaître's solution can help to answer this question. It should be possible to write down the exterior solution in terms of the coordinates in which Lemaître expressed his interior solution. This exterior solution should be identical to the Schwarzschild solution, which, by virtue of Birkhoff's theorem, is unique. In Lemaître's coordinates, there cannot be any gap between the collapsing sphere and the Schwarzschild singularity. In other words, the Schwarzschild solution has to be valid in this no man's land. The basic idea here is simply to deal with the interior and exterior solution in a *single* coordinate system. Even though one does not explicitly find this train of thought in Lemaître's paper, it was probably the guiding idea for him.[69]

Let us see how Lemaître technically works out his idea. Lemaître now had to look for the expression of the Schwarzschild solution in his own system of coordinates. He found that he had to perform the coordinate transformation

$$r^{3/2} = \sqrt{\frac{Gm}{\lambda c^2}} \sinh\left(\sqrt{\frac{3\lambda c^2}{4}}(\tau - \chi)\right) \tag{4}$$

to obtain the Schwarzschild line element in his own coordinates:

$$\mathrm{d}s^2 = c^2\,\mathrm{d}\tau^2 - \left(\frac{\lambda c^2}{3} r^2 + \frac{2Gm}{r}\right)\mathrm{d}\chi^2 - r^2(\mathrm{d}\theta^2 + \sin^2\theta\,\mathrm{d}\phi^2). \tag{5}$$

Hence, Lemaître had found a form of the Schwarzschild line element that explicitly shows that the only singularity of the solution is at $r = 0$. There is no singularity at $r = 2Gm/c^2$. At the same time, Lemaître had generalized his result to the case with a nonvanishing cosmological event.

Finally, Lemaître showed that by performing the transformation

$$c\,\mathrm{d}t = c\,\mathrm{d}\tau + \frac{\sqrt{\frac{2Gm}{rc^2} + \frac{\lambda r^2}{3}}}{1 - \frac{2Gm}{rc^2} - \frac{\lambda r^2}{3}}\,\mathrm{d}r, \tag{6}$$

one recovers the "classical" form of the Schwarzschild solution written in Droste's coordinates:[70]

$$\mathrm{d}s^2 = \left(1 - \frac{2Gm}{rc^2} - \tfrac{1}{3}\lambda r^2\right)c^2\,\mathrm{d}t^2 - \frac{\mathrm{d}r^2}{1 - \frac{2Gm}{rc^2} - \frac{1}{3}\lambda r^2} - r^2(\mathrm{d}\theta^2 + \sin^2\theta\,\mathrm{d}\phi^2).$$

For $\lambda = 0$, the transformation (6) is integrable, and (4) reduces to

$$r^{3/2} = \tfrac{3}{2}\sqrt{2Gm}(\tau - \chi).$$

Robertson's form of the Schwarzschild line element,

$$\mathrm{d}s^2 = c^2\,\mathrm{d}\tau^2 - \frac{\rho}{r}\,\mathrm{d}\rho^2 - r^2(\mathrm{d}\theta^2 + \sin^2\theta\,\mathrm{d}\phi^2), \tag{7}$$

is obtained by setting $\lambda = 0$ in Lemaître's line element (5) and by using the coordinate transformation

$$\chi = \tfrac{2}{3}\rho^{3/2}.$$

As we will see later on, Lemaître Robertson met in the early 1930s, and most likely Lemaître had a direct influence on Robertson's understanding of the Schwarzschild solution.[71] Robertson's remarks on the issue in Toronto in the late 1930s in turn aroused Synge's interest.

Thus, Lemaître explicitly showed, in a very nice way, that the so-called Schwarzschild singularity was not a singularity at all, even though it would continue to be called that for decades. Furthermore, our analysis shows that Lemaître's result was much more than just a mathematical derivation. Not only was Lemaître aware of the fictitious character of the Schwarzschild singularity before he had demonstrated it mathematically, he was also aware—and this point deserves to be stressed—of what was responsible for this tricky problem. As he put it right at the beginning of his demonstration, "we will show that the singularity of the field is not real but the result of using a coordinate system in which the field is static."[72] The important word here is *static*. Lemaître's insight has to be at least partly understood as coming out of his understanding of the dynamical/static character of the De Sitter solution that he worked out in 1925. Such insight was rare and would not return until years later. In fact, it would only return in Finkelstein 1958 and in Kruskal 1960, but the dynamical character of the Schwarzschild solution inside the horizon would take much more time to become truly accepted. In short, Lemaître's mathematical demonstration was backed up by a deep conceptual understanding of some of the fundamental features of Einstein's theory.

6. The Reception of Lemaître's 1932 Article

In 1934, Richard Tolman, about to publish his book *Relativity, Thermodynamics, and Cosmology*, wrote a short article, in which he studied the "Effect of Inhomogeneity on Cosmological Models" (Tolman 1934b). In this article, Tolman referred to Lemaître's dust solution (Lemaître 1932), which he rederived using Dingle's formulae. He then applied the dust solution to "distorted" uniform models of the universe and showed that disturbances from originally uniform distributions will in general increase with time. It is no coincidence that Tolman was aware of Lemaître's 1932 paper. Tolman had worked with Lemaître during the latter's 1932–1933 trip to the United States. As Deprit recalls it, Lemaître left Europe in August 1932 to go to Montreal. As usual, he had a very busy schedule. He participated in a solar eclipse expedition and went to M.I.T. to work with Vallarta on cosmic rays. Afterward, he gave a seminar at Princeton at the invitation of Robertson, and he spent the winter in California where he gave two seminars, one on the expanding universe and one on cosmic rays as fossils of the Big Bang

(Deprit 1984, pp. 373–375). He also met Hubble and worked with Tolman for two months at the California Institute for Technology. It is very likely that during this collaboration Lemaître explained his new solution to Tolman. In all likelihood, Lemaître had written his 1932 paper and had sent it off for publication before he traveled to the United States.

Tolman's paper attracted far more attention than Lemaître's. In fact, Lemaître's 1932 pressure-free solution is usually attributed to Tolman or to Bondi, sometimes to Datt, but never to Lemaître. Only the demonstration of the nonsingular character of the Schwarzschild solution is—sometimes—attributed to him.[73] Tolman's paper stimulated Synge to write an interesting paper, entitled "On the Expansion or Contraction of a Symmetrical Cloud under the Influence of Gravity" (Synge 1934b). In this paper, Synge "do[es] not pursue the method of Tolman but suggest[s] another point of view, which has much to recommend it on the score of mathematical simplicity" (Synge 1934b, p. 635). The mathematical tools he used are indeed remarkable: invariant equations, congruences, Riemannian curvature, and last but not least, the equation of geodesic deviation, an early application of one of Synge's favorite tools in the context of general relativity.[74]

Synge was especially interested in the equation that controls the expansion or contraction of a small cloud of particles. The equation he arrived at, in terms of the radius of the sphere, expressed in Droste's coordinates, is in fact the equation for the radial orbit of a particle moving in a Schwarzschild field with cosmological constant. Moreover, the equation is exactly the same as the corresponding Newtonian equation, with Newtonian absolute time replaced by proper time. It is easily seen that this equation is not singular on the Schwarzschild singularity and that the cloud collapses up to its center. Synge—following his contemporaries—did not mention this feature.[75] He carefully studied the evolution of the cloud, though, and concluded that there can be no oscillation of the cloud: its radius can only have one extremum. More precisely, the evolution depends on the value of the quantity $(3mG/\lambda c^2)^{1/3}$. Depending on the value of this quantity, the cloud either expands steadily and indefinitely or collapses into a point. This result shows in a very simple way that—at least in the pressure-free case—a collapse beyond the Schwarzschild singularity is possible. I find it strange that this useful paper has almost never been cited.[76] Synge himself did not cite it in his fundamental paper on the "Gravitational Field of a Particle," the first paper on the topology of the Schwarzschild field (Synge 1950)[77] after the Einstein and Rosen article (Einstein and Rosen 1935). This is very unfortunate, because these early calculations might have been very helpful, for instance, to Oppenheimer and Snyder. Had Synge forgotten about his 1934 paper in 1950? Or did he think the result was not worth mentioning in the context of the topological problems he addressed in his 1950 paper?

Synge did give Lemaître's line element (5) in his 1950 paper, for which he referred to Lemaître's 1932 paper,[78] as well as Robertson's line element (7). Synge acknowledged "Professor G.C. McVittie for recalling this reference" (Synge 1950, p. 84), and also mentioned that "Professor H.P. Robertson had drawn attention to [Lemaître's article] in a lecture in Toronto in 1939" (Synge 1950, p. 84). I came across a letter that Synge wrote to Robertson in late 1938.[79] In this letter, Synge raises questions about the recent demonstration by Einstein, Infeld, and Hoffmann that the gravitational equations for empty space are sufficient to determine the motion of matter represented as point singularities of the field (Einstein et al. 1938).[80] From this letter, one gathers that Robertson, who had been aware of the apparent character of the Schwarzschild's singularity for quite some time,[81] told Synge about it during a meeting "at State College." Synge, however, was not completely convinced. He added, "I must think more about this, because I am not clear on the point, and I don't think you have published anything on it" (Synge to Robertson, October 31, 1938). This clearly shows how difficult this point was to one of the best relativists of the time—a geometer, moreover—and thus how difficult it must have been to anybody in the field. Some years ago, I wrote to Synge and asked him about his recollection of these issues. He did not remember much about it, but he told me that he "never thought much about very dense concentration of matter" and that "perhaps [his 1950 paper] was better than [he had] regarded it for 38 years" (Synge to Eisenstaedt, May 16, 1988). This seems to indicate that despite his extensive work on the problem and despite the importance of the questions he raised, Synge was not satisfied with what he had accomplished in these matters. This attitude, I may add, is very typical of Synge.

In 1936, Paul Drumaux, in an otherwise quite interesting paper (Drumaux 1936), contested Lemaître's views concerning the apparent character of the Schwarzschild singularity (Eisenstaedt 1987, pp. 316–317). It seems that Drumaux and Tolman were the first and, at the time, only experts to cite Lemaître's 1932 paper. However, thanks to Tolman and thanks to his 1934 account of Lemaître's solution, this situation would change. I am thinking here in particular of the influence of Lemaître's "dust solution" on one of the most important pieces of work in general relativity, the article published in 1939 by J. Robert Oppenheimer and H. Snyder (Oppenheimer and Snyder 1939). A few months earlier, as is well known, Oppenheimer had published an article in collaboration with G.M. Volkoff (Oppenheimer and Volkoff 1939), in which they studied the gravitational equation for a neutron star. In particular, they showed that for masses greater than $^3/_4$ solar masses there were no static equilibrium solutions. It turns out that both Volkoff and Oppenheimer collaborated with Tolman. In the Tolman papers, one finds some letters that Tolman and Oppenheimer exchanged in late 1938 concerning this work. Tolman and Oppenheimer tried to find a solution "with agreeable properties"[82] to describe the gravitational equilibrium of a neutron star. Tolman's contribution, some "static solutions of Einstein's field equations for spheres of fluid" (Tolman 1939), was published separately in the same issue of *Physical Review* in which Oppenheimer and Snyder's paper appeared.[83] A few months later, Oppenheimer and Snyder published a paper entitled "Continued Gravitational Contraction," in which they calculated the collapse up to the *gravitational radius*[84] of a pressure-free fluid sphere. As they explained in the abstract of their paper, Oppenheimer and Snyder showed that:

> When all thermonuclear sources of energy are exhausted a sufficiently heavy star will collapse... the radius of the star approaches asymptotically its gravitational radius; light from the surface of the star is progressively reddened, and can escape over a progressively narrower range of angles.... The total time of collapse for an observer comoving with the stellar matter is finite... an external observer sees the star asymptotically shrinking to its gravitational radius. (Oppenheimer and Snyder 1939, p. 455)

Oppenheimer and Snyder derived these important results using the Lemaître–Tolman "dust solution."[85] I want to emphasize the importance of Lemaître's solution for Oppenheimer and Snyder's work. As I mentioned before, Lemaître's solution is not only one of the very first[86] general dynamical solutions with spherical symmetry in general relativity, it also allows us to describe the complete evolution of a star, its interior as well as his exterior gravitational field, in a *single* coordinate system. As I have already said, this last aspect, of course, was what allowed Lemaître to demonstrate the fictitious character of the Schwarzschild singularity. This property of the coordinate system is of crucial importance for dealing with the tricky question of the boundary conditions between the interior and exterior solutions on the surface of the star. Oppenheimer and Snyder stated that they followed the "earlier work of Tolman," and they "thank Professor R.C. Tolman and Mr. G. Omer for making this portion of the development available to [them], and for helpful discussions" (Oppenheimer and Snyder 1939, p. 457). Thus, it is clear that Lemaître's solution, through Tolman, played a very important role in Oppenheimer and Snyder's work.

At the end of the 1930s, there was great interest in matters of relativistic astrophysics, in neutron stars, in equilibrium of stars, and in star collapse. Tolman and Zwicky[87] were working in this field at the California Institute of Technology, Baade at Mount Wilson Observatory, and the members of the Oppenheimer team, Snyder and Volkoff, at the University of California at Berkeley. Around the same time, Robertson was working on the interpretation of the Schwarzschild solution and lectured on it. These lectures would be published after his death by his assistant Thomas Noonan (Robertson and Noonan 1968). I analyzed this paper elsewhere (Eisenstaedt 1987, pp. 328–338), and I will just outline my conclusions here. In his lectures, Robertson used the line element (7) which can easily be derived from Lemaître's line element (5). Although we do not have absolute evidence that this is in fact how Robertson got it, I think it is very likely that Robertson borrowed Lemaître's nonsingular Schwarzschild metric, which he had heard about during Lemaître's 1932 visit to Princeton, and used it to derive some of his very interesting results.[88] The detailed analysis of the trajectories in a Schwarzschild field, however, that he offered to his students in the late 1930s, is probably entirely his own (Robertson and Noonan 1968, p. 250). Robertson had a detailed understanding of the behavior of incoming as well as outgoing particles.[89] Thus, some of the points that Oppenheimer and Snyder made in 1939 had already been discussed by Robertson in his lectures. For instance, Robertson had already given the description of a particle falling toward the Schwarzschild radius from the point of view of an observer situated at infinity and in terms of the red shift of light rays emitted during the particle's fall. In Robertson's lectures, one also finds the description in proper time of the particle passing through the Schwarzschild horizon. Even more interesting is the fact that his description implies the necessity of a topological view of the Schwarzschild field. This follows from the lack of symmetry in the behavior of ingoing and outgoing particles that is illustrated in his diagrams. In fact, it seems that many results given in (Finkelstein 1958) had already been known to Robertson.

Robertson's office was situated close to Einstein's in Fine Hall at Princeton. Einstein, like Robertson, was very interested in singularities. I will not discuss Einstein's numerous contributions to the subject, which seem to be totally independent of Lemaître's. However, I do want to quote from the conclusion of a paper Einstein published in 1939, entitled "On a Stationary System with Spherical Symmetry of many Gravitating Masses" (Einstein 1939). Einstein wrote:

> This investigation arose out of discussions the author conducted with Professor H.P. Robertson and with Drs. V. Bargmann and P. Bergmann on the mathematical and physical significance of the Schwarzschild singularity. The problem quite naturally leads to the question, answered by this paper in the negative, as to whether physical models are capable of exhibiting such a singularity. (Einstein 1939, p. 936)

Thus, in spite of the discussions mentioned in this paragraph, Einstein continued to be of the opinion that the Schwarzschild singularity was a real singularity that just happened not to be realized in nature.

7. The Meaning of Lemaître's Work

In the preceding sections, we saw how—thanks in large measure to Tolman—Lemaître's results became well known among experts on general relativity in the years immediately preceding World War II, and how these results provided them with some fundamental tools for the analysis of the Schwarzschild solution. We saw that Lemaître was, in fact, the great forerunner of Oppenheimer and Snyder, providing them with the right tools for their seminal work on star collapse. Moreover, we saw that Lemaître understood the inevitability of collapse to zero volume and the fictitious character of the Schwarzschild singularity, insights that even Oppenheimer and Snyder failed to reach. It would be a mistake, however, to restrict our analysis to Lemaître's technical accomplishments. In this final section, I want to take a step back from the specific results Lemaître obtained and take a look at the general approach he took in arriving at them.

One of the most important characteristics of Lemaître's approach, I think, is the subtle interplay between local and global concerns in his work: the stars and the cosmos, contracting nebulae and the expanding universe, the condensation of a star and the collapse of the universe. In a way, Lemaître was able to describe the local in the global: a star is embedded in the universe, and the Schwarzschild solution is described in the same coordinates as Friedman's solution. This tendency to combine the local and the global, the awareness of the parallels between cosmology and the treatment of an individual star, enabled Lemaître to view things in new and unexpected ways, to look, so to speak, at the Schwarzschild singularity from the interior, or at the universe from the exterior. It was this general approach and his extraordinary facility in delicately manipulating the equations of the universe that enabled Lemaître to shake off the dogma of the impenetrability of the Schwarzschild singularity. One should keep in mind that most relativists at the time were working on problems that were almost classical. They were using the methods of post-Newtonian approximation, constructing and endorsing a *neo-Newtonian* interpretation of general relativity.[90] It fell to cosmologists such as Lemaître to develop the sort of global descriptions we have been looking at in this paper, and a crucial factor, in my opinion, was the sort of freedom that being a cosmologist afforded him, a form of independence vis-à-vis various conformist visions and traditional interpretations. Cosmology, so to speak, provided him with a space for work, for thought, a space of the dimensions of the whole theory. It is above all to this freedom of thought, I feel, that we owe the beautiful results that Lemaître gave us, results that collide head-on with the neo-Newtonian interpretation that would remain dominant for about 30 more years.

Lemaître, one might add, did not seem to be at all intimidated by the cosmological dimensions of the problems he chose to work on. The confidence of youth may have been a factor here, but I would like to venture that his vocation played a role here as well: as a priest he probably felt a closeness to God that may have given him a feeling of freedom in front of Creation. As I said above, Lemaître aimed at combining the global and the local: is there a question more suitable for a priest? To be sure, his answers, it seems, were strictly physical and mathematical. And it seems to me that Lemaître cannot be accused of confusing science and religion. As Kragh put it:

> Lemaître was an epistemic optimist who believed that God would hide nothing for the human mind and that consequently there could be no contradiction between Christian belief and scientific cosmology.... This does not mean that Lemaître's cosmology was designed to fit cosmological views or that he used it in defending such views. (Kragh 1987, p. 133)

To conclude this paper, I want to look somewhat more closely at an aspect I touched upon above, namely the small but important role cosmology played in the early history of general relativity. During the years 1925–1955, i.e., during what I have called the "low water mark" period of general relativity (Eisenstaedt 1986 and 1989b), a "neo-Newtonian" interpretation dominated the field. In those circumstances, cosmology was the only place where one could genuinely think about relativity.[91] As Bargmann put it:

> For many reasons, the history of general relativity (from 1920 to 1960) has been much less spectacular. The one field on which it had a decisive and most stimulating influence is cosmology. Its influence on the rest of physics, however, has been slight, notwithstanding the profound changes in our fundamental concepts which it had brought about.[92]

Likewise, in the conclusion of his textbook, Tolman, in trying to justify his interest in cosmology, gave an interesting analysis of the influence of cosmology on general relativity. In cosmology, he wrote, one expresses one's "natural interest and intellectual pleasure" in developing mathematical assumptions "without reference to possible physical applications."[93] Moreover, he felt that the work done in cosmology would "inform," "liberalize," and "illuminate our thinking."[94]

This certainly comes as a surprising declaration from the ultrapositivist America of the 1930s; a declaration that nonetheless has to be taken seriously, I think, since it comes from a scientist who cannot be suspected of blind idealism. Lemaître's work, it seems to me, provides a very good illustration of Tolman's ideas. The fact that he worked in a universe whose structure was not given in advance, but had to be constructed or invented, gave Lemaître a new point of view and a considerable freedom. As I have written elsewhere (Eisenstaedt 1989c), cosmology provided—and continues to provide to this very day—"a space for thought in general relativity."

ACKNOWLEDGMENTS. I wish to thank the Hebrew University of Jerusalem for permission to quote from the unpublished correspondence of Albert Einstein. I also wish to thank the California Institute of Technology for permission to quote from the Papers of Howard P. Robertson and from the Papers of Richard Tolman, and Professor Odon Godart for permission to quote from the Lemaître Archives at Louvain-la-Neuve. Finally, I would like to thank S. Kichenassamy, Michel Janssen, and John Stachel for interesting discussions, comments, and help with the English of this paper.

NOTES

[1] Many articles have been written on the topic; I will cite them as we go along. Right now, I just want to mention the comprehensive discussion of Lemaître's contributions to general relativity by Odon Godart, who worked with Lemaître (Godart 1992). My aim in this paper, however, is rather different from Godart's, and this study is to be understood as complementary to his.

[2] The uniqueness is guaranteed by Birkhoff's theorem; see Goenner 1970.

[3] In fact, they depend upon the very frequently used Droste coordinates, for long improperly called "Schwarzschild's coordinates." In particular, Droste's coordinate time t was generally thought of as the absolute time (in a Newtonian sense). See Eisenstaedt 1982, p. 167.

[4] Lemaître 1932 was first published in *Publication du Laboratoire d'Astronomie et de Géodésie de l'Université de Louvain*, an internal publication, and then reprinted in 1933 in *Les Annales de la Société Scientifique de Bruxelles*, a rather "obscure" journal. Still, if the article is referred to at all, the reference usually is to this 1933 reprint. I will go by the year of the original publication.

[5] Contrary to common belief, the dust solution is neither due to Tolman nor to Bondi and still less to Datt. As we shall see later on, Lemaître 1932 is never even cited except in Tolman 1934b. More on this below.

[6] Accounts of parts of this episode can be found in many books and articles, such as Ellis 1989; Eisenstaedt 1989c; Kerszberg 1989a, 1989b; Merleau-Ponty 1965; North 1965; Smith 1982; Stachel 1979; and Tipler et al. 1980. For a thorough discussion of the technical issues involved, see Rindler 1956, 1977 and Schrödinger 1956.

Kerszberg has given two rather different analyses of the Einstein–De Sitter controversy, one in the Osgood Hill proceedings (Kerszberg 1989a), the other in his book (Kerszberg 1989b). I prefer the former analysis. Kerszberg's book was harshly—but, I think, fairly—criticized in a review by Goenner (Goenner 1991).

[7] Einstein to Schwarzschild, January 9, 1917 (EA 21-561), quoted in Eisenstaedt 1989a, p. 219. For a more extensive discussion of the issue, see Stachel 1979.

[8] "Es scheint mir deshalb, das ihrer Lösung keine physikalische Möglichkeit entspricht." Einstein to De Sitter, March 24, 1917 (EA 20-547).

[9] "Es wäre nach meiner Meinung unbefriedigend, wenn es eine denkbare Welt ohne Materie gäbe. Das $g_{\mu\nu}$-Feld soll vielmehr durch die Materie bedingt sein, ohne dieselbe es nicht bestehen könne. Das ist der Kern dessen, was ich unter der Forderung von der Relativität der Trägheit verstehe" (De Sitter 1917a, p. 1225) and Einstein to De Sitter, March 24, 1917 (EA 20-547). The emphasis is Einstein's.

[10] "Wir werden es als Forderung der Theorie zu bezeichnen haben, dass die [Feld]Gleichungen für alle Punkte im Endlichen gelten. Dies wird nur dann der Fall sein können, wenn sowohl die $g_{\mu\nu}$, wie die zugehörigen kontravarianten $g^{\mu\nu}$ (nebst ihren ersten Ableitungen) stetig und differenzierbar sind; im besonderen darf also die Determinante $g = |g_{\mu\nu}|$ nirgends im Endlichen verschwinden" (Einstein 1918, p. 270). Even now, there is no generally accepted definition of a real singularity. By modern standards, neither the vanishing of g nor a discontinuity of any $g_{\mu\nu}$ constitute a "real" singularity. One way to determine whether a singularity is real or apparent is to maximally extend the space-time under consideration. If the De Sitter space-time is maximally extended, one sees that it does not have any real singularity at all. However, one can say that there will be a physical singularity if the stress-energy tensor becomes infinite somewhere (for example, if the matter density becomes infinite somewhere). Concerning this question, see Tipler et al. 1980, p. 139, where the question of the definition of a singularity is considered.

[11] The question still remains what he meant by a physical distance. Curiously enough, this fundamental concept was not yet clearly defined in general relativity, and it became an object of dispute between Einstein and De Sitter how to define it (Tipler et al. 1980, p. 100). For a discussion of the difficulties surrounding the definition of physical concepts in general relativity, see Eisenstaedt 1986, 1989b.

[12] "Ferner is die Stetigkeitsbedingung für die $g_{\mu\nu}$ und $g^{\mu\nu}$ nicht so aufzufassen, dass es eine Koordinatenwahl geben müsse, bei welcher ihr im ganzen Raum Genüge geleistet wird. Es muss offenbar nur gefordert werden, dass es für die Umgebung eines jeden Punktes eine Koordinatenwahl gibt, bei welcher für diese Umgebung der Stetigkeitsbedingung genügt wird..." (Einstein 1918, p. 270).

[13] In 1917, Hilbert also proposed a definition of regularity compatible with covariance, but he stipulated that one should get it through an "invertible and one-to-one" transformation (Eisenstaedt 1982, p. 172). It is to be emphasized that Einstein's definition is more suitable in this context since he did not specify the nature of the allowed transformations. The class of transformations allowed by Hilbert is too narrow. If, on some part of space-time, one has a regular line element and one asks which coordinate transformations preserve regularity, Hilbert's stipulations are perfectly acceptable and are, in fact, at the origin of the idea of "admissible coordinates" (see Lichnerowicz 1955, p. 5). If, however, on some part of space-time

one has a nonregular line element, Hilbert's one-to-one coordinate transformations will always leave us with a nonregular line element and nothing is gained. This is what happens, for instance, when we go from Droste's system of coordinates (Eisenstaedt 1982, p. 168) to an isotropic system of coordinates. In order to get to a well-behaved coordinate system—a coordinate system in which the line element is, say, as regular as possible—one should be allowed, at first, to perform a coordinate transformation that is not necessarily one-to-one. This is the case, for example, for the Eddington–Finkelstein or the Lemaître coordinate transformations. The idea is to introduce a new system of coordinates and it does not matter how you introduce it as long as the new line element is a solution of Einstein's field equations. Afterwards, in order to preserve regularity, one must only use one-to-one—in fact, C^2 piecewise (Lichnerowicz 1955, p. 5)—coordinate transformations. The question is also linked to the question for which region of space-time the new expression of the solution is valid and to the problem of extension. Concerning these questions, see also Tipler et al. 1980).

[14] "diese Einschränkung der Stetigkeitsforderung ergibt sich naturgemäss aus der allgemeinen Kovarianz der [Feld]Gleichungen" (Einstein 1918, p. 271).

[15] This form of the line element was first given in De Sitter 1917b, p. 230. It is the line element in a Schwarzschild-like static frame of reference, which can readily be seen by introducing the coordinate $r = a \sin c$.

[16] "und zwar scheint es sich hier um eine Unstetigkeit zu handeln, die durch keine Koordinatenwahl beseitigt werden kann" (Einstein 1918, p. 271).

[17] "Bis zum Beweise des Gegenteils ist also anzunehmen, dass die De Sittersche Lösung in der im Endlichen gelegenen Fläche $\chi = \pi/2$ eine echte Singularität aufweist" (Einstein 1918, p. 271). It is to be remarked that Einstein calls discontinuity (*Unstetigkeit*) what can be eliminated; if it cannot, it comes to be a singularity (*Singularität*).

[18] "Dort wird—wie in unmittelbarer Nähe eines gravitierenden Massenpunktes—die Komponente g_{44} des Gravitationspotential zu null" (Einstein 1918, p. 272).

[19] "Das De Sittersche System dürfte also keineswegs dem Falle einer materielosen Welt, sondern vielmehr dem Falle einer Welt entsprechen, deren Materie ganz in der Fläche $\chi = \pi/2$ konzentriert ist" (Einstein 1918, p. 272).

[20] "Es darf also kein $g_{\mu\nu}$-Feld, d. h. kein Raum-Zeit-Kontinuum, möglich sein ohne Materie, welche es erzeugt" (Einstein 1918, p. 271).

[21] Lichnerowicz 1939. See also Einstein 1941 and Einstein and Pauli 1943. For a discussion of this point, see Tipler et al. 1980, p. 108.

[22] With the obvious exception of Minkowski space-time. Lichnerowicz's theorem, of course, does not apply to the De Sitter solution, since this solution is not asymptotically flat.

[23] "Sie haben volkommen recht." Einstein to Klein, June 20, 1918 (EA 14-408). I owe this reference to John Stachel.

[24] "Es existiert tatsächlich eine singularitätsfreie Lösung der Gravitationsgleichungen ohne Materie. Aber diese Welt dürfte als physikalische Möglichkeit keinesfalls in Betracht kommen." Einstein to Klein, June 20, 1918 (EA 14-408).

[25] "Alle diese Resultate sind in voller Übereinstimmung mit De Sitters eigenen Angaben. Sie widersprechen aber dem Einwande, den Einstein in seiner Mitteilung vom März 1918 gegen De Sitter erhob und den dann Weyl in seinem Buche, sowie neuerdings in einem besonderen Aufsatz in der physikalischen Zeitschrift durch ausfürhrliche Rechnungen gestützt hat" (Klein 1918b, p. 422). This is the first time that Klein explicitly cites Weyl's work on the mass-horizon problem (Weyl 1918, 1919). See also Weyl to Einstein, May 19, 1918 (EA 24-036).

[26] This very curious story is treated at length in Kerszberg 1989b. Unfortunately, Kerszberg's analysis—as in many other parts of his book, see Goenner 1991—is problematic. Thus, in the relevant chapter (chapter 4), Kerszberg does not refer to Einstein 1918, which is analyzed in half a page in the preceding chapter (Kerszberg 1989b, p. 207), nor to Einstein's criticism of De Sitter, nor to the correspondence between Einstein and De Sitter. Meanwhile, Kerszberg discusses some crucial points that have their very origin in Einstein 1918, such as "the physical nature of the mass horizon" (Kerszberg 1989b, pp. 262–266), and cites the related correspondence between Einstein and Klein (Kerzsberg 1989b, pp. 266–275).

[27] "Zum mindesten am Horizont müssen sich Massen befinden" (Weyl 1918, p. 225). This passage was only modified in the fifth edition (Weyl 1923).

[28] De Sitter 1918, p. 1309. On this point, see Tipler et al. 1980, p. 100. There is an error in the date given for De Sitter's article in virtually all secondary literature. De Sitter's "Further Remarks..." (De Sitter 1918) was published in 1918, not in 1917. The error probably stems from a typo in the heading of the article, which stated that it was communicated in the meeting of April 26, 1917. This should have been 1918.

[29] This point is discussed in Eisenstaedt 1982, pp. 172–173.

[30] "eine etwaige Singularität des Linienelement [kann] nur von dem benutzten Koordinatensystem herrühren" (Lanczos 1922b, p. 540).

[31] The standard form of the homogeneous universe is sometimes called FLRW (Friedman–Lemaître–Robertson–Walker) after the main cosmologists who successively worked out the cosmological solution. The standard and general form of its line element can be written as

$$ds^2 = c^2\,d\tau^2 - R^2(\tau)\big(dr^2 + f^2(r)(d\theta^2 + \sin^2\theta\,d\phi^2)\big),$$

where $f(r)$ is a function which is $\sinh r$ in the elliptical case ($k = -1$); $\sin r$ in the spherical case ($k = +1$); and simply r in the Euclidean case ($k = 0$). See, e.g., Ellis 1989, p. 368.

[32] These contributions are of two different kinds. Not only did Eddington make a number of important technical contributions, which I will refer to as we proceed; he also put forward opinions, ideas, and images having to do with the formation of physical concepts in his days, i.e., he also much contributed to what I would like to call the ideology of technical relativity, or the heuristics of the field. For a discussion of Eddington's more technical contributions to relativity, see Stachel 1986.

[33] In fact, Eddington gave the central idea of Klein's embedding of the De Sitter line element (a restriction of two dimensions: the time and the radial coordinate) without citing Klein and Lanczos (Eddington 1923, p. 164).

[34] This solution was first published by Weyl (1918, pp. 225–226), although his paper contained an error. In the same year, the solution was independently published by Kottler (1918, p. 443), and a whole set of spherically symmetric static solutions with a cosmological constant was published by Bauer (1918). The solution is also reported in a letter from Weyl to Einstein of May 19, 1918 (EA 24-036).

[35] See Eisenstaedt 1989a, pp. 220–221.

[36] Eddington's view is so close to Lanczos' that one suspects that Eddington actually read Lanczos 1922a. To be sure, there is no independent evidence for this suspicion. If the suspicion is correct, Eddington may also have gotten Klein's transformation of the De Sitter line element, for which he does not give a reference, from Lanczos' paper.

[37] See Note 13.

[38] Of course, Eddington's transformation of coordinates is nonregular on and inside the Schwarzschild singularity; worse, as it is proportional to $\ln(r - m)$, it is not even defined for $r \leq m$. All of this was not clear to Eddington, who was even unconscious of having discovered a well-behaved system of coordinates at $r = 2m$; see Eisenstaedt 1982, note 82, and Note 13 above. In any case, Eddington was not the first scientist to discover a regular line element at the place of the Schwarzschild singularity; as I have shown elsewhere, Paul Painlevé and Allvar Gullstrand also exhibited such line elements. Of course, like Eddington, Painlevé and Gullstrand were unaware of having discovered a well-behaved system of coordinates (Eisenstaedt 1982, pp. 173–179). For more on these technical questions concerning the Schwarzschild solution, see Rindler 1977, pp. 149–165.

[39] For biographical information on Georges Lemaître, see Deprit 1984; Godart 1984; and Kragh 1987, p. 116–117; see also Godart and Heller 1979 and Godart 1992. Deprit 1984 has the best bibliography.

[40] It was reprinted in the same year in *Publication du Laboratoire d'Astronomie et de Géodésie de l'Université de Louvain.* Moreover, a short abstract of the paper was published (Lemaître 1925b). The paper has been analyzed by several authors; see Ellis 1989, p. 373; Godart 1992, p. 438; and Kragh 1987, p. 119.

[41] This feature can also be read off from the form of the De Sitter line element given by Lanczos. Lanczos did not draw attention to it. He only wanted to show that the De Sitter solution was regular and that the Einstein–Weyl mass-horizon interpretation was in error. Notice that Lemaître's space-time is geodesically incomplete: only half of the De Sitter hyperboloid is covered. On this point, see Ellis 1989, p. 373.

[42] Eddington was of the exact same opinion, and Lemaître, in fact, quoted from Eddington here (Eddington 1923, p. 161).

[43] Kerszberg has claimed that "Lanczos is the author that has directly influenced Lemaître" (Kerszberg 1986, p. 84), a claim contested by Kragh on grounds that it "seems to lack documentation" (Kragh 1987, p. 135, note 17). I agree with Kragh. However, as I indicated above, Eddington may well have read Lanczos 1922b, and Eddington certainly had a big influence on Lemaître. Moreover, there is a reference to Lanczos 1922b in Lemaître 1927b, p. 51.

[44] Tipler et al. 1980, p. 103, and elsewhere. The "obscure journal" is *Annales Scientifiques de Bruxelles*, published in French. The "obscurity" of the journal—

and the language!—is sometimes put forward as an explanation of why Lemaître's 1927 work "was ignored for several years" (Tipler et al. 1980, p. 103). I certainly believe there is some truth to that, but one should not forget that Friedman's articles were also ignored for years, even though they were published in *Annalen der Physik*, as were comments on them by Einstein. I believe that the neglect of both Friedman's and Lemaitre's papers has more to do with the fashion—in fact, the ideology—in physics at the time. As always, a paper on a fashionable topic in a fashionable journal is read, even if it is a bad paper, whereas a good paper on a topic that is not in vogue may not be read at all. What determines fashion in physics? How does it develop? These are serious questions. In the case of relativistic cosmology, the topic became quite fashionable in the late 1930s (for good observational reasons). Before that time, apart from the Einstein–De Sitter controversy, it was not. Another factor that may have been responsible for the lack of interest in both Friedman's and Lemaître's articles is that they dealt with a nonstatic solution, while only static solutions were deemed to be physically significant at the time. I will come back to this point below.

[45] This paper, like Lemaître 1925a, has been analyzed by several authors: Ellis 1989, p. 380; Godart 1992, pp. 440–443; and Kragh 1987, pp. 123–125.

[46] Eddington as quoted by Ellis 1989, p. 380.

[47] Kragh writes: "It is remarkable that apparently Lemaître did nothing to make his [1927] theory known outside Belgium" (Kragh 1987, p. 136, note 34). I disagree with Kragh's assessment. Lemaître sent copies of his 1927 article to Eddington in 1930, along with a letter. He participated very actively in discussions and meetings. He had discussions with Einstein on various occasions (in 1927 at the Solvay congress, in 1931 in California, in 1932 and 1933 in Belgium, and in 1935 at Princeton). He arranged to visit De Sitter in 1928. It also seems that he regularly sent copies of his articles to his colleagues. For example, I own a reprint of his 1927 article with an *hommage de l'auteur* to Elie Cartan. In general, I would say that Lemaître was a very dynamic, outspoken, and extroverted scientist. This picture emerges very clearly, for instance, from Deprit's biography (Deprit 1984).

[48] Lemaître to Eddington, draft, early 1930 (Lemaître Archives, Louvain-la-Neuve). The English is Lemaître's, the transcription is mine. Apparently, Lemaître met Einstein while he was attending the Fifth Solvay Conference in October 1927. A. Deprit—who was Lemaître's secretary—tells the story in a slightly different way, but she does not mention her sources (Deprit 1984, p. 371). She probably quoted from "Rencontres avec A. Einstein," the transcript of a radio broadcast in Belgium of April 27, 1957. This is what Lemaître had to say about this meeting with Einstein: "After some favorable technical remarks [concerning his 1927 article], he [Einstein] concluded by saying that from the physical point of view it appeared to him to be completely abominable." (Lemaître Archives, Louvain-la-Neuve, my translation). The reason for quoting Lemaître's 1930 recollection of this meeting with Einstein—which would seem to be more reliable than his recollections nearly 20 years later—is that it indicates that it was not so much Lemaître's results that Einstein found "abominable," but also Friedman's results, i.e., more generally, the existence of nonstatic solutions of his cosmological equations. Einstein would change his mind (Einstein 1931) after Hubble's celebrated article (Hubble 1929).

His extreme reluctance to accept nonstatic solutions before Hubble's results is still poorly understood.

[49] This is not an exact translation of Lemaître 1927b. Some interesting comments on empirical data have been omitted, in particular an analysis of two articles by Hubble and Strömberg. Moreover, important parts of pages 55–56 in the original, as well as a paragraph on page 58, are missing in the translation.

[50] For a discussion of this work, see Peebles 1980, p. 14.

[51] "Es ist damit eine Grenze der Konzentration gegeben über die hinaus eine Kugel inkompressibler Flüssigkeit nicht existieren kann." (Schwarzschild 1916, p. 434).

[52] The early history of the Schwarzschild limit is discussed in Eisenstaedt 1989a, p. 216. The idea is still considered to be respectable; see Misner et al. 1973, pp. 609–612).

[53] Eddington's problem was introduced in Eddington 1923, p. 121 and p. 169. In the 1920s, quite a few papers were published on the problem, e.g., Nuyens 1927 and de Donder 1930.

[54] Lemaître 1927a, p. 3. In fact, Lemaître came to a "paradoxical result," to put it in his own words: he showed that when the invariant density is supposed to be constant, there exists a maximum radius and the pressure remains finite. But what happens if matter is nevertheless added to the sphere? Schwarzschild's way to elude the difficulty in suggesting that "the equations cease to keep their physical meaning" is of course excluded because the pressure is finite (Lemaître 1927a, Summary and p. 27).

[55] It was published in French. In fact, only 19 out of the 101 publications by Lemaître listed in Godart 1984 were written in English, and those are all from before the war.

[56] On this issue, see Lemaître's contribution to the Schilpp volume (Lemaître 1949), Einstein's reply (Einstein 1949, p. 684), and the correspondence between Lemaître and Einstein that resulted from these contributions.

[57] Initially, Lemaître introduced a transverse pressure as well.

[58] These calculations can be found in Godart 1992, p. 446

[59] One may recall that Dingle's calculations of the Christoffel symbols for a "line element of considerable generality" would only be published the following year (Dingle 1933), and Tolman's textbook in which Dingle's calculations were given two years later (Tolman 1934a, pp. 253–257).

[60] "La théorie peut être développée de deux façons différentes: par l'étude de solutions exactes des équations de la gravitation, fournissant des modèles simplifiés ou par le développement approché de la solution de problèmes plus complexes. Il nous paraît utile de ne pas mélanger ces deux méthodes, et dans ce travail nous ne nous occuperons que de solutions mathématiquement exactes" (Lemaître 1932, p. 51).

[61] "la limitation plus sévère du rayon d'une masse donnée introduite par la solution du problème intérieur s'évanouit lorsqu'on n'impose pas à la matière la condition d'être à l'état fluide" (Lemaître 1932, p. 51).

[62] "Dans les applications à l'univers réel la pression est généralement négligeable vis-à-vis de la densité. Dans le cas de l'équilibre nous avons bien dû en tenir compte,

puisque l'étude d'une rupture d'équilibre dépend naturellement de forces minimes, mais pour l'étude de l'expansion de l'univers et le développement de condensations au cours de l'expansion, nous pouvons la négliger" (Lemaître 1932 p. 68).

63 See equations (25), (30), and (31) in Godart 1992, pp. 446–447.

64 Lemaître's article is quoted by Tolman, and Bondi in turn acknowledged Tolman, so Lemaître's priority was never in dispute. The question remains why this important solution was named after Tolman and Bondi and not after Lemaître. This probably has to do with the visibility of the actors: in the 1950s and 1960s, Tolman's textbook was still widely used, and Bondi was a well-known active researcher in the field. Lemaître was not nearly as well known among general relativists. Relativists were not that interested in Lemaître's general way of doing physical cosmology or in his primeval atom hypothesis. The Bondi–Gold steady-state theory, on the other hand, was very much in fashion. The fact that Lemaître wrote in French may also have played a role.

We are dealing here with what I would like to call "structural cosmology": the study of the geometrical structure of the universe. Lemaître's impact in this field certainly has not been as important as his impact on the field of physical cosmology. In a well-documented article, Peebles, a leading physical cosmologist, stresses the importance of Lemaître's 1932 paper. "This paper is remarkable for the freshness and clarity and depth of the ideas," he writes (Peebles 1984, p. 25). Peebles does not address the question of why the paper was hardly ever cited at the time.

65 Lemaître returned to these calculations in two papers he published in 1933 in the proceedings of the Paris Academy of Science (Lemaître 1933a, 1933b). Let me just say that in the following sections Friedman's equations are integrated with the help of Weierstrass's elliptical functions. For a discussion, see Peebles 1984, p. 28.

66 Before Lemaître, Painlevé, Gullstrand, and Eddington exhibited Schwarzschild line-elements with no Schwarzschild singularity (see Note 38).

67 "La singularité du champ de Schwarzschild est donc une singularité fictive, analogue à celle qui se présentait à l'horizon du centre dans la forme originale de l'univers de De Sitter" (Lemaître 1932, p. 82).

68 "Les équations de l'univers de Friedman admettent pour une masse non nulle, des solutions où le rayon de l'univers tend vers zéro. Ceci est en contradiction avec le résultat généralement admis qu'une masse donnée ne peut avoir un rayon plus petit que $2Gm/c^2$" (Lemaître 1932, p. 80). In the following section, Lemaître would prove that Friedman's universe can collapse to zero volume and disappear.

69 There is no explanatory text in Lemaître's paper at this juncture. He simply continues his calculation. Still, I feel that the reconstruction I have given provides the underlying structure of his technical argument.

70 In fact, Schwarzschild's solution in Painlevé's coordinates is an intermediary step between Droste's and Lemaître's. It just comes out in the (r, θ, ϕ, t) coordinate system:

$$ds^2 = \left(1 - \frac{2Gm}{rc^2}\right)c^2\,dt^2 - dr^2 + 2c\sqrt{\frac{2Gm}{rc^2}}\,dr\,dt - r^2(d\theta^2 + \sin^2\theta\,d\phi^2).$$

Clearly, like Eddington's, it is a well-behaved system of coordinates except for $r = 0$.

[71] For a discussion of Robertson's ideas and calculations about the Schwarzschild solution, see Eisenstaedt 1987, chapter 6, entitled "Les silences de Robertson." The line element (7) was also published by Narlikar and Karmarkar in 1946. They called it the "Geodesic Form of Schwarzschild's External Solution" (Narlikar and Karmarkar 1946).

[72] "Nous nous proposons de montrer que la singularité du champ n'est pas réelle et provient simplement de ce qu'on a voulu employer des coordonnées pour lesquelles le champ est statique" (Lemaître 1932, p. 80).

[73] In Misner et al. 1973, for instance, Tolman (1934b) and Datt (1938) are cited for their "analytic solutions for pressure-free collapse" (Misner et al. 1973, p. 859). Axelrad (1964) and Pachner (1966) cite Datt 1938. Lemaître's 1932 paper is cited only for its proof of the nonsingular character of the Schwarzschild horizon (Misner et al. 1973, p. 822). No mention is made of the fact that Lemaître's demonstration was based on the "dust solution." For an account of the question of the gravitational collapse based on the Lemaître solution, see Misner and Sharp 1964, where Bondi 1947 is acknowledged.

[74] Synge had essentially developed the idea of geodesic deviation in the context of Newtonian mechanics. Earlier in 1934, he had written a paper (Synge 1934a) specifically dealing with geodesic deviation in general relativity; see also Synge 1926.

[75] See chapter 5 in Eisenstaedt 1987, entitled "L'impasse (ou les relativistes ont-ils peur de la chute?)."

[76] It is not cited in Misner et al. 1973, for instance.

[77] I will not analyze this important paper here. Synge did include his 1934 paper in the comprehensive bibliography of his textbook *Relativity: The General Theory* (Synge 1960).

[78] In a short paper (Synge 1949), Synge had also referred to Lemaître.

[79] Synge to Robertson, October 31, 1938, from the Papers of Howard P. Robertson, Box 4.47. In this letter, Synge invited Robertson to come to Toronto. Robertson accepted, and it was during this visit that Robertson actually talked to Synge about Lemaître's demonstration. I thank Professor Peter Havas for bringing this letter to my attention.

[80] Concerning this question, see Havas 1989.

[81] See Eisenstaedt 1987, chapter 6.

[82] Tolman to Oppenheimer, November 9, 1938, from the Papers of Richard Tolman, Box 3.20. Tolman showed that the interior Schwarzschild solution has the paradoxical property that the pressure becomes negative when some mass is added after the Schwarzschild limit has been reached.

[83] Tolman cites Lemaître's 1932 paper not for its "dust solution" but for its so-called "layer solution" (Tolman 1939).

[84] The gravitational radius is nothing but the Schwarzschild singularity. Oppenheimer and Snyder still believed that the Schwarzschild singularity was indeed singular. Evidently, they had not read—and certainly did not cite—Lemaître 1932. They simply used Lemaître's solution as given by Tolman, who probably was not interested in Lemaître's use of it in showing that the Schwarzschild singularity is only apparent.

[85] Contrary to what is said in Misner et al. 1973, p. 620, Oppenheimer and Snyder's model is not homogeneous: the pressure is zero everywhere but the density is a function of r and t. They do use in full Lemaître's "dust solution."

[86] That is, if we do not count Friedman's solution.

[87] Zwicky and Baade had predicted the existence of neutron stars in Baade and Zwicky 1934.

[88] In 1935, Robertson was aware of Lemaître's 1932 article. In a letter he wrote to Lemaître on July 19, 1935, he asked him to send a reprint of his 1932 article to Dr. P.Y. Chou who "recently sent [him] a reprint along the same line" (Lemaître Archives, Louvain-la-Neuve). Most likely, this is Chou 1936, which unfortunately I have not been able to get a copy of so far. I thank Professor Peter Havas for this reference.

[89] See Robertson's diagram concerning the "trajectories near the Schwarzschild singularity" (Robertson and Noonan 1968, p. 251); see also Eisenstaedt 1987, p. 333.

[90] Concerning this concept, see Eisenstaedt 1986, p. 149.

[91] I first made this point in Eisenstaedt 1986, and I developed it in Eisenstaedt 1989c.

[92] Quoted in Eisenstaedt 1989c, p. 292.

[93] Quoted in Eisenstaedt 1989c, p. 293.

[94] Ibid.

References

Axelrad, M. (1964). "Une solution complète (intérieur et extérieur) non statique à symétrie sphérique en relativité générale." *Cahiers de Physique* 18: 351–360.

Baade, W. and Zwicky, F. (1934). "Supernovae and Cosmic Rays." *Physical Review D* 45: 138.

Bauer, Hans von (1918). "Kugelsymmetrische Lösungssysteme der Einsteinschen Feldgleichungen der Gravitation für eine ruhende, gravitierende Flüssigkeit mit linearer Zustandsgleichung." *Wiener Berichte* 127: 2141–2227.

Beck, Guido (1925). "Zur Theorie binärer Gravitationsfelder." *Zeitschrift für Physik* 33: 713–728.

Bondi, Hermann (1947). "Spherically Symmetrical Models in General Relativity." *Royal Astronomical Society. Monthly Notices* 107: 410–425.

Chou, P.Y. (1936). "A Relativistic Theory of the Expanding Universe." *Chinese Journal of Physics* 1(3): 1–17.

Datt, B. (1938). "Über eine Klasse von Lösungen der Gravitationsgleichungen der Relativität." *Zeitschrift für Physik* 108: 314–321.

De Donder, Theophile (1921). *La gravifique einsteinienne*. Paris: Gauthier-Villars.

——— (1930). *Application de la gravifique einsteinienne*. Mémorial des Sciences Mathématiques, no. 43. Paris: Gauthier-Villars.

De Sitter, Willem (1917a). "On the Relativity of Inertia. Remarks Concerning Einstein's Latest Hypothesis." *Koninklijke Akademie van Wetenschappen te Amsterdam. Section of Sciences. Proceedings* 19: 1217–1225.

——— (1917b). "On the Curvature of Space." *Koninklijke Akademie van Wetenschappen te Amsterdam. Section of Sciences. Proceedings* 20: 229–242.

——— (1918). "Further Remarks on the Solutions of the Field-Equations of Einstein's Theory of Gravitation." *Koninklijke Akademie van Wetenschappen te Amsterdam. Section of Sciences. Proceedings* 20: 1309–1312.

Deprit, André (1984). "Monsignor Georges Lemaître." In *The Big Bang and Georges Lemaître. Proceedings of a Symposium in Honour of G. Lemaître Fifty Years after His Initiation of Big-Bang Cosmology, Louvain-la-Neuve Belgium, 10–13 October 1983*. André Berger, ed. Boston and Dordrecht: D. Reidel, 1984, pp. 357–392.

Dingle, Herbert (1933). "Values of $T_m{}^n$ and the Christoffel Symbols for a Line-Element of Considerable Generality." *National Academy of Sciences. Proceedings* 19: 559–563.

Drumaux, Paul (1936). "Sur la force gravifique." *Société Scientifique de Bruxelles. Annales* B 56: 5–14.

Eddington, Arthur S. (1920). *Space, Time and Gravitation*. Cambridge: Cambridge University Press.

——— (1923). *The Mathematical Theory of Relativity*. Cambridge: Cambridge University Press.

——— (1930). "On the Instability of Einstein's Spherical World." *Royal Astronomical Society. Monthly Notices* 90: 668–678.

Einstein, Albert (1917). "Kosmologische Betrachtungen zur allgemeinen Relativitätstheorie." *Königlich Preussische Akademie der Wissenschaften* (Berlin). *Sitzungsberichte*: 142–152. Translated as "Cosmological Considerations on the General Theory of Relativity." In Lorentz, Hendrik Antoon et al., *The Principle of Relativity*. Arnold Sommerfeld, ed. W. Perrett and G.B. Jeffery, trans. London: Methuen, 1923; reprint New York: Dover, 1952, pp. 175–188.

——— (1918). "Kritisches zu einer von Herrn De Sitter gegebenen Losung der Gravitationsgleichungen." *Königlich Preussische Akademie der Wissenschaften (Berlin). Sitzungsberichte*: 270–272.

——— (1931). "Zum Kosmologischen Problem der allgeinenen Relativitätstheorie." *Preussische Akademie der Wissenschaften (Berlin). Sitzungsberichte*: 235–237.

——— (1939). "On a Stationary System with Spherical Symmetry of many Gravitating Masses." *Annals of Physics* 40: 922–936.

——— (1941). "Demonstration of the Non-Existence of Gravitational Fields with a Non-Vanishing Total Mass Free of Singularities." *Universidad Nacional de Tucuman. Mathematicas y Ficica Teorica. Revista* 2: 11–15.

——— (1949). "Reply to Criticism." In *Albert Einstein: Philosopher-Scientist*. Paul Arthur Schilpp, ed. Evanston, Illinois: The Library of Living Philosophers, pp. 663–693.

Einstein, Albert, Infeld, Leopold, and Hoffmann, Banesh (1938). "Gravitational Equations and the Problem of Motion." *Annals of Mathematics* 39: 65–100.

Einstein, Albert and Pauli, Wolfgang (1943). "On the Non-Existence of Regular Stationary Solutions of Relativistic Field Equations." *Annals of Mathematics* 44: 131–137.

Einstein, Albert and Rosen, Nathan (1935). "The Particle Problem in the General Theory of Relativity." *Physical Review* 48: 73–77.

—— (1937). "On Gravitational Waves." *Journal of the Franklin Institute* 223: 43–54.

Eisenstaedt, Jean (1982). "Histoire et singularités de la solution de Schwarzschild (1915–1923)." *Archive for History of Exact Sciences* 27: 157–198.

—— (1986). "La relativité générale à l'étiage: 1925–1955." *Archive for History of Exact Sciences* 35: 115–185.

—— (1987). "Trajectoires et impasses de la solution de Schwarzschild." *Archive for History of Exact Sciences* 37: 275–357.

—— (1989a). "The Early Interpretation of the Schwarzschild Solution." In *Einstein and the History of General Relativity. Proceedings of the 1986 Osgood Hill Conference*. Einstein Studies, Vol. 1, Don Howard and J. Stachel, eds. Boston: Birkhäuser, pp. 213–233.

—— (1989b). "The Low Water Mark of General Relativity, 1925–1955." In *Einstein and the History of General Relativity. Proceedings of the 1986 Osgood Hill Conference*. Einstein Studies, Vol. 1, Don Howard and J. Stachel, eds. Boston: Birkhäuser, pp. 277–292.

—— (1989c). "Cosmology: A Space for Thought on General Relativity" In *Foundation of Big Bang Cosmology. Proceedings of the Seminar on the Foundations of Big Bang Cosmology*. F. Walter Meyerstein, ed. Singapore: World Scientific, pp. 271–295.

Ellis, George F.R. (1989). "The Expanding Universe: A History of Cosmology from 1917 to 1960." In *Einstein and the History of General Relativity. Proceedings of the 1986 Osgood Hill Conference*. Einstein Studies, Vol. 1, Don Howard and J. Stachel, eds. Boston: Birkhäuser, pp. 367–431.

Finkelstein, David (1958). "Past–Future Asymmetry of the Gravitational Field of a Point Particle." *Physical Review* 110: 965–967.

Godart, Odon (1984). "The Scientific Work of Georges Lemaître." In *The Big Bang and Georges Lemaître. Proceedings of a Symposium in Honour of G. Lemaître Fifty Years after His Initiation of Big-Bang Cosmology. Louvain-la-Neuve, Belgium. 10–13 October 1983*. André Berger, ed. Boston and Dordrecht: D. Reidel, 1984, pp. 393–397.

—— (1992). "Contribution of Lemaître to General Relativity." In *Studies in the History of General Relativity. Based on the Proceedings of the 2nd International Conference on the History of General Relativity, Luminy, France, 1988*. Einstein Studies, Vol. 3, J. Eisenstaedt and A.J. Kox, eds. Boston: Birkhäuser, pp. 437–452.

Godart, Odon and Heller, M. (1979). "Les relations entre la science et la foi chez Georges Lemaître." *Pontificia Academia delle Scienze. Commentarii*. 3: 1–12.

Goenner, Hubert (1970). "Einstein Tensor and Generalization of Birkhoff's Theorem." *Communications in Mathematical Physics* 16: 34–47.

—— (1991). "Book Review. The Invented Universe: The Einstein–De Sitter Controversy (1916–17) and the Rise of Relativistic Cosmology by P. Kerszberg. Oxford: Oxford Science Publication, Clarendon Press, 1989." *General Relativity and Gravitation* 23: 615–617.

Havas, Peter (1989). "The Early History of the 'Problem of Motion' in General Relativity." In *Einstein and the History of General Relativity. Proceedings of the 1986 Osgood Hill Conference.* Einstein Studies, Vol. 1, Don Howard and J. Stachel, eds. Boston: Birkhäuser, pp. 234–276.

Hubble, E.P. (1929). "A Relation between Distance and Radial Velocity among Extragalactic Nebulae." *National Academy of Sciences. Proceedings* 15: 168–173.

Kerszberg, Pierre (1986). "Le principe de Weyl et l'invention d'une cosmologie non-statique." *Archive for History of Exact Sciences* 35: 1–89.

—— (1989a). "The Einstein–De Sitter Controversy of 1916–1917 and the Rise of Relativistic Cosmology." In *Einstein and the History of General Relativity. Proceedings of the 1986 Osgood Hill Conference.* Einstein Studies, Vol. 1, Don Howard and J. Stachel, eds. Boston: Birkhäuser, pp. 325–366.

—— (1989b). *The Invented Universe: The Einstein–De Sitter Controversy (1916–17) and the Rise of Relativistic Cosmology.* Oxford: Clarendon Press.

Klein, Felix (1918a). "Bemerkungen über die Beziehungen des De Sitter'schen Koordinatensystem B zu der allgemeinen Welt konstanter positiver Krümmung." *Koninklijke Akademie van Wetenschappen te Amsterdam. Wis- en Natuurkundige Afdeeling. Verslagen van de Gewone Vergaderingen* 27 (1918–1919): 488–489. Also in: *Koninklijke Akademie van Wetenschappen te Amsterdam. Proceedings* 21(1918–1919): 614–615.

—— (1918b). "Über die Integralform der Erhaltungssatze und die Theorie der räumlichgeschlossenen Welt." *Königliche Gesellschaft der Wissenschaften zu Göttingen. Nachrichten*: 394–423.

Kottler, F. (1918). "Über die physikalischen Grundlagen der Einsteinschen Gravitationstheorie." *Annalen der Physik* 56: 401–462.

Kragh, Helge (1987). "The Beginning of the World: Georges Lemaître and the Expanding Universe." *Centaurus* 32: 114–139.

Kruskal, M.D. (1960), "Maximal Extension of Schwarzschild Metric." *Physical Review* 119: 1743–1745.

Lanczos, Cornelius (1922a). "Ein vereinfachendes Koordinatensystem für die Einsteinschen Gravitationsgleichungen." *Physikalische Zeitschrift* 23: 537–539.

—— (1922b). "Bemerkung zur De Sitterschen Welt." *Physikalische Zeitschrift* 23: 539–543.

Lemaître, Georges (1924). "The Motion of a Rigid Solid according to the Relativity Principle." *Philosophical Magazine* 48: 164–176.

—— (1925a). "Note on De Sitter's Universe." *Publication du Laboratoire d'Astronomle et de Géodésie de l'Université de Louvain* 2: 37–41. Also in *Journal of Mathematics and Physics* 4: 188–192.

—— (1925b). "Note on De Sitter's Universe." *Physical Review* 25: 903.

—— (1927a). *The Gravitational Field in a Fluid Sphere of Uniform Invariant Density, according to the Theory of Relativity*. Ph.D. Thesis, Massachusetts Institute of Technology.

—— (1927b). "Un univers homogène de masse constante et de rayon croissant, rendant compte de la vitesse radiale des nébuleuses extragalactiques." *Société Scientifique de Bruxelles. Annales A* 47: 49–59.

—— (1931). "A Homogeneous Universe of Constant Mass and Increasing Radius Accounting for the Radial Velocity of Extra-Galactic Nebulae." *Royal Astronomical Society. Monthly Notices* 91: 483–490. [Slightly amended English translation of Lemaître 1927b].

—— (1932). "L'univers en expansion." *Publication du Laboratoire d'Astronomie et de Géodésie de l'Université de Louvain* 9: 171–205. Also in *Société Scientifique de Bruxelles. Annales A* 53 (1933): 51–85.

—— (1933a). "Condensations sphériques dans l'univers en expansion." *Académie des Sciences (Paris). Comptes Rendus* 196: 903–904.

—— (1933b). "La formation des nébuleuses dans l'univers en expansion." *Académie des Sciences (Paris). Comptes Rendus* 196: 1085–1087.

—— (1949). "The Cosmological Constant." In *Albert Einstein: Philosopher-Scientist*. Paul Arthur Schilpp, ed. Evanston, Illinois: The Library of Living Philosophers, pp. 437–456.

Lichnerowicz, André (1939). *Sur certains problèmes globaux relatifs au système des équations d'Einstein*. Paris: Hermann. Also issued as *Problèmes globaux en mécanique relativiste*. Actualités Scientifiques et Industrielles, no. 833. Paris: Hermann.

—— (1955). *Théories relativistes de la gravitation et d'électromagnétisme*. Paris: Masson.

McCrea, W.H., and McVittie, George C. (1931). "The Expanding Universe." *Royal Astronomical Society. Monthly Notices* 92: 7–12.

Merleau-Ponty, Jacques (1965). *Cosmologie du XXème siècle*. Paris: Gallimard.

Misner, Charles W., Thorne, Kip S., and Wheeler, John Archibald (1973). *Gravitation*. New York: Freeman

Misner, Charles W. and Sharp, D.H. (1964). "Relativistic Equations for Adiabatic, Spherically Symmetric Gravitational Collapse." *Physical Review B* 136: 571–576.

Narlikar, V.V. and Karmarkar, K.R. (1946). "Geodesic Form of Schwarzschild's External Solution." *Nature* 108: 515–516.

Nordström, Gunnar (1918). "Berekening voor eenige bijondere gevallen volgens de gravitatietheorie van Einstein." *Koninklijke Akademie van Wetenschappen te Amsterdam. Wis- en Natuurkundige Afdeeling. Verslagen van de Gewone*

Vergaderingen 26 (1918–1919): 1577–1589. English translation, "Calculation of Some Special Cases in Einstein's Theory of Gravitation." *Koninklijke Akademie van Wetenschappen te Amsterdam. Section of Sciences. Proceedings* 21 (1918–1919): 68–79.

North, John D. (1965). *The Measure of the Universe. A History of Modem Cosmology*. Oxford: Clarendon Press.

Nuyens, Maurice (1927). "Solution du Problème d'Eddington." *Académie Royale de Belgique. Classe des Sciences. Bulletin* 13: 440–446.

Oppenheimer, J. Robert and Snyder, H. (1939). "On Continued Gravitational Contraction." *Physical Review* 56: 455–459.

Oppenheimer, J. Robert, and Volkoff, G.M. (1939). "On Massive Neutron Cores." *Physical Review* 55: 374–381.

Pachner, J. (1966). "Contribution to the Problem of a Gravitational Collapse of Matter." *Bulletin of the Astronomical Institutes of Czechoslovakia* 17: 108–111.

Peebles, Phillip J.E. (1980). *The Large-Scale Structure of the Universe*. Princeton: Princeton University Press.

—— (1984). "Impact of Lemaître's Ideas on Modern Cosmology." In *The Big Bang and Georges Lemaître. Proceedings of a Symposium in Honour of G. Lemaître Fifty Years after His Initiation of Big-Bang Cosmology, Louvain-la-Neuve Belgium, 10–13 October 1983*. André Berger, ed. Boston and Dordrecht: D. Reidel, 1984, pp. 23–30.

Rindler, Wolfgang (1956). "Visual Horizons in World Models." *Royal Astronomical Society. Monthly Notices* 116: 662–677.

—— (1977). *Essential Relativity*. 2d. ed. New York: Springer-Verlag.

Robertson, Howard P. and Noonan, Thomas W. (1968). *Relativity and Cosmology*. Philadelphia: W.B. Saunders.

Schrödinger, Erwin (1956). *Expanding Universes*. Cambridge: Cambridge University Press.

Schwarzschild, Karl (1916). "Über das Gravitationsfeld einer Kugel aus inkompressibler Flüssigkeit nach der Einsteinschen Theorie." *Königlich Preussische Akademie der Wissenschaften (Berlin). Sitzungsberichte*: 424–434.

Smith, R.W. (1982). *The Expanding Universe*. Cambridge: Cambridge University Press.

Stachel, John (1979). "The Genesis of General Relativity." In *Einstein Symposium Berlin*. H. Nelkowski, A. Herman, H. Poser, R. Schrader, and R. Seiler, eds. Berlin, Heidelberg and New York: Springer-Verlag, pp. 428–442.

—— (1986). "Eddington and Einstein." In *The Prism of Science*. E. Ullmann-Margarit, ed. Dordrecht and Boston: D. Reidel, pp. 225–250.

Synge, John L. (1926). "On the Geometry of Dynamics." *Royal Society of London. Philosophical Transactions* 226: 31–106.

—— (1934a). "On the Deviation of Geodesics and Null Geodesics, Particularly in Relation to the Properties of Spaces of Constant Curvature and Indefinite Line-Element." *Annals of Mathematics* 35: 705–713.

——— (1934b). "On the Expansion or Contraction of a Symmetrical Cloud under the Influence of Gravity." *National Academy of Sciences. Proceedings* 20: 635–640.

——— (1949). "Gravitational Field of a Particle." *Nature* 164: 148–149.

——— (1950). "The Gravitational Field of a Particle." *Royal Irish Academy* (Dublin). *Proceedings* 53: 83–114.

——— (1960). *Relativity: The General Theory.* Amsterdam: North-Holland.

Tipler, C., Clarke, C.J.S., and Ellis, G.F.R. (1980). "Singularities and Horizons: A Review Article." In *General Relativity and Gravitation.* Vol. 2. A. Held, ed. New York: Plenum Press.

Tolman, Richard C. (1934a). *Relativity, Thermodynamics and Cosmology.* Oxford: Clarendon Press.

——— (1934b). "Effect of Inhomogeneity on Cosmological Models." *National Academy of Sciences. Proceedings* 20:169–176.

——— (1939). "Static Solutions of Einstein's Field Equations for Spheres of Fluid." *Physical Review* 55: 364–373.

Weyl, Hermann (1918). *Raum-Zeit-Materie: Vorlesungen über allgemeine Relativitätstheorie.* Berlin: Julius Springer.

——— (1919). "Über die statischen Kugelsymmetrischen Lösungen von Einsteins 'kosmologischen' Gravitationsgleichungen." *Physikalischer Zeitschrift* 20: 31–34.

——— (1923). *Raum-Zeit-Materie*, 5th ed. Berlin: Julius Springer.

E.A. Milne and the Origins of Modern Cosmology: An Essential Presence

John Urani and George Gale

1. Introduction

When it is remembered at all, E.A. Milne's kinematic relativity is remembered as a quirky, even "oddball," dead-end offshoot of relativistic cosmology. This view is mistaken. Not only is it unfair to Milne, it also presents a completely Whiggish revision of the actual history of early modern cosmology. From the moment of its first appearance on the scene, kinematic relativity was accepted as an alternative to cosmologies based upon the general theory of relativity. Moreover, and perhaps more importantly, Milne's ability to embed his program in solid philosophical foundations shaped the form and content of the debate about the nature of cosmology as a science. His attack upon the "bunkum" of curved and expanding space-time in favor of operational definitions based upon the primitive experience of the passage of time constrained the more effusive proposals of others such as Jeans and Eddington. His vigorous arguments supporting rationalist epistemology and hypothetical-deductive methodology created tolerance for subsequent efforts—steady state cosmology, for example—based upon these philosophical positions. And, ultimately, his theory that all sciences developed naturally from inductive empiricism toward free-standing axiomatization forcefully shaped the self-conception of cosmologists about their own practices.

In addition to these more general, philosophical contributions to modern cosmology, Milne's work contributed to one particularly important substantive development: the Robertson–Walker metric. It is well known that A.G. Walker and H.P. Robertson developed, independently, the space-time metric that bears their names. What is not so well known is the important role played by E.A. Milne in this development. Milne's influence upon Walker is direct: Walker was Milne's student and colleague. As Walker himself has recently noted,

> [With] regard to the so-called Robertson–Walker metric, it was just a perfectly natural kind of thing when one deals with the ideas of Milne using, you might say, a complete symmetry in the space. When you have the kind of group of motions that is required by Milne's kinematics, the natural thing is to look at it as a group, a Lie group, and the metric drops out of it because the Lie group has got a natural metric associated with it—and that is the Robertson–Walker metric. (Walker 1990)

Robertson's case is more complicated. Robertson was clearly—and self-admittedly—attracted by Milne's methods and philosophical foundations. Perhaps of equal importance is that Robertson's mathematical development of the groundwork of his version of the metric is directly and explicitly fashioned as a counterpart to Milne's own development of central themes in space-time frameworks. Thus, in the cases of both Walker and Robertson, Milne's work provided the point of departure for the genesis of today's standard cosmological metric. Analyses of published works, personal correspondence, and interviews will be used to demonstrate these points.

2. Beginnings

Milne had no cosmological theory to begin with. There is no doubt about that whatever. When the cosmology community gathered on September 29, 1931 to agree that it had reached consensus about what later came to be called the Big Bang model, Milne attended, but played no role save that of an astrophysicist (Dingle 1931; Milne 1931). Indeed, it was as an astrophysicist that Milne had become highly regarded, especially for his work on the internal constitution of stars (Milne 1929a). So far was cosmology from Milne's interest that even in his wide-ranging position paper "On the Aims of Mathematical Physics"—written for his inaugural as Rouse Professor of Mathematics at Oxford—not even the slightest whiff of cosmology was evinced (Milne 1929b). As we shall soon see, astrophysicist Milne's interest in cosmology was not to be piqued until seven months after the September consensus meeting, and then only by an especially singular event. But in the meantime, a community had coalesced solidly around a particular version of cosmology.

Modern theoretical cosmology began in 1917, when Einstein proposed a model based upon his general theory of relativity (GTR), a proposal soon countered by De Sitter's competing model (De Sitter 1931a). The problem, however, was that neither model seemed very likely to have much of a future:

Einstein's model predicted that the universe was full but static; De Sitter's proposed a universe that was but Pickwickianly "static," and this solely in virtue of its emptiness. The observed universe of course was neither full nor empty. When Hubble announced his red shift findings 12 years later, it was apparent that something different needed proposing, namely, a solution that Eddington called intermediate, due to its falling as it must mid-way between De Sitter's emptiness and Einstein's fullness (Eddington 1931). As is now well known, Lemaître had described just such a model two years earlier, in 1927, and had sent it to Eddington, who read it and promptly forgot all about it (McVittie 1987). In early 1931 an embarrassed Eddington, by now fully refreshed in memory, announced Lemaître's results via a translation of the paper in question (Lemaître 1931). The stage was thereby set for the September 29 special session of the British Association devoted to the topic "The Evolution of the Universe." Present at that session were all the main participants—De Sitter, Dingle, Eddington, Jeans, McCrea, McVittie, and as noted earlier, Milne. Consensus formed around what came to be called relativistic cosmology (RC). RC held that the universe began in a static Einstein state, suffered an indeterminate period of "stagnation," a state terminated by one of McCrea and McVittie's "condensations," this latter soon followed by a period of expansion as described by Lemaître. It was believed that near its end the universe would enter the De Sitter state, with galaxies spread out so thinly that a virtual emptiness would result (De Sitter 1931b).

Involved in accepting RC as the consensus model was implicit acceptance of the philosophical commitments of GTR. Metaphysically, GTR was committed to belief in four-dimensional space-time; and so, thereby, were the proponents of RC. The general philosophy underlying this relationship might fairly be called *theoretical realism*, the view that commitment to a theory implies commitment to the theory's realities as well. Hence, since GTR was committed to the reality of a four-dimensional space-time, four-dimensional space-time is real for anyone committed to GTR. This metaphysical stance was most certainly the one adopted by all the major figures on the RC side.[1]

Epistemologically, the situation was not nearly so clear. Initially, it seemed to involve commitment at least to a vague sort of *inductive empiricism*. According to this view, successful hypotheses based upon direct observational evidence are generalized into laws of nature. While their general allegiance to this view is evident, it is not terribly clear to what degree or to what specificity the major figures were committed (De Sitter 1931b; Dingle 1933; Tolman 1932). Clarity did not enter the epistemological situation until several years later, when the battle between Milne, Dingle, and nearly every other practicing cosmologist was well and truly joined. At the start it was metaphysics, and metaphysics alone, that awakened Milne from his astrophysical slumbers, catapulting him against the metaphysical bulwarks of the RC community.

3. Milne's Theory

3.1 Philosophy and Physics

The strength of these bulwarks was revealed in a loud debate that broke out in the pages of *The Times* seven months after the BA meeting, in May 1932 (Jeans 1932). During the exchange, Sir James Jeans took on all comers, defending with vigor and polemical skill the reality of an expanding, curved, four-dimensional space-time. Since it was this exchange that catapulted Milne into cosmology, it is worth looking into in some detail.

3.2 Metaphysical Motivations

The exchange began on May 14, when Stephen Coleridge of The Ford, Chobham, asked

> Sir James Jeans says "the Universe is expanding." What does he mean by "the Universe"?... Then we are told that space "must necessarily curve back on itself." This means nothing unless space is something quite inconceivable to the human mind. Space being manifestly infinite cannot curve; a thing without limit can have no shape.... "The Universe is doubling its dimensions once every 1,300,000,000 years." What is doubling? Into what is, whatever it is, doubling? (Jeans 1932, May 14, 1932)

Jeans responded in immediately recognizable theoretical realist fashion:

> But a flat map of this kind [Mercators projection the earth] does not correspond to reality.... Geography tells us that only a curved and finite representation of the earth's surface can be true to Nature, and present-day science conjectures that the same is true of space. Just as we can only make the countries of the world fit properly together on a globe, so we can only fit the parts of space properly together in a finite, curved whole. It is not a matter of common sense or the reverse, but of interpreting ascertained facts of Nature. (Jeans 1932, May 14, 1932)

On May 23, Jeans replied to yet another sally. This time, his answer took on a bit of stridency, which perhaps is what ultimately piqued Milne's response. Note, in particular, Jeans indictment of common sense. The respondents, he said,

> want to impose the newly discovered properties of space onto an out-of-date concept of the nature of space which is not flexible enough to receive them. We must make our views of space conform with the ascertained properties of space, and not attempt the reverse process. When the scientific and common sense views clash, the latter must obviously yield to the former, since science has knowledge of all the facts known to the man-in-the-street, and a host of others as well. The man-in-the-street may nevertheless prefer to retain his old common sense view of space; it will serve for his everyday requirements. (Jeans 1932, May 23, 1932)

Although there are obvious difficulties in understanding precisely what Jeans here meant by "newly discovered properties of space," there is no gainsaying his enthusiasm for the completely up-to-date space of GTR. Milne read *The Times*, and, within 10 short days,[2] had his own response to Jeans' plumping for cosmological modernity: a new, and highly original theory of relativistic cosmology, motivated in large part by his disagreement with the theoretical realism of Jeans and his RC colleagues.[3] Milne himself was in absolutely no doubt about the philosophical foundations of his proposal. "I don't know," he told his brother Geoff at the time, "whether you understand how very deep it goes—it destroys at one swoop much of the recent much-advertised work of Einstein, Jeans & Eddington." We can only surmise here, of course, that an element of the "recent much-advertised work" would include Jeans' letters to the editor. But this surmise is grounded a bit more firmly as Milne continued his comments to his brother, leading them directly into philosophical territory. His theory, he claimed, "gives the only satisfactory (*philosophically* satisfactory) picture of the universe & of the content of reality which I am acquainted with" (Milne 1932b, August 10, 1932; emphasis added). In metaphysical particulars, he went on to say, his theory "destroys time and space as legitimate objective conceptions & brings the light of cold reasoning into the fantastic medleys of thought created by J & E" (Milne 1932b, August 10, 1932).[4]

Milne first aired his theory at a colloquium at Wadham College, Oxford, on June 7, 1932. A synopsis of this paper appeared almost immediately in *Nature*, July 2, 1932 (Milne, 1932a). Since we will save discussion of the physical motivations and details of the theory until the next section, at this point we need only note that Milne led his presentation with an attack upon the metaphysics of RC and the claim that his theory was simpler.

Milne's claimed virtues for his new theory go beyond mere simplicity—it is metaphysically superior: "This *common-sense* explanation" he asserts, "has many advantages *in addition* to that of rendering unnecessary the introduction of a curved 'space' and a non-static metric" (Milne 1932a, ibid., emphasis added). Milne actually made two important points here. First, his new explanation was a common sense explanation, a point directed unerringly at the precise target presented by Jeans' comment that relativity theory was scientific and not common sense.[5] Secondly, if Milne's new theory "rendered unnecessary" curved space and expanding space, then, needless to say, there could be no question about their reality. Thus was introduced Milne's metaphysical minimalism, but his minimalist metaphysics was not an unprincipled one. Near the end of the brief article, Milne told why he could not abide the "fantastic medleys" of relativistic space:

> We cannot observe space. We observe point-events.... It seems best to avoid the phrase "the structure of space" or of "space-time" and consider simply the structure of the hyper-complex of world-lines which can be reconstructed from our observations. (Milne 1932a, p. 10)

Space, as far as Milne was concerned, was not an observational object. Point-events are, and from them, "we can recognize the continued existence of material objects" (Milne 1932a, ibid.). But space is not a material object, period. What Milne was introducing into cosmological theory was a rigorous connection between observation and metaphysical commitments. Rather than the loose connection between commitment to theory and commitment to metaphysics as evidenced in the theoretical realism of proponents of RC, Milne wanted to argue that a strict observationalist criterion should precede any metaphysical commitments to, e.g., spaces, times, or even space-time. Throughout the ensuing debate, Milne was never to relinquish his hold upon this view.

Although Milne did not go any further to develop his observationalist criterion, he had discussed it earlier in other places. In one of his first major papers on stellar interiors, he laid out a rigorous observationalist metaphysical criterion:

> Contemporary physics makes progress by discarding ideas corresponding to quantities which experience shows can never be observed. The "velocity of a system through the aether" proves to be an "unobservable"; consequently a dynamics must be constructed which avoids this concept.... [T]he physical content of the assertion which results from a piece of mathematico-physical reasoning must be a relation between observables only.... Theory is enriched by being pruned of unnecessary assumptions. (Milne 1929a)

Geometry, for Milne, was theory. Thus, if a piece of geometrical reasoning is to acquire any metaphysical significance, it can do so solely by being linked to observation, to experience. He said this very precisely in a letter to Walker written a year or so after his *Nature* paper appeared:

> You can only get physical results from geometrical formulation of problems by stating explicitly the bridge relating the geometrical embodiment to experience. It is essential to start with experience. The general theory of relativity is capable of criticism in the regard. (Milne 1933–1951)

The strongest and clearest statement in a scientific context of this philosophical point was made in a letter written to *Observatory* at the same time as the above to Walker.[6] Milne started out with the claim that "assignments of distances, epoches, densities and so on are at bottom all conventions, and we do not get much further by discussing conventions" (Milne 1934a, p. 25). So far, no problem; these comments are quite consistent with what we've just seen regarding the link between geometry and physics. Milne continued on to make an astonishing statement. "But what are independent of conventions," he asserted, "are the observed phenomena. *Theories differ simply and solely when their predictions as to phenomena differ*" (Milne 1934a, ibid., emphasis added). If this claim were generally accepted as true, then there would be no possibility for two theories to differ formally. Clearly, Milne was indicating a breathtakingly strong commitment to observationalist metaphysics. Just what this commitment came to in practice was laid out in a few simple, straightforward sentences:

> Now what Dr. Hubble, Dr. Shapley, and their co-workers actually observe may be described as follows. A certain area on a photographic plate is taken, representing a certain solid angle in the sky, and attention is fixed on a number of small nebulous patches and their spectra. For each patch its Doppler shift s and apparent brightness b are measured, and the patches are counted.... Every solution of the cosmological problem, every world-model, predicts in principle the smoothed-out values to be expected for $f(s_0, t)$ for a given patch, and the brightness and distribution functions $b(s, t)$ and $n(s, t)$ for different patches. *Two theories differ when their predictions of these functions differ.* This method of comparison avoids all reference to distance-assignments, world-geometry, schemes of projection or the like. (Milne 1934a, p. 25, emphasis added)

Having forcefully stated his observationalist position, Milne now took out after his detractors. First came Eddington, Jeans, and their ilk: "The neo-relativists have been so busy with discussing geometry that they have overlooked the necessity for discussing what should be observed" (Milne, 1934a, p. 26). Moreover, Milne declared with a certain waspishness, "it is greatly to be wished that the expounders of all world-models, static or expanding, would eliminate their differing geometries and reduce them to statements concerning the observables $f(s_0, t)$, $b(s, t)$ and $n(s, t)\,ds$" (Milne 1934a, ibid.). Eddington himself comes in for a certain amount of fire, too: "Sir Arthur Eddington asks me to project. But the boot is on the other leg; I have no intention of projecting. The non-Euclidean geometers must state what they expect to observe" (Milne 1934a, ibid.). Milne concluded with the reminder—as if one were needed—that he wrote the letter, "not to advocate the kinematic model, but to point out that geometry is irrelevant in the ultimate comparison of theories, and that comparison may be and should be made purely in terms of observables" (Milne 1934a, ibid.).

His heroic interpretation of the relationship between physics—real, natural-world-based observational physics—and geometry was to haunt Milne throughout his life. He began his own cosmological research during the initial explosion of enthusiasm for relativity theory, with its concomitant exploitation of new geometrical ideas. But Milne never did infer from the success of GTR its truth. This point is made simply and elegantly by Walker and Whitrow in their unpublished obituary:

> At that time the relativistic models of the expanding universe were still fairly new, and their success taken to support the belief that the essential structure of the universe is geometrical—that physics is subordinate to geometry. This idea, however, contradicted all that Milne had come to believe about the nature of physical phenomena and the relation between physics and mathematics. (Walker and Whitrow 1951)

It is clear that Milne's attack upon RC was based upon an explicit and deep-seated philosophical position: a metaphysical commitment to the observable world and to the reality of that world alone. What is now needed is an explanation of exactly how Milne brought this commitment to fruition in his cosmological theorizing. We turn now to that topic.

3.3 Metaphysics Begets Physics

In the first published version of his theory, the June 7 synopsis in *Nature*, Milne came directly to the physical point. Although the received cosmology accounted for the observations, namely, that "the most distant nebulae appear to be receding from us, and the velocity of recession is proportional to the distance," the theory had certain problems, among the most difficult of which figure the points "that it has been impossible to explain why 'space' is expanding and not contracting" and "that at the present moment it is impossible to determine the algebraic sign of the curvature of 'space'" (Milne 1932a, p. 9).[8] Given these problems, Milne proposed an alternative explanation:

> A much simpler explanation of the facts may be obtained as follows. The explanation abandons the curvature of space and the notion of expanding space, and regards the observed motions of distant nebulae as their actual motions in Euclidean space. (Milne 1932a, ibid.)

The upshot of Milne's approach was this: Even as he rejected the curved space and expanding space concepts arising in relativistic cosmology, he maintained special relativity as the primary mathematical basis for his "simpler explanation of the facts" of astronomical observation. Making full, and in fact ingenious, use of special-relativistic methods in constructing related statistical and hydrodynamic cosmological models, Milne studiously avoided not only tensor analysis but also covariant notation in most of his published work. The theory almost immediately came to be known as *kinematical relativity*, a curious label, oddly akin to the use of the misnomer *general relativity* to describe Einstein's theory of gravitation.

Although special relativity was undeniably a cornerstone of Milne's mathematical methodology, he also generally avoided speaking of space-time as a four-dimensional real continuum, preferring to discuss the motion of matter in Euclidean space. Milne also assailed RC for its claimed use of global time, but in his own models he routinely used Lorentz transformations that were applied globally to all points in his flat, Lorentzian space-time. In short, Milne's cosmology is difficult, even annoying, to read. This is probably not an attempt to obfuscate but rather a need on Milne's part to reiterate and accentuate his rejection of geometric methods even while, at the same time, he was constructing a practical and accessible theory. Yet, even given the inherent annoyances in Milne's presentation, it is manifestly clear that Milne's colleagues were simply bowled over, once the masterful completeness and elegant simplicity of Milne's constructions were finally appreciated.

In its earliest formulation (Milne 1932a), KR was based on two postulates, the first being the constancy (Lorentz invariance) of the speed of light. This is tantamount to accepting special relativity fully, which is not at all unusual given that Milne, by default, had formulated his cosmology in flat Lorentzian space-time. Milne's second postulate asserted that two observers in relative motion should have identical views of the universe (both local and inferred global views). The notion behind this postulate is usually described as related to Einstein's extended principle of relativity or Robertson's principle of uniformity (Robertson 1933). A third major assumption, a postulate for Milne's statistical cosmology (Milne 1933), is the constancy along particle world lines (*à la* Liouville) of the distribution of points in phase space. This additional restriction was seen by Milne to be necessary to obtain spatial distributions of particles in his spatially unconfined system. Massive amounts of computation (Milne 1933) led Milne to the inevitable Boltzmann equation that enforced the necessary phase space density requirement.

The key element of Milne's cosmological models is, however, the shrewd physical observation that an initially concentrated system of particles with negligible collision losses and sufficient kinetic energy to overcome binding (gravitational) forces will naturally sort itself out into a correlated spatial-velocity distribution as (global) time progresses to asymptotic extremes. The particle swarm will ultimately approach a Hubble-like state: essentially all velocities are radial, with the faster particles arrayed at greater distances from the initial concentration according to a distribution $r = v/t$ for fixed t. Milne noted the variability of the proportionality factor $(1/t)$ as a significant difference between KR and RC. Milne also noted the clear time-reversibility of this inevitable sorting phenomenon, while leaving untouched the question of the beginning of things as represented by the initial concentration (Milne 1933).

Having found this highly plausible velocity-sorting mechanism, Milne then applied the requirement of local Lorentz invariance to the distribution function, $f(u, v, w)$, of particle velocities at an arbitrary spatial point. Here u, v, and w are not-too-convenient Cartesian velocity components. Requiring Lorentz invariance of the local particle number $f(u, v, w)\,du\,dv\,dw$ in an infinitesimal volume $dV = du\,dv\,dw$ immediately led Milne distribution law,

$$f(u, v, w) = \frac{Bc}{(c^2 - u^2 - v^2 - w^2)^2},$$

where B is the local particle density at some fixed time. This velocity distribution function, singular as speeds approach c, indicated to Milne that the universe could contain objects with speeds approaching the speed of light. By 1931, apparent recession speeds had already been identified to be as high as $c/15$ within the known limited viewing range of Mt. Wilson, a fact that Milne took to be quite encouraging for his model (Milne 1932a).

Following a busy leave of absence at the Einstein Institute at Potsdam (late fall and winter 1932), Milne published the results of extensive work that he had completed there on statistical and hydrodynamic cosmological models (Milne 1933). The resulting 95-page tome covered virtually every aspect of KR. With the exception of some philosophically driven results having to do with time scales and clock graduations (to be briefly noted in the final section of this paper), Milne's later versions of KR did not differ substantially from that published here. Of particular interest was that both models led to identical velocity distribution functions, the aforementioned $f(u, v, w)$, and to the same particle spatial distribution function,

$$\rho = \frac{Dt}{c^3(t^2 - r^2/c^2)^2},$$

with D constant and $\rho = 0$ for $r > ct$. The particle density distribution then exhibits a feature in addition to the velocity-sorting mechanism that produces the Hubble-like velocity–distance relation for any given large t-value, namely, a clear size for the matter universe.

The surprising singularity of the particle density function as $r \to ct$ for $t > 0$, while consistent with the velocity distribution function's singularity as speeds approach c, immediately raises the suspicion that Milne's model might possess a catastrophic Olbers phenomenon. A diverging density at $r = ct$ could imply an infinite radiant flux upon interior points, $r < ct$, if the particle density represents luminous matter.

Milne treated this problem with his typical thoroughness (Milne 1933), showing that the divergent particle density was, in fact, more than compensated for by the Doppler red shift of spectra of any luminous matter near $r = ct$, which must then have $v \to c$.

Despite Milne's thoroughness, KR was still somewhat susceptible to questions concerning the specific role of gravity in the dynamics of the matter in the universe. These questions were to be addressed by Milne and many others over the next several years, e.g., Dingle 1933; Kermack and McCrea 1933; McVittie 1933; Robertson 1933; and McCrea and Milne 1934, to cite only a very few. The attention that KR received was truly remarkable. Milne's initial formulation of a kinematical theory of cosmology must be considered as spectacular for both its physical insight and its comprehensive mathematical formulation. Even more significant was its immediate inspirational effect on the cosmological community, influencing the work of McCrea (Kermack and McCrea 1933), McVittie (1933), and especially the foundational work by Walker (1937) and Robertson (1935). We turn now to discussion of this reception of KR.

4. Reception of Kinematic Relativity

Milne introduced his theory in June of 1932. During the next eight years, 70 papers related to the theory in one way or another were published in physics, mathematics, and philosophy journals. This constitutes the vast majority of cosmologically oriented papers published during the period. Probably the major reason underlying the strength of this response to Milne lies in the fact that his proposals were not simply physical ones, but rather were physical proposals presented in conjunction with a full-blown philosophy: metaphysics, epistemology, and philosophy of science. Hence, in order to understand the reception of Milne's theory correctly, we must necessarily understand not only the reception of its physical aspects but of its philosophical aspects as well. We begin with the former.

4.1 Reception of KR's Physical Aspects

From the very first, Milne's theory was taken seriously. Most perceived it as an ingenious rival to RC. McVittie's comments are typical. Following his judgment that "the basic idea of this theory is both simple and elegant," McVittie goes on to note the difficulty of choosing between the rival theories:

> Both theories are in accordance with observation and it seems impossible to decide definitely for or against either so long as the phenomenon of the recession of the nebulae, in isolation from all other phenomena, is to be the only criterion. (McVittie 1933, p. 534)

Whittaker, after remarking that Milne's "working out of the new concepts has the unmistakable stamp of originality and power," concludes that "the cleavage between relativistic cosmology and kinematical relativity is fundamental" (Whittaker, 1935, p. 180). Dingle, who responded along with Robertson to Milne's first extended presentation of the theory (Milne 1933), was no less positive:

> In a recent number of *Zeit. für Astrophysik*, there appeared an exceedingly ingenious paper by Professor Milne.... [W]e can hardly avoid regarding it as an alternative theory [to RC,] ... the article is therefore of fundamental importance and demands careful and critical examination." (Dingle 1933, p. 167)

Robertson's response was more complicated. In the first place, he did not view Milne's theory as an alternative to RC. Rather, he claimed, either Milne's theory was the "kinematical preliminary to the dynamical problem ... completely solved [by RC]" or it is a

> special case of the alternate theory in which the influence of matter on the structure of the universe is considered negligible. The gravitational extension of Milne's solution. from this latter point of view, leads to an expanding universe which differs inappreciably from it, unless the density of all matter in the universe exceeds Hubble's lower limit by at least a hundred-fold. (Robertson 1933, p. 153)

Robertson was never to swerve from this interpretation of the purely physical aspects of KR; he quite simply found them unattractive. Another aspect of Milne's work caught Robertson's eye right from the start, however, generating an interest that was to deepen over the next two years, eventually leading him to produce his version of what we now know as the Robertson–Walker metric. The aspect of Milne's work Robertson found attractive was the use of purely *operational* methods to define theoretical primitives, whereby Milne satisfied his metaphysical desires to avoid unobservable entities in theorizing. We shall examine this part of the story in the next section.

Yet, even though Milne's theory received a generally widespread welcome, not everything was rosy. A major difficulty nearly everyone confronted while trying to assess KR involved its mathematical expression. As we saw earlier, Milne had refused to frame his theory in the usual differential geometric form, choosing instead to stick to algebra. This irritated his colleagues no little bit, with the result that they refused essentially *en masse* to respect Milne's decision to avoid geometry.[9] In nearly every discussion of KR, the first move was to attempt to translate Milne's algebra into some sort of geometrical analogue. Often the result of this translation was the conclusion that KR was equivalent to an RC model with $k = -1$ (cf., for example, Kermack and McCrea 1933 and Walker 1935).[10] Sometimes, however, more intriguing possibilities were suggested, as when first Walker and then Hosokawa suggested that KR exhibited Finslerian geometry (Walker 1934, 1935; Hosokawa 1938).

As these examples show, despite the generally positive tone, a major theme in the reception of KR's physical aspects was the difficulty of deciding, in the very first place, how it was to be assessed. Eddington's comments sum up the case well: "Most of his critics have occupied themselves with the question, not whether Milne's theory is right, but whether it differs from current relativity cosmology" (Eddington 1935, p. 635). Some aspects of Milne's proposals received a much less ambiguous reception: most, Robertson and Walker in particular, found Milne's operationalism attractive.

4.2 Reception of KR's Operationalism

Although Milne had argued in favor of the operational definition of theoretical primitives in his earliest papers, it was not until his article Milne 1934b that he worked out in full detail how the procedure would work. Interestingly enough, the procedure was presented in an essay entitled "Some Points in the Philosophy of Physics: Time, Evolution, and Creation," published in the journal *Philosophy*. In exquisite detail Milne described how an observer, starting simply from his own "definite temporal experience" of before and after, may build up the operations necessary to carry out cosmological observations. Equipment used in the procedure consists of some arbitrarily running clock, whose only function is to put the sequence of events at the observer in association with the real numbers, and some sort of signaling device.[11] The signalling device acts as an "echo-locator," whose pulse—light or otherwise—is timed at its moments of origin and echo-reception.[12] Dingle's later description and assessment of this aspect of KR provides a useful summary of the situation:

> kinematical relativity rests on the postulate that the whole of physical law must be deducible from the characteristics of our awareness of the passage of time; accordingly, the only measuring instruments to be used are clocks.... Distances as well as times are measurable in terms of the same clock readings. On this simple basis Milne constructed a highly elaborate system of cosmology expressed primarily in purely mathematical terms. This was a magnificent achievement. (Dingle 1951)

It was Robertson, however, who was to pay the first tribute to Milne's inspiration.

4.3 Robertson–Walker Metric

4.3.1 Robertson's operational inspiration. Robertson's first close contact with Milne's system occurred during the winter of 1933. In early November of 1932, Milne had submitted his first extended description of his theory and its philosophical context to *Zeit. für Astrophysik.* The article, a 95-page opus entitled "World-Structure and the Expansion of the Universe," was published the following July (Milne 1933); Robertson's reply appeared in the next number of *ZfA* (Robertson 1933).[13] Robertson's response to Milne's theory, as we have already seen, focused for the most part on showing that KR was no rival to RC but more like a special case of it. Other efforts included analysis of Milne's "extended" principle of relativity (which will be discussed below), and repulsing Milne's attacks on the orthodox RC metaphysics of expanding and curved real space. Conspicuous by its absence is any mention—on either man's part—of operationalist procedures. Milne had not coupled his critique of real space to any replacement; Robertson hadn't commented on the fact. At the time of his next interaction with Milne's theory, however, Robertson not only commented upon Milne's operationalism, he adopted it. And in so doing, he produced the work he is remembered for.

In a remarkable series of three papers entitled "Kinematics and World Structure" (Robertson 1935, 1936d, 1936c), Robertson responded to Milne's issuing "a challenge which cannot be ignored," namely, determining "to what extent, then, must a strictly operational attack on the problem, with the aid of the 'cosmological principle,' necessarily lead to Milne's conclusion?" (Robertson 1936b, pp. 65–66). Robertson's goal in the three papers was stated succinctly: "We propose to analyze the general problem *ab initio*, using the operational methodology throughout and avoiding what Milne chooses to call the 'conceptual terms' of the general theory of relativity" (Robertson, 1935, p. 287).[14] The operational methodology was straight from Milne: nebulae were replaced by "fundamental particle observers," each of whom would be "equipped with a clock, a theodolite, and apparatus for sending and receiving light-signals—these latter considered as corpuscular impulses" (Robertson 1935, p. 285). According to Milne's procedure, "briefly stated, the operational viewpoint restricts the observations of each of these fundamental observers to such as can be made on events on his own world-line with the aid of these instruments" (Robertson 1935, ibid.).

But not only had Milne set Robertson the general challenge "of determining the most general kinematical background suitable for an idealized universe in which the cosmological principle[15] holds," he had set him another, quite specific task. Robertson introduced the task by noting that Milne had initiated "a promising attack on the problem of collinear observers suffering relative acceleration," but failed "to extend the results there obtained to the full three-dimensional problem" (Robertson 1936b, p. 61). Having said this, Robertson now issues the fateful promissory note:

> As I hope to show elsewhere in this *Journal*, Milne's failure to obtain the solution of the problem in which the observers are relatively accelerated is attributable to his imposition of restrictions foreign to the general nature of the program, and that on avoiding them one is led directly to the invariant theory of that general line element on which the general relativistic theory of cosmology is based. (Robertson 1936b, p. 62)[16]

Robertson here identified the most significant weakness that Milne's KR embodies in trying to be a comprehensive theory of "world structure." Despite Milne's insistence that his theory was grounded solely in empirical methodology, we must infer that by "world structure" Milne meant a self-evident topology for the universe (three-dimensional real space with a time axis adjoined), a geometry of the universe (flat Lorentzian geometry), an obvious invariance group for that geometry (the proper Lorentz group applied on a global basis) along with physically motivated "principles" such as a cosmological principle and a Hubble expansion principle.

Milne's immediate embracing of the above bedrock mathematical assumptions, which to him seemed obvious or at least more sensible than some general Lorentzian manifold structure of relativistic cosmology, in fact deprived the theory of sufficient tools to enforce consistency between its mathematical foundations and its physical assumptions and predictions. It is most likely that this is the point Robertson made above when he attributed Milne's failure to solve the problem of accelerated observers to "his imposition of restrictions foreign to the general nature of the program" (Robertson 1936b, p. 62). Robertson's first move was to avoid them.

Milne's use of global Cartesian coordinates, and Lorentz transformations thereof, made it essentially impossible for him to deal consistently with arbitrarily moving observers. Yet, as Robertson pointed out (Robertson 1936b), in the end Milne was able to make at least a certain amount of headway in treating collinear observers with relative accelerations. Milne recognized that for a pair (or discrete array) of such observers, general invertible (nonlinear) coordinate transformations were necessary in order to enforce a cosmological principle. In fact, a group structure was implied for the transformation functions relating observations made by pairs of observers. Milne's immediate difficulty was that of trying to impose Lorentz invariance on a coordinate-based speed of light, a property he was potentially willing to abandon. Robertson and Walker, on the other hand, since each was firmly committed to a differential-geometric formulation of cosmology, were both able to see that the key to a consistent treatment of accelerating observers, especially within the context of a cosmological theory with a "uniformity" requirement, was to relax Milne's global Lorentz transformation requirements.

Robertson, the consummate general relativist, exploited Milne's collinear observers' group requirements. He passed to a continuum of observers, thereby inducing a Lie group structure on the operationally based transformation functions between the observers. He then showed that simple local generator properties (forcing isotropy), in league with Milne's cosmological principle (forcing homogeneity), together mandated, via the Helmholtz–Lie theorem, that his (Robertson's) space-time carry a local metric structure of constant spatial curvature. (In this case, since scalar curvature equals sectional curvature, no confusion results concerning which curvature is intended.) The great power of Robertson's general mathematical arguments is evident in his masterful derivation of the metric form with which his name came to be associated. What has hardly been appreciated to this point is that the original spark of interest in kinematical methodology—which inevitably led to their independent discovery of the metric—was kindled in Robertson and in Walker by the foundational work of Milne. Robertson's familiarity with the work of Milne was strictly through the professional literature, which makes Robertson's inspiration by Milne's writings all the more surprising since Milne's published work was so generally difficult to endure by those who had embraced the newer tensor methods.

As we shall see shortly, Milne, through personal contact as well as professional collaboration, had a profound influence on Walker's work in cosmology and in the general treatment of observers. This influence, we shall find, was even more significant and productive in Walker's case than it had proved to be in Robertson's.

4.3.2 Walker's personal inspiration. A.G. Walker first encountered Milne during his last spring as an Oxford undergraduate, in 1931, when he took one of Milne's seminars. The following autumn, Walker went to work on his Ph.D. under Whittaker. From Whittaker he received a thorough grounding in mathematics and relativity, especially in differential geometry (Walker 1990). Walker soon proved to be quite a talent. Even though he was just a young research student, Walker's physical sophistication and mathematical depth were amply displayed in his first publication, a discussion of general relativistic observers (Walker, 1932). In that paper, the problem of assigning coordinates to events in some open tubular neighborhood of an arbitrary observer moving along a smooth time-like curve is treated using orthonormal tetrads along with their associated (frame-bundle) connection. This work is among the first in the mathematical physics literature to employ Cartan moving-frame methods. Later work in this area, in which he generalized Milne's kinematical methods to general relativity (Walker 1940), led Walker to ideas that anticipated the subsequent efforts by many others on the treatment of torsional space-times. Because of this anticipation, Walker's name came to be associated with what we now call the Fermi–Walker transport formalism. But all of this lay in the future.

Walker finished his thesis in early 1933. Then, in late spring, he decided to return to Oxford and work with Milne. Milne was only too pleased to acquire the cooperation of such a talented young mathematician:

> I have a number of potential ideas on gravitation which I do not know how to work out, and on which your own thinking would be valuable. There are firstly a number of problems arising out of my paper [most probably this is Milne 1933]—though I have of course gone a good way lately with some of them. Still there are many left over.... I am no expert on differential geometry, but I saw your thesis and admired it. (Milne 1933–1951, June 12, 1933)[17]

Walker went that autumn to Oxford, where, in addition to working further on topics from his thesis, he began a long-term, comprehensive investigation of KR and its implications for generally relativistic observers and cosmological models. Notably, and probably in deference to Milne, essentially all of Walker's published work on KR is in Cartesian coordinate notation,[18] thereby avoiding the overt appearance of grafting KR onto generally relativistic cosmology but, without doubt, at the same time accomplishing the task.

An example of Walker's investigation of some of the intrinsic properties of KR is found in a 1934 paper (Walker 1934) on least action in KR. Walker began the discussion with a restatement of one of Milne's fundamental proposals, namely, that the theoretician began his work in the face of a choice. According to this view, one may either first specify a geometry and subsequently infer the physics intrinsic to that geometry, or alternatively one may first specify physical laws and subsequently determine within which geometric setting(s) such laws would consistently reside. Walker argued that Milne's KR represents an example of the first option, while GR represents an example of the second.

After this beginning discussion, Walker, again following Milne, then reiterated the possibilities for constructing particle equations of motion:

(1) Modify the intrinsic line element $\mathrm{d}s$, thereby altering the implied geometry and adopting the new line element as the integrand in a stationary integral for particle trajectories;

(2) Retain the intrinsic geometrical line element $\mathrm{d}s$, but use a weighted action integral integrand, $W\,\mathrm{d}s$, to obtain particle trajectories.

Attributing the first choice to general relativity, Walker then pursued the second possibility in a KR setting and inferred that such a structure implied an underlying Finsler[19] basis for the basic Lorentzian (Gaussian) geometry of Milne's KR. Soon afterward, in January of 1935, Walker published his first investigations of the formal comparison of KR and generally relativistic cosmology (Walker 1935). Here Walker considered the gravitational properties that Milne's fundamental particle congruence must have when embedded in a general-relativistic metric formalism with an appended cosmological principle. Relying solely upon Milne's idealized operational method of local time-keeping for distance assignments, Walker found that only the $k = -1$, open homogeneous isotropic universes, were consistent with both Milne's particle distribution and his use of global Lorentz transformations. Walker also found that a global time variable must exist in the general setting in order for the chronological operationalism of KR to be implemented. Milne adamantly denied that KR implied a basic, underlying global measurable time. The saving qualifier, of course, was the term "measurable," this because special relativity in flat R^4 possesses a global time variable as an example of a stably causal Lorentzian manifold.

While the aforementioned papers by Walker constitute, on their own, an impressive body of work in special and general relativity, the paper that caused Walker's name to be permanently affixed to the important class of cosmological space-times of constant spatial curvature (the Robertson–

Walker cases) was yet to come. This paper was entitled "On Milne's Theory of World Structure" (Walker 1936).

Walker had largely completed his work on the implications of Milne's operationalism for generally relativistic cosmologies satisfying a cosmological principle, which constituted the groundwork for this paper, by the summer of 1935. He had in fact lectured on various aspects of the subject on several occasions some time before Robertson's parallel work applying Milne's kinematical methods appeared in a three-part series during late 1935 and early 1936 (Robertson 1936c, 1936d, 1935). At Milne's urging, Walker presented his results in several stages of development, the ultimate presentation occurring before the London Mathematical Society in June of 1935; the Society's *Proceedings* appeared early in 1936.[20]

Walker's approach was to generalize Milne's arguments and kinematical methods to determine the requirements that a fully relativistic cosmological metric must satisfy to support a cosmological principle. Realizing that Milne's global Lorentz transformations could not generally be applied in general relativity, Walker, like Robertson, made extensive use of local (Lie) groups of motions, both translational and rotational (on level 3-surfaces of a global time function). The requirements of isotropy and homogeneity on level time surfaces produced Killing equations which, taken together with null geodesic requirements adapted from Milne's time-keeping methods of distance measurement, restricted the metric on the space-time to be of the Robertson–Walker form.

In addition, Walker demonstrated the consistency of an isotropy criterion coupled with a cosmological principle and a least-action principle for the fundamental particles of such a theory. Finally, Walker showed that Milne's fundamental particle distributions were consistently classified as $k = -1$ space-times because the least-action principle implied for such cases does not necessarily lead to gravitational geodesic equations.

4.4 Outcomes

Walker was to continue his work with Milne for several more years, ever remaining a staunch defender of his inspirational mentor and life-long friend.[21] Robertson, on the other hand, never again wrote about the kinematical aspects of cosmology.[22] Yet the two men, each in his own way inspired by Milne, will be forever linked by the space-time they separately found together.

5. The Reception Sours: Philosophical Controversy

Milne's book appeared in 1935. In substance, the book represented not much that was new; rather, ideas expressed in his earlier papers merely received fuller, and in some cases sharper, expression. During the long run-up to the book's appearance, however, an idea occurred to Milne that was to shape his thoughts, and their reception, throughout the rest of his life. Unfortunately, it is this idea and its consequences that have endured until today, altering, indeed, destructively deforming, the opinion we hold of him and his work. This idea was new to Milne. As he notes to Chandra, "I have not developed these ideas in the book—they have only occurred to me since I finished it" (Milne and Chandrasekhar 1929, September 27, 1934). Two weeks earlier, as the idea first occurred to him, he had written to Walker, "I begin now to see that my theory is an approach to the ideal of deducing so-called laws of nature without any appeal to experience whatever—merely the embodiment of the compatibility or self-consistency of different observations by different observers of the same phenomena." Then, prophetically, he goes on to say, "I shall develop this idea later in some lecture or another" (Milne 1933–1951, September 12, 1934). The "some lecture or another" soon became a series of paragraphs, then a paper, several papers, a huge controversy, and in the end a virtual obsession, separating Milne from his colleagues and their now flourishing research program in cosmology. Yet, although we must say that, for Milne, on the whole, the results of his obsession were negative, it would be unfair to leave it there, for it is also true that at least some consequences of cosmological practice at large were quite positive.[23] Details of the situation are so rich, and the historical dynamic so complex, that it would be quite impossible to attempt anything other than a cursory survey here. Even such a survey provides insight into the richness of the situation, however. As always, it is Milne's complicated admixture of physics and philosophy that generates the problem. We have already noted the role played in KR's earlier reception by some of the physical virtues of Milne's KR scheme as begot by his philosophical drive toward metaphysical minimalism and its operationalist manifestation. KR's later reception, on the other hand, was driven by yet another kind of philosophical drive, namely, Milne's penchant for a mathematician's epistemology–methodology.

5.1 Milne's Epistemology–Methodology: Axioms, Models, Deductions, and Experience

Milne's new idea was a polymerization of previously disconnected strands of his thought. The first strand was *axiomatization*. From his earliest days, Milne had been committed to the goal that physics, including astrophysics and cosmology, should become an axiomatized theory.[24] Apparently, this commitment derived from an early reading of Whitehead and

Russell (Crowther 1970, p. 12). Taken by itself, this commitment would be quite uncontroversial. The second of Milne's early commitments, however, *was* controversial, at least among his astrophysical–cosmological colleagues. Milne believed strongly, unswervingly, in a particular mode of doing theoretical physics, a mode we today would call *hypothetico-deductive*. He described his belief—three years prior to his KR proposal—in no uncertain terms:

> It is the prime business, then, of theoretical astrophysics to suggest not one hypothesis in any given field, but many. The duty of the theoretical astrophysicist is to construct models, and rigorously infer their properties.... The peculiar contribution [that] the theoretical astrophysicist can make to his science is a set of models constructed on as many different plans as he can conceive, with a corresponding set of consequences. It is of little importance in the first instance whether the models reproduce nature or not. (Milne 1929b, p. 26)

Models thus generated are to be tested against the facts. In contrast to his colleagues, whom Milne disparaged as Baconian inductivists, facts are used to *test* theoretical models, not to generate them: "What are large collections of facts for? To make theories *from*, says Bacon; to try ready-made theories *by*, says the history of discovery" (Milne 1935, pp. 125–126). Milne believed that there were essentially no constraints upon the creativity of the theoretician imagining models; since, ultimately, the consequences of the models would be compared to observable facts, and constraint would enter only at a late stage of the process (Milne 1929b, p. 19).

The third philosophical strand leading to Milne's new idea can only be called *rationalism*, that is, the view that the source of theoretical concepts is within the mind, as opposed to within experience. Clearly, Milne's view on this developed in counterpoint to his hypothetico-deductivism. If, as he believed, the theoretician was unconstrained in his creation of models, then pure reason could serve as a source of models. Milne stated his view in no uncertain terms: "It is, in fact, possible to *derive* the laws of dynamics rationally... without recourse to experience" (Milne 1937, p. 324).

In the end the three philosophical strands—axiomatization, hypothetico-deductivism, and rationalism—came together in a coherent view of theory construction, a view that Milne vigorously opposed to the views of his colleagues:

> Now the methods of theoretical physics seem to be reducible to two species, the method of starting with concepts and the method of starting with things observed.... When a subject is developed from concepts the concepts play the part of the terms occurring in the axioms

> of geometry.... The concepts are undefined save as being governed by propositions of which they are subjects. (Milne 1934b, p. 19)

The concepts themselves, the sources of the axioms, come from intuitive ideas:

> That is, in short, to use only such brute facts, such irreducible facts, as are of the intuitive sort or do not rest on the questionable principle of induction, and thus to appeal to no empirical "laws of nature" of a quantitative kind. (Milne 1940, p. 132)

Milne's cosmological principle was an instance of such an "irreducible fact." Any observer, Milne believed, must observe exactly the same universe. Moreover, any observer must begin, according to the operationalist metaphysics, with his own internal sense of the passage of time. From these two "facts," all the rest—KR—followed. Milne described his first realization of the consequences of his new synthesis of the three philosophical strands in a letter to Chandra. "I believe," he said, "I have established kinematic theorems of the same validity as the theorems of pure geometry. The only appeal to experience is the existence of a temporal sequence for the individual, necessary in order to introduce time" (Milne and Chandrasekhar 1929, September 27, 1934). On the basis of this experience is constructed the first set of axioms describing a model universe, which is then compared with the facts. Further work should result in a reduction of the number of axioms. This, according to Milne, is the normal process of science, a process that, in the end, raises a fascinating question:

> The tendency of all scientific theory is to reduce the axiomatic basis, to deduce more and more phenomena from fewer and fewer statements of general principles. When will this process stop? Can we reduce the axiomatic basis to zero? My work strongly suggests that we can.... (Milne and Chandrasekhar 1929, September 27, 1934)

Here we see expressed the ultimate Holy Grail of rationalism: to generate a theory of the universe from the merest wisp of formal requirements, to show that "laws of nature are... but inevitable general relations following from the condition of the compatible observation by different observers" (Milne and Chandrasekhar 1929, September 27, 1934). Similarly, as Milne wrote to his brother Geoff, "I have to explain how probably all 'laws of nature' are not fiats, but the conditions of creation itself" (Milne 1932b, April 10,1936).

Milne's colleagues reacted violently to his new philosophy, especially to his postulation of the cosmological principle. He was surprised at the fuss: although "this was the most natural condition in the world to impose," he was "still amazed at the outcry it caused" (Milne 1944, p. 128). It seemed to him completely intuitive that there was only one universe, and that all systems of observations of it must be compatible. Moreover, having granted this, and the deductive consequences which follow, "their study and subsequent comparison with the world of nature was a perfectly legitimate procedure" (Milne, 1944, p. 128). Yet, because of this, "I was accused of abandoning the scientific method, of imposing a form on the world" (Milne 1944, p. 128).

Dingle led the opposition. The major bout occurred as an exchange in the pages of *Nature* and ultimately involved essentially all the first-ranked scientists in England. From his first encounter with KR, Dingle had objected to Milne's hypothetico-deductivism and presagings of rationalism. Milne's cosmological principle, Dingle held, got things wrong right from the start. There is, he said, a

> fundamental distinction between Milne's principle and the generally acknowledged principles of world structure, such as the principle of relativity and the laws of thermodynamics; namely, that the former [Milne's principle] requires the events of nature to conform to it, whereas the latter are abstractions which are true (or false) whatever the events of nature are. (Dingle, 1933, p. 178)

Science, good science, legitimate science, according to Dingle, begins with observations and only later, much later, is abstracted into a law or principle of world structure.[25] Milne's hypothetico-deductivism not only got things the wrong way around, it violated scientific method:

> The spirit of relativity is simply a reaffirmation of Newton's principle of induction from phenomena.... Milne approaches the problems of physics in precisely the opposite way. He starts, not with phenomena, but with a hypothetical smoothed-out universe which must obey an arbitrary principle.... It would seem that the general course of Milne's theory is at variance with the fundamental principles of scientific method. (Dingle 1933, p. 178)

Dingle's view on this question never changed in content, but the tone became more strident. In his 1937 *Nature* article "Modern Aristotelianism," Dingle sallied forth once more from his bastion within inductive empiricism. According to him, 17th century science held that "the first step in the study of Nature should be sense observation, no general principles being admitted which are not derived by induction therefrom" (Dingle 1937, p. 784). Opposed to this was "Aristotelianism... the doctrine that Nature is the visible working-out of general principles known to the human mind apart from sense perception." Although Dingle obviously confused Aristotle with Descartes, his point is clear enough: Milne, and other moderns of his ilk such as Eddington and Dirac, were, as he later said, "traitors" to the virtues of the Galilean scientific method (Dingle 1937, p. 385). The issue is sharply defined: "The issue between Galileans and Aristotelians is still sharply defined: ... Should we deduce particular conclusions from *a priori* general principles or derive general principles from observations?" Milne, especially, is guilty of creating, via his cosmological principle, an imaginary world. But

> the position must not be misunderstood. We are not dealing here with legitimate imagination transcending the temporary limits of exact demonstration.... Instead of the induction of principles from phenomena we are given a pseudoscience of invertebrate cosmythology. (Dingle 1937, p. 786)

In the end, Dingle concludes, "the question presented to us now is whether the *foundation* of science shall be observation or invention."

Dingle's polemic ignited an instant outburst of clangor. Indeed, the outburst was so immediate, so vigorous, and so widespread, that *Nature* had no recourse but to publish an entire supplement to contain the response (Milne et al. 1937). Virtually the entire population of first-rank British scientists joined the battle over the philosophical credentials of induction vs. deduction, of empiricism vs. rationalism. In general, it is safe to say that the result of this debate was a breaching of the dogmatic foundations enshrining inductive empiricism as a basis for cosmology. Ten years later, Bondi and the other steady-staters were to capitalize on this point (Bondi 1948).

6. Conclusion

Modern cosmology's first decade was marked by a curious mix of confident success and vigorous debate. In a fashion often seen during the initial events of a new science's emergence, the controversial elements of modern cosmology's genesis exhibit a distinct philosophical tone. For the most part, E.A. Milne's determined commitment to operational metaphysics, rationalist epistemology, and a hypothetico-deductive, axiomatic methodology both drove the debate and shaped its eventual outcome. Chandra, in an appraisal given many decades later, justly summed up Milne's contribution to these formative years (Chandrasekhar 1990). Milne, he noted, demonstrated clearly that RC was not necessarily the only theory that could explain the cosmological observations. This fact alone would be worthy of our respect. Yet this was by no means his only contribution. Of perhaps equal value was his devising a simple kinematic model capable of representing the basic details of the expanding system of nebulae. Milne's model will always remain a basic element of expanding universe theories. Even more, as we hope our account demonstrates, we must admit our debt to Milne as one who inspired others, particularly Robertson and Walker, in their own contributions to our present outlook on the universe.

NOTES

[1] Besides himself, Whittaker includes in this group "Einstein, De Sitter, Friedman, Lemaître, Weyl, Eddington, Robertson and others" (Whittaker 1935, p. 179).

[2] During which, as he remarked to his brother Geoff, "I was definitely visited by 10 days of inspiration—it was like the flinging aside of a curtain" (Milne 1932b, August 10,1932).

[3] Walker and Whitrow, in their unpublished obituary, clearly note, first, that Milne came late to cosmology, and, second, that it was the stimulus of Jeans' exchange that had done it: "It was however, not until early in May, 1932, that Milne was provoked and stimulated by letters, published in *The Times*, on the subject of the curvature of space" (Walker and Whitrow 1951).

[4] The letter's context (and orthographic style) strongly suggest that "J & E" here refers to Jeans and Eddington, whom Milne quite often ran together with only the "&" to keep them apart.

[5] Jeans states: "when the scientific and commonsense views clash, the latter must obviously yield to the former, since science has knowledge of all the facts known to the man-in-the-street, and a host of others as well. The man-in-the-street may nevertheless prefer to retain his old commonsense view of space; it will serve for his everyday requirements" (Jeans 1932, May 23, 1932). According to the criteria Jeans gives here, Milne is not only a "man-in-the-street," but also one who attempts to serve his everyday requirements *and* those of the most contemporary science as well!

[6] As will be evident, Milne's statements in this letter are particularly hard-nosed and uncompromising. We think that this severity can be accounted for simply by noticing that Milne is responding to some of Eddington's criticism of an earlier Milne R.A.S. presentation, criticism to which Milne had had "no opportunity of replying" at the time (Milne 1934a, p. 24).

[7] One obvious explanation for this astonishing view is tied directly to Milne's hypothetico-deductivism. Although we will save major discussion of his methodology until a later section of this essay, it is useful to note here that Milne simply didn't care about the truth or falsity of any given theoretical hypothesis. He viewed them all indifferently as "models," whose only role was to imply observational phenomena. This point was stated with admirable precision in his Oxford inaugural address "The Aims of Mathematical Physics: "It is the prime business, then, of theoretical astrophysics to suggest not one hypothesis in any given field, but many. The duty of the theoretical astrophysicist is to construct models, and rigorously infer their properties.... The peculiar contribution which the theoretical astrophysicist can make to his science is a set of models constructed on as many different plans as he can conceive, with a corresponding set of consequences. It is of little importance in

the first instance whether the models reproduce nature or not. A model which fails to reproduce nature is really more valuable than one which does—it at least shows what nature is not like, whilst a successful model leaves it an open question which of its characteristics is responsible for its success" (Milne 1929b, p. 26). Milne's falsificationist position here should also be noted.

[8] Milne's use of shudder quotes around "space" is entirely germane to our argument above, namely, due to its entraining a metaphysics he abhorred, Milne is reluctant to even use the term.

[9] Robertson hated it: "But undoubtedly the greatest drawback in the exposition, for the mathematically expert and inexpert alike, will be found in the cumbrousness and obscureness of the mathematical parts. Instead of applying fundamental results long established by invariant and group theory, Milne plows through pages of complicated analyses, involving mainly functional and differential equations, which are in effect but roundabout proofs, for the specific cases in point, of these well-known results" (Robertson 1936b, p. 65).

[10] Milne always objected strenuously to these attempts. In particular he denied that, since he held that geometry was conventional, he had in any sense chosen his system to be $k = -1$: "I hope you don't mind my objecting to your statements that I *choose* $k = -1$ (made in various places in your writings) but I honestly cannot see that I ever do any choosing. In my line of attack, starting from purely physical (not merely logical) considerations, alternatives simply do not occur. My work can be considered as either resting on the assumption of the Lorentz formulae, or going behind these & establishing them from light-signalling" (Milne 1933–1951, April 9, 1945). Here we see again Milne's view that geometry depends upon physics, and not vice versa.

[11] Although he never departed from the basics of this account, Milne continued to sharpen his procedure throughout the next decade. His ultimate versions appear in Milne 1940 and Milne 1941.

[12] This method will be familiar to anyone who has been caught speeding. Bondi is quite firm in awarding Milne credit for discovery of this method (Bondi 1988).

[13] It is clear from the reception dates that Robertson saw a preprint of Milne's paper. What is not clear is where he got it. Perhaps *ZfA* sent it to him. Examination of Robertson's papers should clear this up.

[14] Milne from the earliest distinguished between concepts and experiences. Among the fundamental notions of physics, Milne especially emphasized the conceptual status of "space" and "space-time." Both, for him, were constructs of the most abstract sort (cf., e.g., Milne 1933, p. 31). Indeed, "space-time is a concept of which we have no experience, a mathematical invention, useful solely for correlating experiences" (Milne 1934b, p. 26).

[15] It's not entirely clear who named the principle. Robertson apparently thought Milne had named it, and indeed thought well of the name itself: "The uniformity postulate, which Milne fittingly calls 'the cosmological principle,' asserts that the description of the whole system, as given by A in terms of his immediate measurements, is to be identical with the description given by any other fundamental observer A' in terms of his own measurements" (Robertson 1935, p. 285).

[16] Robertson here was speaking in his review of Milne's *World Structure*. Through a fluke of publication, the review was published later than the first member of Robertson's "Kinematics and World Structure" series.

[17] He certainly had seen the thesis, since, as he later admitted to Walker, he had been on a committee reviewing it for an award!

[18] Walker confessed that he and Milne often disagreed, and strongly, over Milne's refusal to adopt geometry as his vehicle: "He always preferred algebraic expression" (Walker 1990).

[19] Walker's suggestion was picked up by Hosokawa, who promptly generalized it, showing that such a relationship held in all Finslerian geometries (Hosokawa 1938). Unfortunately, Hosokawa's work apparently never penetrated the later controversy.

[20] Walker apparently was a bit slow in getting the work into print. On November 28, 1935, Milne wrote to Walker, "I should be interested in your views of Robertson's paper in the new number of the *Astrophysical Journal*. Does it partly anticipate your work on cosmology that you gave an account of at our colloquium last year? I think you ought to get this written up at once and published." The colloquium here mentioned had occurred the preceding April 28. Milne had set McCrea up for Walker's presentation: "It is probable, or at least possible, that McCrea will be up in Oxford next Tuesday, and if so I shall bring him to the colloquium at which you are to speak. I hope you will not mind this. I have no doubt that you will have some good stuff for him to think about. Make it as hard, abstract, and provocative as you can!" (May 22, 1935). Walker was obviously a great success: "McCrea was very complimentary about your paper last night—said he greatly admired the power of your methods" (May 29, 1935). Milne encouraged Walker at the same time to present his work at the upcoming autumn British Association meeting: "By all means read a paper at the Norwich B.A. Whitrow has been asked similarly. It will do hardened relativists good to see your method of building up a universe. I expect that that is the one [the paper] you will choose" (April 20, 1935).

[21] Yet, in terms that might sound strange to the ear of an American-educated physical scientist, Walker strongly denies ever having any but a purely mathematical interest in Milne's (or for that matter, anyone else's) cosmology. He was always a mathematician, Walker says. And it is obvious when talking to him that, for him, relativity theory is a mathematical theory, period (Walker 1990).

[22] And Milne never forgot that Robertson had written about the kinematical aspects of cosmology! Although he at first thought that Robertson's work encroached upon Walker's (November 28, 1935), Milne soon came to believe that it encroached upon his own as well. (McCrea obviously agreed with both assessments: Referring to Robertson's work, he remarked, "Actually a similar conclusion had been previously announced by A.G. Walker, though details of his work have not yet been published.... It is only fair to recognize that this paper is a modified discussion of work already described by Milne in his book *World-Structure* and elsewhere" [McCrea 1936, p. 203]. This verdict is important, since it came as an editorial in *Observatory*, of which McCrea was then the editor.) Ultimately, however, what galled Milne was not only that Robertson seemed to have adopted his operationalist methods without enough credit being given but that Robertson (to Milne's mind) had slighted Milne's book in a review (Robertson 1936b). Many years later, a still

smarting Milne wrote to Chandra, "I have no objections to E.L. Hill's review at all, though it is not a very favourable one But it is not irritating or depreciatory, as HPR's was.... I met HPR several times over here during the war and I remained perfectly friendly. But I always thought that he owed me an apology for running down *World Structure* and then with very little acknowledgement trying to make a development of the same ideas" (October 29, 1949).

[23] Since these consequences will be developed at length in another essay, we will provide here only the briefest of mentions. Milne's view of the natural evolution of physical theories from an inductive-empiricist origin to stand-alone axiomatization was adopted by cosmologists as essentially an official story of their community's development (cf., especially, McCrea 1939 and Whittaker 1941). Additionally, Milne's firm avowal of the right of a physical scientist to use rationalist sources—the second "species" of physics' methods (Milne 1934b, p. 19)—led straightaway into Bondi's cosmological methodology (cf., for example, Bondi 1948).

[24] Walker recently noted how Milne's enthusiasm and commitment to axiomatization had ignited in him, Walker, a similar life-long commitment (Walker 1990).

[25] Dingle was always suspicious of mathematical theorists. For him, mathematics had no role to play in discovery, and indeed functioned solely to effect economy of thought about the observations. He and Milne were thereby completely at odds on this question (Whitrow 1990).

References

Bondi, Hermann (1948). "Review of Cosmology." *Monthly Notices of Royal Astronomical Society* 108: 104–120.

—— (1988). Interview with George Gale, November 12, 1988. Unpublished.

Chandrasekhar, Subrahmanyan (1990). *Truth and Beauty—Aesthetics and Motivations in Science*. Chicago: University of Chicago Press.

Crowther, Jeffrey G. (1970). *Fifty Years with Science*. London: Barrie & Jenkins.

De Sitter, Willem (1931a). "The Expanding Universe." *Scientia* 49: 1–10.

—— (1931b). "Contribution." In *The Evolution of the Universe*. Herbert Dingle, ed. London: *Nature*, October 24, pp. 706–709.

Dingle, Herbert (ed.) (1931). *The Evolution of the Universe*. London: special supplement to *Nature*, October 24.

—— (1933). "On E.A. Milne's Theory of World Structure and the Expansion of the Universe." *Zeitschrift für Astrophysik* 7: 176–179.

—— (1937). "Modern Aristotelianism." *Nature* 138: 784–786.

—— (1951). "Letter to O.U.P." February 27, 1951. Referee's report on Milne's *Cosmology and the Christian Idea of God*. Bodeleian Modern Manuscript Collection, Oxford. Unpublished.

Eddington, Arthur S. (1931). "The Expansion of the Universe." *Monthly Notices of Royal Astronomical Society* 91: 412–416.

—— (1935). "Review of *Relativity, Gravitation and World Structure*." *Nature* 135: 635–636.

Hosokawa, Taneka (1938). "Finslerian Wave Geometry and Milne's World-Structure." *Hiroshima University Journal of Science* 8: 249.

Jeans, Sir James (1932). "The Universe: Space Finite and Expanding: Sir James Jeans to His Inquirers." *The Times*, May 14, 18, 23, 26, 27.

Kermack, William O. and McCrea, William H. (1933). "On Milne's Theory of World Structure." *Monthly Notices of Royal Astronomical Society* 93: 519–529.

Lemaître, Georges (1931). "The Expanding Universe." *Monthly Notices of Royal Astronomical Society* 91: 490–501.

McCrea, William H. (1936). "Kinematics and World-Structure." *Observatory* 59: 202–204.

——— (1939). "The Evolution of Theories of Space-Time and Mechanics." *Philosophy of Science* 6: 137–162.

McCrea, William H. and Milne, Edward A. (1934). "Newtonian Universes and the Curvature of Space." *Quarterly Journal of Mathematics (Oxford)* 5: 63–70.

McVittie, George C. (1933). "Milne's Theory of the Expansion of the Universe." *Nature* 131: 533–534.

——— (1987). "An Anglo-Scottish University Education." In *The Making of Physicists.* Rajkumari Williamson, ed. Bristol: Hilger, pp. 66–70.

Milne, Edward A. (1929a). "The Masses, Luminosities, and Effective Temperatures of the Stars." *Monthly Notices of Royal Astronomical Society* 90: 17–54.

——— (1929b). *The Aims of Mathematical Physics.* November 19, 1929. Bodelian Modern Manuscript Collection, Oxford. Unpublished.

——— (1931). "Contribution." In *The Evolution of the Universe.* Herbert Dingle, ed. London: *Nature*, October 24, pp. 715–718.

——— (1932a). "World Structure and the Expansion of the Universe." *Nature* 130: 9–10.

——— (1932b). Correspondence with Geoffrey Milne. Bodelian Modern Manuscript Collection, Oxford. Unpublished.

——— (1933). "World-Structure and the Expansion of the Universe." *Zeitschrift für Astrophysik* 6: 1–95.

——— (1933–1951). Correspondence with A.G. Walker. Unpublished; Prof. Walker holds the correspondence.

——— (1934a). "World-Models and the World-Picture." *Observatory* 57: 24–27.

——— (1934b). "Some Points in the Philosophy of Physics: Time, Evolution and Creation." *Philosophy* 9: 19–38.

——— (1935). "Reviews, Reviewers and Reviewed." *Observatory* April: 124–126.

——— (1937). "Kinematics, Dynamics, and the Scale of Time." *Proceedings Royal Society (London)* A158: 324–329.

——— (1940). "Cosmological Theories." *Astrophysical Journal* 91: 129–158.

——— (1941). "Remarks on the Philosophical Status of Physics." *Philosophy* 16: 356–370.

——— (1944). "The President's Address." *Monthly Notices of Royal Astronomical Society* 104: 120–136.

Milne, Edward A. and Chandrasekhar, Subrahmanyan (1929). Correspondence with S. Chandrasekhar, 1929–1949. Bodelian Modern Manuscript Collection, Oxford. Unpublished.

Milne, Edward A. et al. (1937). "On the Origins of Laws of Nature." *Nature* 139: 997–1007.

Robertson, Howard P. (1933). "On E.A. Milne's Theory of World Structure." *Zeitschrift für Astrophysik* 7: 152–162.

——— (1935). "Kinematics and World-Structure." *Astrophysical Journal* 82: 284–301.

——— (1936a). "An Interpretation of Page's 'New Relativity.'" *Physical Review* 49: 755–760.

——— (1936b). "Review of Milne's *Relativity, Gravitation and World Structure*." *Astrophysical Journal* 83: 61–66.

——— (1936c). "Kinematics and World-Structure, II." *Astrophysical Journal* 83: 187–201.

——— (1936d). "Kinematics and World-Structure, III." *Astrophysical Journal* 83: 257–271.

Tolman, Richard C. (1932). "Models of the Physical Universe." *Science* 75: 367–373.

Walker, Arthur G. (1932). "Relative Observers." *Proceedings Royal Society (Edinburgh)* 52: IV: 345–353.

——— (1934). "The Principle of Least Action in Milne's Kinematical Relativity." *Proceedings Royal Society (London)* 147A: 478–490.

——— (1935). "On the Formal Comparison of Milne's Kinematical System with the Systems of General Relativity." *Monthly Notices Royal Astronomical Society* 95: 263–269.

——— (1936). "On Milne's Theory of World Structure." *Proceedings London Mathematical Society* 42: 90–107.

——— (1940). "Relativistic Mechanics." *Proceedings London Mathematical Society* 46: 113–153.

——— (1990). Interview with George Gale, January 10, 1990. Unpublished.

Walker, Arthur G. and Whitrow, Gerald J. (1951). *Milne Obituary*. Bodelian Modern Manuscript Collection, Oxford. Unpublished.

Whitrow, Gerald J. (1990). Interview with George Gale, January 5, 1990. Unpublished.

Whittaker, Edmund T. (1935). "Review of *Relativity, Gravitation, and World Structure*." *Observatory* 58: 179–188.

——— (1941). "Some Disputed Questions in the Philosophy of the Physical Sciences." *Proceedings Royal Society (Edinburgh)* 61: 160–175.

Contributors

Silvio Bergia, Dipartamento di Fisica, Università degli Studi di Bologna, Via Irnerio 46, 40126 Bologna, Italy

Carlo Cattani, Dipartamento di Matematica, Università degli Studi di Roma, La Sapienza, Piazzale Aldo Moro, 2, I-00185 Rome, Italy

Michelangelo De Maria, Dipartimento di Fisica, Università degli Studi di Roma La Sapienza, Piazzale Aldo Moro, 2, I-00185 Rome, Italy

John Earman, Department of History and Philosophy of Science, University of Pittsburgh, Pittsburgh, PA 15260, U.S.A.

Jean Eisenstaedt, Laboratoire de Physique Théorique, Institut H. Poincaré, 11, rue P. et M. Curie, 75231 Paris Cedex 05, France

George Gale, Department of Philosophy, University of Missouri–Kansas City, 222 Cockefair Hall, Kansas City, MO 64110-2499, U.S.A.

Hubert Goenner, Institut für Theoretische Physik, Universität Göttingen, Bunsenstrasse 9, D-3400 Göttingen, Germany

Gennady Gorelik, Dibner Institute for the History of Science and Technology, 38 Memorial Drive, Cambridge, MA 02139, U.S.A.

Peter Havas, Department of Physics, Temple University, Barton Hall 009-00, Philadelphia, PA 19122, U.S.A.

Don Howard, Department of Philosophy, University of Kentucky, 1415 Patterson Office Tower, Lexington, KY 40506-0027, U.S.A.

Michel Janssen, Department of History and Philosophy of Science, University of Pittsburgh, Pittsburgh, PA 15260, U.S.A.

S. Kichenassamy, Laboratoire de Physique Théorique, Institut Henri Poincaré, 11 rue P. et M. Curie, 75231 Paris Cedex 05, France

A.J. Kox, Collected Papers of Albert Einstein, Boston University, 745 Commonwealth Ave., Boston, MA 02215, U.S.A., and Institute of Theoretical Physics, University of Amsterdam, Valckeniertstraat 65, 1018 XE, Amsterdam, Netherlands.

John D. Norton, Department of History and Philosophy of Science, University of Pittsburgh, Pittsburgh, PA 15260, U.S.A.

Karin Reich, Ehrenhalde 23, 7000 Stuttgart 1, Germany

John Urani, Department of Physics, University of Missouri–Kansas City, Kansas City, MO 64110-2499, U.S.A.

Kameshwar C. Wali, Department of Physics, Syracuse University, 201 Physics Building, Syracuse, NY 13244-1130, U.S.A.

Index

GPSR Compliance

The European Union's (EU) General Product Safety Regulation (GPSR) is a set of rules that requires consumer products to be safe and our obligations to ensure this.

If you have any concerns about our products, you can contact us on ProductSafety@springernature.com

In case Publisher is established outside the EU, the EU authorized representative is:

Springer Nature Customer Service Center GmbH
Europaplatz 3
69115 Heidelberg, Germany

Batch number: 10725800

Printed by Printforce, the Netherlands